Advances in Global Change Research

Volume 49

More information about this series at http://www.springer.com/series/5588

Robert J. Nicholls • Richard J. Dawson
Sophie A. Day (née Nicholson-Cole)
Editors

Broad Scale Coastal Simulation

New Techniques to Understand and Manage Shorelines in the Third Millennium

Springer

Editors
Robert J. Nicholls
Tyndall Centre for Climate Change Research, Faculty of Engineering and the Environment
University of Southampton
Southampton, UK

Richard J. Dawson
School of Civil Engineering and Geosciences, Tyndall Centre for Climate Change Research
Newcastle University
Newcastle upon Tyne, UK

Sophie A. Day (née Nicholson-Cole)
Tyndall Centre for Climate Change Research, Faculty of Engineering and the Environment
University of Southampton
Southampton, UK

ISSN 1574-0919 ISSN 2215-1621 (electronic)
Advances in Global Change Research
ISBN 978-94-024-0427-2 ISBN 978-94-007-5258-0 (eBook)
DOI 10.1007/978-94-007-5258-0

Springer Dordrecht Heidelberg New York London

Softcover reprint of the hardcover 1st edition 2015

Printed on acid-free paper

Springer Science+Business Media B.V. Dordrecht is part of Springer Science+Business Media (www.springer.com)

Foreword

Global change, including climate change, presents one of the greatest challenges facing the world today. It is affecting food, water and human security as well as biodiversity and ecosystems and their services, with developing countries and poor people being the most vulnerable. Nowhere are the pressures of global change more evident than along our coastlines, where an ever-increasing number of people live. Hundreds of millions of people are threatened by rising sea levels and storm surges, especially those living in deltaic areas, such as Bangladesh; low-lying small island states, such as Kiribati; and cities sinking due to excessive groundwater extraction, such as Bangkok. Even rich cities such as New York are not immune to flooding as witnessed with Hurricane Sandy.

The systems approach that the Tyndall Centre researchers describe in this book has developed an appropriate conceptual framework for integrated coastal assessment at the scales of coastal management. It has enabled them to answer previously intractable questions about how climate change interacts with changing coastal systems. Whilst the work focussed upon North Norfolk in the East of England, the methods and insights are generic and potentially transferable to other coastlines around the UK and the world.

Coasts are complex systems that evolve over a range of time and space scales. Marine climate, beaches, cliffs, farmland, urban areas, erosion and flood protection infrastructure change over timescales of decades, so we need to understand the implications of those changes and start to prepare for them to avoid being burdened with long-term problems. This research recognised that coastlines are embedded within wider socio-economic and environmental systems and sought to understand the implications of climate, socio-economics and coastal management policy as drivers of long-term change. The Tyndall Coastal Simulator simulates the interactions between these processes in order to develop scenarios of change at spatial scales of relevance to coastal decision-makers.

Coastal governance is also a key issue as it is complex and involves multiple actors. By mapping out the process of long-term change, the Tyndall Coastal Simulator can help assess the potential effectiveness of policy instruments and their implications in terms of a range of different issues, including erosion, flooding and biodiversity.

Central to the success of the Tyndall Centre's holistic and policy-relevant approach has been the transdisciplinary nature of the research. This engaged a range of relevant stakeholders, including national and local government, the private sector and members of the public, in co-developing the research programme in order to address the full range of societal issues and concerns. This research showed that integrated modelling, as delivered by the Tyndall Coastal Simulator, can help to bring different stakeholders together in order to develop common understanding of processes and consequences of long-term change. That collective understanding is essential if society is to manage coastal change rather than become its victims.

There is no doubt that this work has broken new ground in terms of transdisciplinary climate change research by demonstrating that individual academic disciplines can advance the state of their art whilst importantly addressing questions of relevance to society that cut across these disciplines. The importance of the work was recognised by industry when an earlier synthesis paper (Dawson et al. 2009, included in Chap. 1 of this book) won the Lloyd's Science of Risk Prize in 2012.

Whilst significant advances have been made, inevitably a host of challenges and scientific uncertainties remain, and so I am pleased to see the Tyndall team reflect on these issues with practitioners and international experts in the final two chapters of this book. Not only does this offer a manifesto for an international research agenda but also a framework/demonstration of how interdisciplinary research should be conducted in international programmes such as the recently launched ICSU/ISSC Future Earth programme. That, however, is for the future. The purpose of this book is to summarise the results from eight years of Tyndall Centre research, which has significantly advanced our understanding of climate change and coastal systems.

Tyndall Centre
University of East Anglia
Norwich, UK
Bob Watson

Monash University
Melbourne, VIC, Australia
Louis Matheson

Preface

Coastal scientists, engineers and policy makers around the world are increasingly recognising the challenge of sustainable coastal management in the third millennium. Long-term geomorphological, climatic and socio-economic changes are influencing coastal systems at unprecedented spatial scales and over extended timeframes – with profound implications for people, coastal infrastructure and settlements, biodiversity, ecosystem services and governance of the coastal zone.

Coastal researchers and decision-makers are presently ill-equipped to deal with the problems emerging from multiple drivers of change across multiple coastal sectors. This reflects that the coast is a linked system, and any change in one area or sector may influence the impacts for other areas or sectors. An integrated systems-based approach that seeks to represent the interactions between different issues within the coastal zone is fundamental to understanding the impact of global change on coastlines and to assist the sustainable management of our shorelines over the twenty-first century.

In 2000, the Tyndall Centre for Climate Change Research, an interdisciplinary consortium of engineers, scientists and social scientists, was established in the UK. This provided a unique platform to develop a coastal research programme with a major focus on an integrated assessment – this became known as the Tyndall Coastal Simulator – and is now reported in full, for the first time, in this book. An earlier synthesis paper from this research by Dawson et al. (2009) (included within Chap. 1 of this book) won the Lloyd's Science of Risk Research Prize for Climate Change in 2012.

However, there are already a number of books on coastal management, so why might we need another one? Whilst this is certainly a valid question, this book is not a handbook for design, nor is it a compendium of methods that cover every aspect of coastal systems or a compilation of case studies with differing aims. Rather it is a perspective on integrated assessment as applied to coastal problems, which represents a topic where there is an important gap in the literature. This book is structured as follows: Chapter 1 expands upon the challenges of sustainable coastal management and provides an overview of the integrated assessment for the core case study site in

North Norfolk in England. Chapters 2 and 3 describe how global changes in marine climate may alter future sea level, storm surge and offshore wave conditions around the British Isles and especially North Norfolk. Similarly, Chap. 4 quantifies the effects of global and local socio-economic drivers on changes to land use and development in the East of England. The following five chapters consider a range of impacts associated with the climatic and socio-economic changes considered above. Chapter 5 analyses broadscale geomorphic change in the region and Chap. 6 analyses habitat degradation and loss. Chapter 9 brings together the cliff erosion and flood modelling in Chaps. 7 and 8, respectively, to evaluate future erosion and flood risks in terms of expected economic damages under different climatic, socio-economic and management scenarios. Central to the success of this programme was comprehensive engagement with stakeholders. Chapters 10 and 11 describe the visualisation techniques and graphical user interface used to present results in an interactive manner to stakeholders, whilst Chap. 12 considers current decision-making processes and how the management of transitions in the coastal zone could be more adaptive and equitable and effectively implemented. The final two chapters of the book reflect on the key findings of the research, describe a general framework for transferring the Tyndall Coastal Simulator and identify future research priorities for integrated assessment in the coastal zone. This includes reflection on the range of coastal problems across the globe.

Each chapter could be read in isolation, but distinctive to other books on coastal management, each chapter also contributes to the wider integrated assessment. Throughout the book, we reflect on the process of integrating information on the different environmental, social and economic dimensions of coastal management.

Taking a systems perspective of the natural, physical and social environment at a scale that is relevant to livelihoods and the economy has enabled us to analyse how the coastal system as a whole might evolve in a changing physical and socio-economic environment. The application of the Tyndall Coastal Simulator to North Norfolk, UK, demonstrates that it is now feasible to explore long-term integrated projections of coastal processes such as geomorphology, flood risk and land use change, greatly increasing the evidence base available for coastal management decisions. Moreover, the methods and integrated assessment framework are transferable to other coastal areas.

The integrated assessment presented here has highlighted a number of the opportunities, challenges and trade-offs and the need for a long-term perspective on coastal policy in order to allow adaptation to coastal change to occur, for example, the difficulties faced by coastal managers, who in reducing the risk of erosion may actually enhance flood risk (or the cost and viability of mitigating this risk) at sites within the same coastal system. Such results were captured within the Tyndall Coastal Simulator interface allowing the technical results to be accessible to a wide range of stakeholders.

It is now clear that the management of any coastline and the governance structures upon which that management depends need to reflect the connectivity between the various coastal features that comprise the natural and human coastal system and consequential trade-offs in management policy. Furthermore, the Norfolk analysis relates the technical aspects of coastal change to the present, and often emotive, debate around long-term shoreline management – in particular it strengthens the argument for a change in the widespread historic management approach of increasing lengths of "hold the line" towards allowing as much of the coastline as possible to return to a more natural and dynamic configuration, including the associated sediment supply from eroding coasts. Inevitably, this raises a number of fundamental questions from stakeholders, which we have explored through using results from the Tyndall Coastal Simulator, about how to address the concerns of directly and indirectly affected landowners and householders to facilitate this fundamental change in management approach.

More generally, the work presented here shows that there is great potential for coastal stakeholders to develop improved understanding of coastal futures and for decisions to be based on a stronger evidence base. However, our integrated analysis exposed the magnitude of many uncertainties about coastal futures, and so in the context of adaptation, although the broadscale coastal simulation of the type presented here can provide a rich evidence base, it should be regularly reassessed, debated and reviewed as part of an ongoing process to reflect improving knowledge and changing priorities. Thus, we believe the Tyndall Coastal Simulator and tools like it have the potential to provide a platform for the longer-term adaptation process.

Southampton, UK — Robert J. Nicholls
Newcastle upon Tyne, UK — Richard J. Dawson
Southampton, UK — Sophie A. Day (née Nicholson-Cole)

Acknowledgements

The majority of this work was supported by the Tyndall Centre for Climate Change Research, which was funded by the Natural Environment Research Council (NERC), the Engineering and Physical Sciences Research Council (EPSRC) and the Economic and Social Research Council (ESRC). The methodological development of the risk analysis benefitted from additional support through the Flood Risk Management Research Consortium which was funded by EPSRC in partnership with the DEFRA/EA Joint Research Programme on Flood and Coastal Defence, UK Water Industry Research (UKWIR), Office of Public Works (OPW) (Ireland) and the Rivers Agency (Northern Ireland). In addition, the habitat and stakeholder analysis was supported through the Biodiversity Requires Adaptation in North-West Europe under a changing climate (BRANCH) Project which was funded by EU INTERREG IIIB funds. Richard Dawson's EPSRC fellowship (EP/H003630/1) provided him the flexibility and space to work on this.

For the first phase of work, the Tyndall Centre was led by Professor Mike Hulme (now at King's College London) and Professor John Schellnhuber (currently director of the Potsdam Institute for Climate Impact Research) who both provided practical support and developed an exciting vision for interdisciplinary research. Professor Andrew R. Watkinson (University of East Anglia), who led the coastal research programme during its first phase and latterly the Tyndall Centre, is thanked not just for these roles but for his regular advice and critical review throughout the production of this book.

Stakeholder engagement and collaboration has been central to the Tyndall Centre's philosophy. We are extremely grateful to the many individuals, organisations and coastal communities who contributed their time, data and expertise to provide valuable feedback on the research and insights into the case study location. It is impossible to name them all, but we extend particular thanks to Steve Hayman, Gary Watson and Julian Wright from the Environment Agency; Peter Frew and Robert Young from North Norfolk District Council; Malcolm Kerby, a resident of Happisburgh and founder of the Coastal Concern Action Group; and Clive Stockton, also of Happisburgh and landlord of the Hill House.

The analysis presented here would not have been possible were it not for access to data collected and managed by the Environment Agency, Natural England, North Norfolk District Council, the Met Office, the Ordnance Survey, the British Geological Survey, the British Oceanographic Data Centre (BODC) and the Centre d'Etudes Techniques Maritimes et Fluviales (CETMEF) for their Atlas Numérique d'Etats de Mer Océanique et Côtier.

The aerial photographs that appear in this book were taken by Mike Page and the historical images were provided by Norfolk County Council's Library and Information Service. Others were taken by Sophie Day and Richard Dawson during the course of their research.

The inspiration from this book stemmed from the Dawson et al. (2009) paper that was published in Climatic Change. A number of co-authors of this earlier paper did not contribute directly to this book, but we are grateful for their earlier contribution: Julie Richards (ABPmer), Jianguo Zhou (University of Liverpool), Steve Pearson (British Geological Survey), Jon Rees (NERC and British Geological Survey) and Paul Bates (University of Bristol). However, were it not for Margaret Deignan at Springer inspiring us into action, this book may not have happened, and we also thank Takeesha Moerland-Torpey and the rest of the publishing team at Springer for their guidance and patience.

David Walker and Andrew R. Watkinson also provided useful reviews of some of the earlier chapter drafts. Finally, we are grateful for Susan Hanson at Southampton University for reviewing the final book chapters and sparing us many blushes.

Contents

Contributors

Peter Balson Formerly of British Geological Survey, Kingsley Dunham Centre, Nottingham, UK

Iain Brown The James Hutton Institute, Aberdeen, UK

Jessica Bryson (née Milligan) Geography and the Lived Environment Research Institute, School of Geosciences, University of Edinburgh, Edinburgh, UK

Nicolas Chini Tyndall Centre for Climate Change Research, School of Mechanical, Aerospace and Civil Engineering, University of Manchester, Manchester, UK

Richard J. Dawson School of Civil Engineering and Geosciences, Tyndall Centre for Climate Change Research, Newcastle University, Newcastle upon Tyne, UK

Sophie A. Day (née Nicholson-Cole) Tyndall Centre for Climate Change Research, Faculty of Engineering and the Environment, University of Southampton, Southampton, UK

Mark Dickson School of Environment, The University of Auckland, Auckland, New Zealand

Corentin M. Fontaine Namur Centre for Complex Systems and Research Group in Sustainable Development, Department of Geography, University of Namur, Namur, Belgium

Jon French Coastal and Estuarine Research Unit, Department of Geography, University College London, London, UK

Peter Frew Coastal Strategy for North Norfolk District Council, Coastal Management Consultant, Ely, UK

Sarah Gardiner Faculty of Engineering and the Environment, University of Southampton, Southampton, UK

Jim W. Hall Environmental Change Institute, Oxford University Centre for the Environment, University of Oxford, Oxford, UK

Susan Hanson Faculty of Engineering and the Environment, University of Southampton, Southampton, UK

Tom Howard Storm Surges, Hadley Centre, Met office, Exeter, UK

Simon Jude Centre for Environmental Risks and Futures, Department of Environmental Science and Technology, School of Applied Sciences, Cranfield University, Bedford, UK

Sotiris Koukoulas Department of Geography, University of the Aegean, Mytilene, Greece

Qiuhua Liang School of Civil Engineering and Geosciences, University of Newcastle, Newcastle upon Tyne, UK

Jason Lowe Department of Meteorology, Met Office, University of Reading, Earley Gate, UK

Department of Meteorology, University of Reading, Reading, UK

Nobuo Mimura Institute for Global Change Adaptation Science, Ibaraki University, Mito, Ibaraki Japan

Mustafa Mokrech Environmental Institute of Houston, School of Science and Computer Engineering, University of Houston Clear Lake, Houston, TX, USA

Robert J. Nicholls Tyndall Centre for Climate Change Research, Faculty of Engineering and the Environment, University of Southampton, Southampton, UK

Leonard Nurse Centre for Resource Management and Environmental Studies (CERMES), Faculty of Science and Technology, University of the West Indies (UWI), Barbados

Melissa Nursey-Bray Geography, Environment, Population, School of Social Sciences, University of Adelaide, North Terrace, Adelaide, South Australia, Australia

Tim O'Riordan School of Environmental Sciences, University of East Anglia, Norwich, Norfolk, UK

Munsur Rahman Institute of Water and Flood Management (IWFM) Bangladesh, University of Engineering and Technology (BUET), Dhaka, Bangladesh

Mark D.A. Rounsevell Institute of Geography and the Lived Environment, School of Geosciences, University of Edinburgh, Edinburgh, UK

Tom Spencer Cambridge Coastal Research Unit, Department of Geography, University of Cambridge, Cambridge, UK

Peter Stansby Tyndall Centre for Climate Change Research, School of Mechanical, Aerospace and Civil Engineering, University of Manchester, Manchester, UK

William J. Sutherland Conservation Science Group, Department of Zoology, University of Cambridge, Cambridge, UK

Owen Tarrant Flooding and Communities, Environment Agency, Bristol, UK

James Thomas Experience Prototyping Technician, School of Design, Northumbria University, Newcastle upon Tyne, UK

Mike Walkden WSP, Keble House, Southernhay Gardens, Exeter, UK

WSP Group, Exeter, UK

David Walker School of Civil, Environmental and Mining Engineering, The University of Adelaide, Adelaide, SA, Australia

Andrew R. Watkinson School of Environmental Sciences, University of East Anglia, Norwich, UK

Kathleen D. White Global and Climate Change, Institute for Water Resources, US Army Corps of Engineers, Portland, OR, USA

Judith Wolf Marine Systems Modelling Group, National Oceanography Centre, Liverpool, UK

Xing Zheng Wu Department of Applied Mathematics, School of Applied Science, University of Science and Technology, Beijing, P.R. China

Robert Young Coast and Community Partnerships, North Norfolk District Council, Norfolk, UK

Barbara Zanuttigh DICAM, University of Bologna, Bologna, Italy

Acronyms and Abbreviations

A1B SRES	A1B Special Report on Emission Scenarios
ABM	Agent-Based Model
ANEMOC database	Atlas Numérique d'Etats de Mer Océanique et Côtier database
AR4	IPCC Fourth Assessment Report
AR5	IPCC Fifth Assessment Report
ASMITA	Model describing inlet-coast morphological interaction described by Stive et al. (1998)
CHaMPS	Coastal Habitat Management Plans
CS3	Continental Shelf model grid (~12km resolution) developed for UK tide-surge forecasting
DEFRA	Department for Environment, Food and Rural Affairs (England)
DELFT3D	Hydro-informatics system developed by Deltares
DPSIR	Driver-Pressure-State-Impact-Response is a causal framework for describing the interactions between society and the environment (see, e.g. Maxim et al. 2009)
ECMWF	European Centre for Medium-Range Weather Forecasts
ERA40	ECMWF 40-year reanalysis data
EU	European Union
EurOtop	An assessment manual for Wave Overtopping of Sea Defences and Related Structures
FDGIA	Flood Defence Grant in Aid
FP7	7th Framework Programme of the European Union-funded research
GEV distribution	Generalised extreme value distribution
GCM	General circulation model
GHG	Greenhouse gas
GIA	Glacial isostatic adjustment

GIS	Geographic information systems
HadCM3	Hadley Centre Coupled Model, version 3
HadRM3	Hadley Centre Regional Climate Model, version 3
HAT	Highest astronomical tide
IA	Integrated assessment
iCOASST	Research project 'Integrating Coastal Sediment Systems'
IPCC	Intergovernmental Panel on Climate Change
LAT	Lowest astronomical tide
LiDAR	Light detection and ranging
LISFLOOD-FP	The 2D inundation model originally created by Bates and De Roo (2000) and applied to coastal flood modelling by Bates et al. (2005)
LUC	Land use change
MCA	Multi-criteria analysis
MEECE	Marine Ecosystem Evolution in a Changing Environment
Met Office	UK Meteorological Office
MHWN	Mean high water neap
MHWS	Mean high water spring
MHW	Mean high water
MIKE	Hydro-informatics system developed by DHI
MSL	Mean sea level
MUSCL	Monotonic Upstream-Centered Schemes for Conservation Laws
NAO	North Atlantic Oscillation
NewChan	A finite volume 2D flood model developed by Liang (2008) that employs a Godunov-type scheme to capture different types of shallow-flow hydrodynamics over uneven bed bathymetry
NOC	National Oceanography Centre (UK)
NNDC	North Norfolk District Council
ODN	Ordnance Datum Newlyn
OpenMI	Open Modelling Interface (http://www.openmi.org/)
POLCOMS	Proudman Oceanographic Laboratory Coastal Ocean Modelling System
POLCS3	POLCOMS on CS3 grid
RCM	Regional climate model
SAC	Special Areas of Conservation
SCAPE	Soft Cliff And Platform Evolution modelling tool (Walkden and Hall 2005)
SISYPHE	Near-bed mass sediment conservation solver
SLP	Sea-level pressure
SLR	Sea-level rise
SMP	Shoreline Management Plan
SMP6	Shoreline Management Plan 6 (Weybourne to Lowestoft)
SPA	Special Protection Areas

SRES	Special Report on Emissions Scenarios
SSSI	Special Sites of Scientific Interest
SWAN	Wave action conservation equation solver
SWH	Significant wave height
TAW	The Technical Advisory Committee for Flood Defence in the Netherlands
TE2100	The Thames Estuary 2100 Project
TELEMAC	Hydro-informatics system developed by EDF
TELEMAC-2D	Depth-averaged version of TELEMAC
TOMAWAC	Wave action conservation equation solver
UKCIP02	UK Climate Impact Projections 2002
UKCP09	UK Climate Projections 2002
WAM	Wave Model, referring to the first third-generation wave model
WAMCS3	WAM on CS3 grid

References

Bates, P. D., & De Roo, A. P. J. (2000). A simple raster-based model for floodplain inundation. *Journal of Hydrology*, *236*, 54–77.

Bates, P. D., Dawson, R. J., Hall, J. W., Horritt, M. S., Nicholls, R. J., Wicks, J., & Hassane, M. A. A. M. (2005). Simplified two-dimensional modelling of coastal flooding for risk assessment and planning. *Coastal Engineering*, *52*(9), 793–810.

Liang, Q. (2008). Simulation of shallow flows in nonuniform open channels. *Journal of Fluids Engineering*, *130*(1), 011205. Doi:10.1115/1.2829593.

Maxim, L., Spangenberg, J., & O'Connor, M. (2009). The DPSIR framework for biodiversity assessment. *Ecological Economics*, *69*(1), 12–23.

Stive, M. J. F. Z. B., Wang, M., Capobianco, P. R., & Buijsman, M. C. (1998). Morphodynamics of a tidal lagoon and the adjacent coast. In J. Dronkers, & M. Scheffers (Eds), *Physics of estuaries and coastal seas* (pp. 397–407). Rotterdam: Balkema.

Walkden, M. J. A., & Hall, J. W. (2005). A predictive mesoscale model of the erosion and profile development of soft rock shores. *Coastal Engineering, 52*, 535–563.

Chapter 1
The Challenge for Coastal Management During the Third Millennium

Richard J. Dawson, Robert J. Nicholls, and Sophie A. Day (née Nicholson-Cole)

Abstract Coastal planners and managers face a wide range of challenges around the world during the twenty-first century. These include geomorphological, climatic, and socio-economic drivers of change, their interaction and the societal and governance issues that they raised. The interplay between these challenges motivated the Tyndall Centre for Climate Research's Coastal Research Programme.

Assessing the impacts of multiple scale drivers and possible responses is problematic, yet failure to do so can limit the utility of the analysis – or lead to undesirable outcomes that are a consequence of thinking too narrowly about a problem. To define and analyse coastal problems in a comprehensive manner, an interdisciplinary team of researchers was assembled. Embracing natural, social and engineering sciences, this team engaged in the development of an integrated assessment, called the Tyndall Coastal Simulator, which was applied and demonstrated in East Anglia in the UK. However, the approach could be applied widely, as discussed at the end of the book.

The Tyndall Coastal Simulator provides a platform to integrate the diverse knowledge and methods developed as part of the integrated assessment process in a meaningful and accessible way. This chapter reviews the need for the Tyndall Coastal Simulator, outlines the main aims of the simulator work and defines its unique contribution to broadscale coastal simulation. It gives an overview of the integrated assessment structure, scenario framework and case study locations adopted in this work, especially North Norfolk. The study site was chosen as an exemplar of the challenges facing coastal stakeholders due to its long history of erosion and flooding and the fact that it is going through a major transition in coastal management strategy. Under this strategy, a number of currently protected clifftop communities will lose their defences causing widespread concern. Hence, this

R.J. Dawson (✉)
School of Civil Engineering and Geosciences, Tyndall Centre for Climate Change Research, Newcastle University, Newcastle upon Tyne, NE1 7RU, UK
e-mail: richard.dawson@newcastle.ac.uk

R.J. Nicholls • S.A. Day (née Nicholson-Cole)
Tyndall Centre for Climate Change Research, Faculty of Engineering and the Environment, University of Southampton, Southampton SO17 1BJ, UK
e-mail: r.j.nicholls@soton.ac.uk; day_soph@yahoo.co.uk

R.J. Nicholls et al. (eds.), *Broad Scale Coastal Simulation*, Advances in Global Change Research 49, DOI 10.1007/978-94-007-5258-0_1

provides a good study site to develop transferable lessons on the analysis of coastal change and hazards, as well as the issue of managing transitions which will be essential under climate change.

This book builds upon, and significantly extends, work reported in an earlier paper by Dawson et al. (Climatic Change 95:249–288, 2009), included as an Appendix to this chapter, by providing a complete record of methods, results and analysis from the Tyndall Coastal Simulator as well as reflections on broadscale coastal simulation from British and International practitioners and researchers on this 10-year-research effort.

Keywords Integrated assessment • Coastal change • Shoreline management • North Norfolk • Tyndall Coastal Simulator

1.1 Introduction

1.1.1 Background

This book presents a comprehensive description of an analysis of the coastal zone at a scale suitable to support strategic coastal management. It is distinctive because it looks over broad spatial scales and extended timescales and considers multiple and sometimes competing coastal priorities and drivers. The book demonstrates the first quantification of the relationship between erosion risks and flood risks. The chapters in this book demonstrate the application of a range of relevant methods using real case studies in England and synthesise the lessons learnt from this work at both national and international levels, drawing on the expertise of scientists and policymakers working in the UK and abroad (Box 1.1).

Box 1.1 What Does This Book Offer?

- Understanding the implications of global drivers of change for local coastal communities.
- An analysis of the relationships between different dimensions of coastal change, integrating across natural, engineering and social sciences.
- Presentation of a methodology for coastal integrated assessment – each chapter provides an additional contribution to the overall coastal integrated assessment.
- National and international reflection of lessons learnt, transferability and a research agenda for integrated assessment in coastal areas.

Long-term management of the coastal zone is challenging for a number of reasons. Coastal zones attract settlements, are ideal for a range of economic activities and contain important natural habitats that provide a range of ecosystem services. As sea-level rise and possibly intensified storms could increase the incidence of flooding and erosion and degrade coastal ecosystems, all these activities are vulnerable to climatic and other pressures unless appropriate coastal management policies are implemented. Change in the coastal zone is fraught with uncertainty and requires understanding long-term climatic, socio-economic and marine processes. These play out over a range of scales, with global drivers of change interacting with local activities. It is essential to assess this uncertainty in order to improve understanding of impending choices. Hence, it supports informed long-term coastal management policies that can adapt to growing observations and understanding, as well as evolving societal priorities. Moreover, coastal systems are highly interconnected, and any changes in one area or sector may influence impacts elsewhere. In complex situations such as this, an integrated assessment (IA) is required to understand the impacts of change and the implications of different management choices.

The Tyndall Coastal Simulator provides regional impact assessments of climate and socio-economic futures under various management options, with a regional application in the coastal zone of Norfolk, East Anglia, which quantifies for the first time how coastal erosion and coastal flood risk are strongly linked. This has important implications for shoreline management planning both in the UK and more widely. The book covers the development and application of a framework for coastal modelling (including coupling a range of models to describe various processes such as morphodynamics, surge, erosion, flooding and land use change) and visualisation techniques and interface development for the results. It acts as a step-by-step guide through the process of integrated assessment of coastal areas at a broad scale suitable for strategic coastal management and demonstrates the application of these methods using real case studies. It also considers the social science dimension and documents the process of engagement with a range of relevant stakeholders, institutions and the public who are responsible for and affected by coastal management decisions. This stakeholder dimension is an important element of the research and the final methods. The specific research content of this book is set in a broader UK and international context to illustrate the wide-ranging application of the methods used and lessons learned.

This book describes research undertaken by the Tyndall Centre for Climate Change Research on their Coastal Simulator from 2000 to 2012, as part of a major programme of work concerned with sustainable coasts (see Box 1.2). There were two phases of this work: Phase 1 from 2000 to 2006 and Phase 2 after 2006. Whereas Phase 1 necessarily focused on developing an understanding of coastal integrated assessment through the development of interdisciplinary partnerships, Phase 2 was able to extend the depth and breadth of the assessment. Where necessary, these phases are distinguished within the text.

Box 1.2 Emergence of the Tyndall Centre Coastal Programme

In 2000, in recognition of the need for more integrated responses to climate change, UK research councils funded an interdisciplinary research consortium, the Tyndall Centre for Climate Change Research, comprising a number of the UK's leading climate research institutions. This centre was funded by UK research councils until 2010 and continues today. When the Tyndall Centre for Climate Change Research was established, the importance of coastal issues in the context of responding to climate change was recognised in a Coastal Research Programme.[1] This research addressed these challenges at a scale that is relevant to coastal management, including considering the interrelated issues of erosion, flood risk management and habitat change, as well as linking this to stakeholder engagement and coastal governance implications.[2] While the starting point was climate change and sea-level rise, it rapidly became apparent that all the relevant drivers of coastal change and risk need to be considered to produce analyses of interest and relevance to coastal managers and policymakers. Further, integration across these different knowledge domains represents a major challenge, and integration has to be built into the process from the beginning if it is to be achieved. This book is based on this research into integrated assessment and its application to coasts.

[1] This continues as the Cities and Coasts theme of the Tyndall Centre (http://www.tyndall.ac.uk/research/cities-and-coasts).

[2] Other important coastal management issues such as water quality are pertinent to the UK and reflected on in a broader context in Chap. 14.

1.1.2 Overview of Book Structure

Figure 1.1 provides an overview of the book structure in the context of the Tyndall Coastal Simulator. This chapter begins by considering the need for developing integrated responses to coastal management in the third millennium. Subsequently, a case study site in North Norfolk on the East of England, which is used throughout the book, is introduced before outlining the conceptual framework for our IA application in North Norfolk – the Tyndall Coastal Simulator. The North Norfolk case study provides a common demonstration site for individual components and crucially acts as a unifying real world case study. The study site was chosen as an exemplar of the challenges facing coastal stakeholders due to its long history of erosion and flooding and the fact that it is going through a major transition in coastal management strategy as part of Shoreline Management Planning (Sect. 1.3).

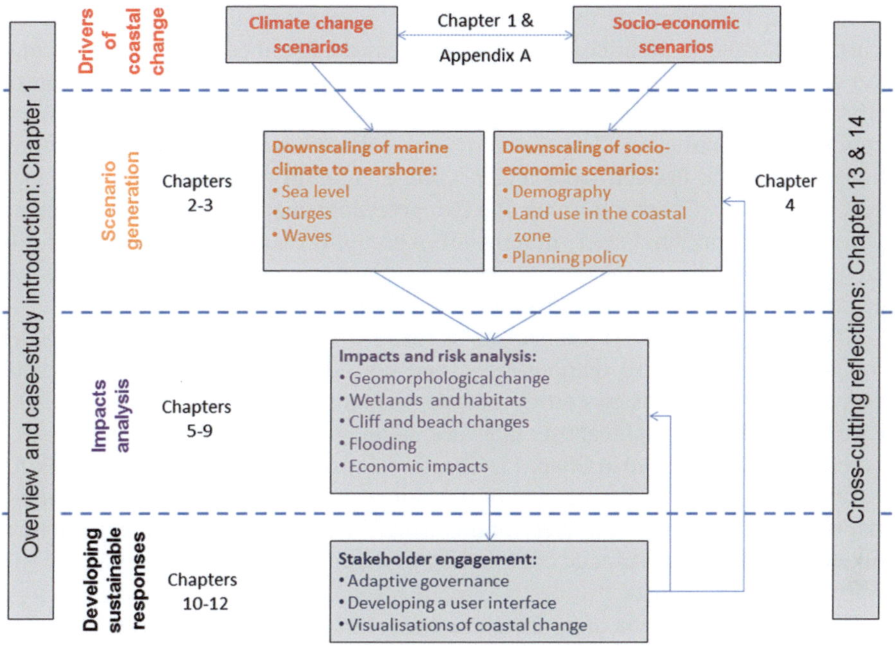

Fig. 1.1 The structure of this book

Under current proposals, many clifftop communities with coastal defences will not have them replaced as they deteriorate. At Happisburgh, a village in North Norfolk, the defences have already been removed. This is causing an ongoing concern and debate in the region. Hence, North Norfolk provides a good study site to develop transferable lessons on the analysis of coastal change and hazards, as well as the issue of managing transitions which will be essential under climate and other pervasive environmental changes.

Following this introduction, Chap. 2 describes how global changes in marine climate may alter future sea level, storm surge and offshore wave conditions around the British Isles and especially North Norfolk. These changes in offshore conditions are used to drive a wave transformation model, described in Chap. 3, to understand the impact of these changes on storm surges and waves at the shoreline in North Norfolk. In both these chapters, sensitivities to key uncertainties such as climate model parameterization and the potential evolution of offshore sandbanks are explored. While Chaps. 2 and 3 downscale climate drivers to the spatial scale of the North Norfolk case study, Chap. 4 quantifies the effects of global and local socio-economic drivers on changes to residential and nonresidential development in East Anglia, with a focus on the coastal zone.

Chapters 5, 6, 7, 8, and 9 consider a range of impacts and risks associated with the climatic and socio-economic changes considered above. Chapter 5 uses an innovative behavioural systems approach to elicit qualitative insights about geomorphic behaviour of the North Norfolk case study site and neighbouring coastline. Chapter 6 describes a method to assess the likelihood of habitat degradation and loss as a result of coastal change. Long-term changes to the position of cliffs and beach volumes, under different climate scenarios and management regimes, are quantified in Chap. 7 while Chap. 8 describes the inundation modelling used to determine flooded areas under a range of storm surge events and defence breach scenarios. Chapter 9 brings the two preceding chapters together to evaluate future erosion and flood risks, in terms of expected economic damages, under different climatic and socio-economic scenarios. Furthermore, the effects of different management policies on these risks are compared.

Chapters 10, 11, and 12 reflect on a range of issues relevant to stakeholder engagement. Chapter 10 demonstrates, and considers the utility of, a range of visualisation techniques to convey coastal change. The process of developing a bespoke GIS-based graphical user interface, to integrate the results and engage with stakeholders, is described in Chap. 11. Chapter 12 considers current decision-making processes and how the management of transitions in the coastal zone could be more adaptive, equitable and effectively implemented. This chapter also highlights how stakeholders provided a crucial contribution to shaping the research programme and contributing to its impact.

Chapter 13 draws upon input from international authors to reflect on this body of research and consider the potential for broadscale coastal simulation around the world and the transferability of the Tyndall Coastal Simulator. Chapter 14 provides a summary of the key findings of the work, proposes a generic framework for integrated assessment in the coastal zone, including some key issues to consider, and outlines a manifesto for future research and application.

1.2 Coastal Management in the Third Millennium: The Need for an Integrated Response

Coastal management in the third millennium is increasingly characterised by large-scale, long-term policy conundrums. These usually involve multiple, often additive, risks and uncertainties and frequently relate to the prioritisation of economic, social, political or environmental objectives that can clash with differing public values. This standoff can paralyse the decision-making process. Analysis of the drivers and processes of change, and their associated uncertainties, requires an integrated approach to understand how risks interact and evolve through time. Furthermore, it is essential to bring different stakeholders together in order to develop a collective understanding of these processes and their consequences of long-term change as well as identify and reach a collective view with regard to suitable options for coastal management.

1.2.1 Drivers of Coastal Change

Interest and concern about the future of coastal areas is widespread (Valiela 2006). Coastal areas are a major focus for humanity, often with high population densities, growing urban areas and economic importance (Dasgupta et al. 2009; Lichter et al. 2011; Foresight 2011). They also host important coastal habitats and deliver a range of ecosystem services (Agardy et al. 2005; Nicholls et al. 2007). Climate change, and especially sea-level rise through the twenty-first century, is one major source of concern, but other drivers of change such as socio-economic change (e.g. urbanisation) or modifications of sediment budgets by various means are also important (Crossland et al. 2005; Wong et al. 2014). The major issues at the global scale are summarised in Box 1.3.

About 10 % of the total global population live within 10-metre elevation of sea level (Lichter et al. 2011). This is forecast to continue to grow rapidly through the twenty-first century mainly in coastal urban areas – of the world's 19 megacities (with a population of over ten million), 14 are situated in the coastal zone (UN Habitat 2008). Across the 136 large port cities considered by Nicholls et al. (2008a), the collective 2005 population of those cities could grow to 1.2 billion by the 2070s and the number of coastal megacities is estimated to increase from 14 to 31. Urbanisation is also occurring widely in the smaller coastal cities.

Concern about coastal areas is widespread and expressed by the activities of many international organisations. The Intergovernmental Panel on Climate Change (IPCC) has typically included a coastal assessment chapter in its Working Group II reports to date (Nicholls et al. 2007). The United Nations Environment Programme and the United Nations Development Programme are both focusing on coastal adaptation to climate change. The Intergovernmental Oceanographic Commission (IOC)

Box 1.3 Key Coastal Trends at the Global Scale

- Population
  - Growing coastal population (double global trends)
  - Urbanising coastal zone (new residents are urban)
  - Increasing tourism, recreation and retirement
- Subsiding, densely populated deltas, especially in urban areas
- Globalisation of trade and international shipping routes
- Degrading coastal habitats and declining ecosystem services' cause
- Increasingly costly coastal disasters
- Climate change and sea-level rise
- A reactive approach to adaptation

is conducting extensive work and guidance on responding to climatic and geologic coastal hazards. The International Geosphere-Biosphere Programme Land-Ocean Interactions in the Coastal Zone (LOICZ) is a dedicated international science project looking at coastal zones. The World Bank and regional development banks are also increasingly considering coastal issues including assessments of the Economics of Adaptation to Climate Change and the future of coastal cities in Asia (e.g. World Bank 2010; Westphal et al. 2013).

1.2.2 Adapting to Coastal Change

Adaptation to environmental change is not a new phenomenon. Throughout history, people and societies have adapted their environments to cope with change and reduce environmental risks (van Koningsveld et al. 2008). Accordingly, when coastal changes have adverse consequences, human beings will adapt to those following a range of possible strategies.

As Klein et al. (2001) note, adaptation may take place in both planned and autonomous contexts. The former is a result of deliberate policy decisions and based on an awareness that coastal conditions have changed or are about to change and that action is required to return to or maintain a desired state, while the latter does not constitute a policy response to coastal stimuli, but is triggered by spontaneous changes in natural and human systems such as market-induced responses. Planned adaptation can be classified in a number of ways (e.g. Adger et al. 2007; Ford et al. 2013). The widely followed approach developed by the Intergovernmental Panel on Climate Change (IPCC) is focused on the adaptation measures themselves (e.g. Wong et al. 2014):

- *(Planned) Retreat* – all natural processes are allowed to occur, and human impacts are minimised by pulling back from the coast via land use planning, development controls etc.
- *Accommodation* – all natural system effects are allowed to occur, and human impacts are minimised by adjusting human use of the coastal zone via flood resilience measures, warning systems, insurance (providing the capacity to repair and rebuild) etc.
- *Protection* – natural system effects are controlled by soft or hard engineering (e.g. seawalls, breakwaters, nourished beaches and dunes), reducing human impacts in the zone that would be impacted without protection.

A critical and overarching challenge of climate change is how and when to adapt in the face of scientific evidence: What are the thresholds for action? How are the impacts and consequences valued? And how should adaptation be implemented (Adger et al. 2009)? The adaptation challenge also requires not just incremental but also transformational changes (Moser et al. 2012). Planning for this kind of change presents a significant challenge (e.g. because of uncertainties about climate change risks and adaptation benefits) and is further discussed in Chap. 14. Kates et al. (2012) suggest that to date adaptation to human-induced climate change has largely been incremental. In some places and for some systems, however, future vulnerabilities and risks may be so significant that they require transformational rather than incremental adaptation.

Throughout most of the twentieth century, coastal adaptation in the UK comprised a choice between protection and no action of any kind. Over time, the length of defended coast increased, especially in response to extreme events such as the 1953 storm surge, after which most of the current defences in Norfolk were constructed (see e.g. Fig. 14.3a–c). This response was rather piecemeal and ad hoc and the indirect effects of protection were ignored. For many decades, it seemed the UK was on a course toward the universal protection of its shoreline. Over the last two decades, coastal governance in the UK has evolved significantly with the development of more strategic assessments such as Shoreline Management Plans (SMPs, explained further in Sect. 1.3), structured flood-risk analyses (e.g. Hall et al. 2003; Thorne et al. 2007) and the development of improved flooding guidance for land use planning (e.g. UK Planning Policy Statement 25). A big change during this time is a wider acceptance that universal protection is neither economically feasible nor environmentally desirable in the long term. However, this remains controversial and the shift in thinking is not yet accepted by all.

This book recognises that adaptation to coastal change is more than a set of technical measures. It is a process that involves awareness, design and planning, implementation and monitoring and evaluation (Klein et al. 2001; Dawson 2007; Hay 2009; Linham and Nicholls 2010). In practice, adaptation is a governance issue involving processes through which environment and society interact (Adger et al. 2009). Regardless of the location, a vast number of different interests are at stake when the future of the coast is being considered. Knowing what or who these are and conducting inclusive processes in the planning and management of future coasts are major challenges. One aspect of the Tyndall Centre Coastal Programme's research looked specifically at the challenges of adaptive governance for the coastal zone (Chap. 12).

1.2.3 The Need for Integration

As already noted, coasts are subjected to a range of climatic and non-climatic drivers, onshore and offshore, that operate at local to global scales (Sect. 1.2.1). Changes to climatic conditions can alter the exposure of the coast to extreme weather events. Modifications to land use alter exposure to erosion and flooding and may directly and indirectly influence coastal habitats and ecosystems. In particular, urban development is important as this not only increases our exposure to damage and hence raises risks, but also influences the politics of the issue and the outcome of decision analysis tools such as cost-benefit analysis. Human intervention on the coast can disrupt coastal processes, in particular modification of sediment transport regimes, for example, as a result of defence works to protect cliffs from erosion, or indirectly by dams on rivers flowing into the sea.

Coastal management requires integrated thinking that encompasses the range of coastal functions and relevant drivers. An integrated systems approach seeks to represent the interactions between different coastal functions and objectives. This integrated approach helps us better understand the legacy of past choices and the potential direct and indirect consequences of decisions and to develop portfolios of measures that aim to address a number of different challenges in a synergistic way. At its most radical, it causes us to consider the potential for fundamental transitions in management approach and delivery and how these might be achieved.

In this book, the focus is on changes that are important in the sites that were specifically considered in the Tyndall research: changing marine climate (sea level, surges and waves), the effect of coastal engineering structures and the expansion of the built environment. This is accomplished via developing an appropriate conceptual framework for analysis that describes how these factors interact and hence defines a series of analytical steps and linkages. This framework, and the associated modelling and analysis, is referred to as an IA (Box 1.4). This is then used to develop appropriate scenarios (or plausible futures) which in turn are used to drive the models to explore and understand the resulting change across a range of uncertainty. Importantly, this exercise was conducted with a distributed model where different analytical components are developed by experts at different universities and laboratories.

The role of scale is critical, e.g. from global (climate mitigation and earth system analysis) to local (coastal adaptation and management). While this book pertains specifically to the UK context and concentrates largely on a particular case study area, the approach has much broader relevance and is transferable as considered in Chaps. 13 and 14. As Misdorp (2011) points out, although specific problem identification in each country or region has to be considered unique, there is considerable commonality in the types of problems encountered and the potential measures to be applied to deliver sustainable coastal management.

Having understood how things might change, possible responses can be considered, including different types of hard and soft protection and changes in management approach, via retreat, such as defence abandonment and managed realignment. The uncertainties surrounding future socio-economic, demographic and climate changes may be large, but by exploring the range of possible futures, it is possible to identify options that are more robust to uncertainties. Typically, this suggests a strategy of adaptive management that starts with a broad trajectory, but the details of how that policy response develops will depend on a combination of observations (monitoring), model outputs and other evidences as society 'learns' more about the future climate change and resulting coastal evolution (e.g. Penning-Rowsell et al. 2013; Ranger et al. 2013).

As the physical drivers of coastal change operate at different scales, so do those that could be considered 'socio-economic'. Past and present decisions about the management and use of the coastal zone – development, defence, land use, land value and so on – all have an impact on the current state of any coastal system and the likely nature of impending coastal change. Further, they often constrain our management choices into the future.

Box 1.4 What Is Integrated Assessment?

As Kelly et al. (2013) note, 'effective environmental management requires an understanding of the interactions between policy choices and complex social, economic, technical and environmental processes and elated aims'. Systems with interacting drivers such as coastal areas have emergent properties (unexpected outcomes and consequences) which are a challenge for policymakers and their expectation of linear systems. Taking an integrated assessment (IA) approach enables us to take a long-term view and reframe the questions that are asked so as to link global, regional and local scales and their interactions in the context of future coastal planning. IA gives a more complete picture about how issues may evolve than is possible when taking a more conventionally sectoral view of the problems associated with a changing coast. Additionally, IA gives us the opportunity to expand our focus beyond the outputs offered by the models incorporated within it, so as to more holistically see how these interact with people and their locality.

However, putting these insights together into an IA that helps to inform decision making is a significant challenge that had defied researchers before the Tyndall Coastal Simulator. There are technical reasons for this as the scales and model elements are not easily linked – their integration needs to be planned, designed and centrally coordinated. The process requires a multidisciplinary team and takes significant time to implement and these both have costs. Moreover, there are practical challenges of interpreting IA model outputs and assimilating their complex model-based evidence into decision-making processes (see Fig. 1.2). However, the benefits of an IA for long-term coastal management, as will be shown in the following chapters, provide new insights and tools for policy analysis that were hitherto unavailable and provide a large potential for improving decision making in the future. Given the complexity of interactions and the large range of possible futures and decision options, it is hard to see how system-scale policy analysis of long-term change could be conducted without the support of computer-based tools.

(continued)

Box 1.4 (continued)

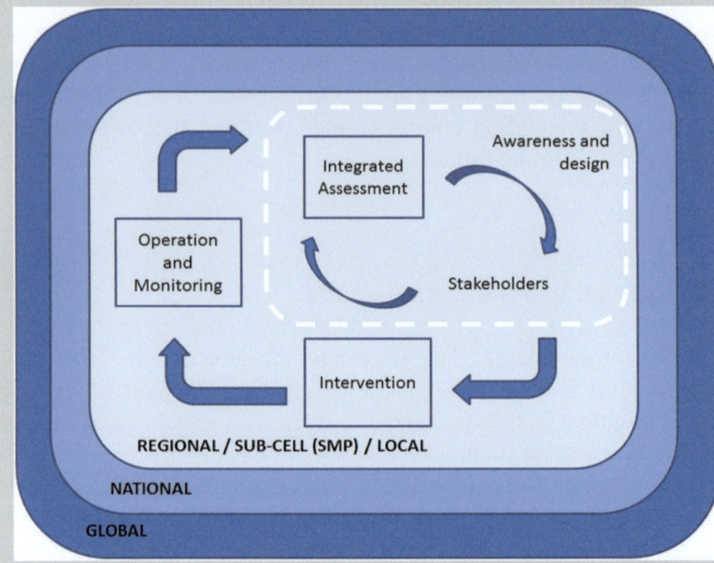

Fig. 1.2 An illustration of the role of the IA process in coastal management, including modelling, monitoring, intervention/implementation and engagement processes

It should be noted that IA is often associated with global assessments, examples of which include the Community Integrated Assessment Model (Warren et al. 2008), although another work (Hall et al. 2009; Walsh et al. 2011) addresses smaller or urban scales. In this book, the focus is on local assessments in a coastal context.

1.3 Coastal Governance Arrangements in England and Wales

The Environment Agency has an overview of the English coastline which enables decisions about coastal management to be made in a joined-up manner (Natural Resources Wales, the Scottish Environmental Protection Agency and the Northern Ireland Environment Agency provide similar roles in Wales, Scotland and Northern Ireland, respectively). In partnership with local authorities and other coastal bodies, the Environment Agency can help provide a consistent approach to managing coastal risks. Coastal groups, including representatives of the Environment Agency and maritime local authorities, prepare *Shoreline Management Plans* (SMPs), which are the principal strategic coastal planning and policy mechanism that exists in England and Wales (Leafe et al. 1998; DEFRA 2011; Nicholls et al. 2013). SMPs undergo public consultation during their preparation.

For the purposes of shoreline management, the England and Wales coast has been divided into lengths which are relatively self-contained as far as movement of beach sediment is concerned (Motyka and Brampton 1993). These boundaries have been reviewed and currently there are 22 SMPs in England and Wales (Fig. 1.3). The Tyndall Coastal Simulator is most concerned with SMP6 which is introduced fully in Sect. 1.4.

Fig. 1.3 A map of the 22 second-generation Shoreline Management Plans around the coast of England and Wales.

SMPs are non-statutory, high-level documents that consider the entire coast. The aim of an SMP is to provide a strategic assessment of the risks associated with coastal processes and present a long-term policy framework to reduce these risks to people and the developed, historic and natural environment in a sustainable manner (DEFRA 2010). They provide estimates of how the coast is likely to change over the next 100 years, taking into account the future implementation of coastal policies, geology, likely impacts of climate change and the existing condition of the coast including coastal defences. Further to providing a strategic direction for coastal management and informing decisions on investment, SMPs also inform regional and local spatial planning strategies (e.g. in England, Local Development Frameworks at District level) to ensure that inappropriate development does not take place in areas that are at risk of flooding or erosion either now or in the future.

In practice, each SMP is further divided into smaller, distinct management units based on land use (and value). Based on the most up to date knowledge available, a strategic management policy for each unit is identified to (1) 'advance the existing defence line', (2) 'hold the existing defence line', (3) 'managed realignment' or (4) 'no active intervention'. These are applied over three timescales, or epochs of 25, 50 and 100 years into the future. Nationally, a reduction in the length of existing coastline to be held at the existing defence line is apparent in later epochs, as this is seen to be the best long-term approach in many locations. The first generation of SMPs were completed in the 1990s. Following a review of their strengths and weaknesses, updated guidance on SMPs has been published (DEFRA 2011) to promote the preparation of the second round SMPs and ensure a more consistent evidence base, engagement of stakeholders throughout the process (but particularly in objective setting and selection of preferred options) and seek adoption of the plans by the relevant authorities.

1.4 North Norfolk – An Exemplar of Coastal Complexity

Our flagship case study site is a 72 km length of coastline and hinterland in North Norfolk, UK (see Figs. 1.4 and 1.5), that corresponds to SMP6 (formerly SMP 3b). Some material on coastal habitats is explored on the Solent coast in Hampshire in Southern England. The Norfolk case study is introduced fully in this section as a coherent reference point for the book.

SMP6 has a long history of erosion and flooding: climate change exacerbates both these risks. As a designated 'coastal subcell', it is reasonably self-contained in terms of beach-grade sediment interactions with neighbouring coasts. The north between Weybourne and Happisburgh comprises retreating chalk and till cliffs, whereas much of the southern area from just south of Happisburgh to Great Yarmouth is low lying and at risk of coastal flooding. In the extensive coastal lowlands of the Norfolk Broads, sea flooding can penetrate 10–20 km inland (Fig. 1.4a).

There are a series of towns and villages along the coast, infrastructure of national significance and important, often designated, ecosystems and habitats. The region has a complex legacy of coastal engineering since the latter half of the nineteenth century. However, most defence construction followed the devastation caused by the North Sea storm surge in 1953 (McRobie et al. 2005; Lumbroso and Vinet 2011). See, for example, Figs. 1.4b and 14.3a–c.

These alterations to the natural coastal system have led to an array of coastal problems, such as declining beach volumes. Long-term viability of these post-1953 defences as they reach the end of their life, and changing coastal management approaches including a contentious set of recommendations (East Anglian Coastal Group 2012) to allow more cliff retreat from Happisburgh to Weybourne, makes the site ideally suited to an IA.

1.4.1 Offshore Bathymetry

The bathymetry associated with SMP6 (Fig. 1.6) has prominent long narrow sandbanks. These sandbanks have been widely surveyed and their mobility has been studied (Reeve et al. 2001; Park and Vincent 2007; Horrillo-Caraballo and Reeve 2008). Studies suggest that the residual sediment pathways around the sandbanks located off Great Yarmouth show a complex pattern with the seaward side of the banks having a northward transport and vice versa (Park and Vincent 2007; Horrillo-Caraballo and Reeve 2008). Clockwise residual circulation has been observed for the offshore sandbanks that are not connected to the shore (Collins et al. 1995), with bottom friction and Coriolis acceleration responsible for this circulation (Huthnance 1982). However, the interconnection between the sandbanks, as well as the connection to the shore, remains poorly understood.

Fig. 1.4 (**a**) The Norfolk Broads, with the North Sea in the background. (**b**) Sea Palling and the off-shore reefs which are designed to prevent breaching and resulting flooding of the Norfolk Broads

Fig. 1.4 (continued)

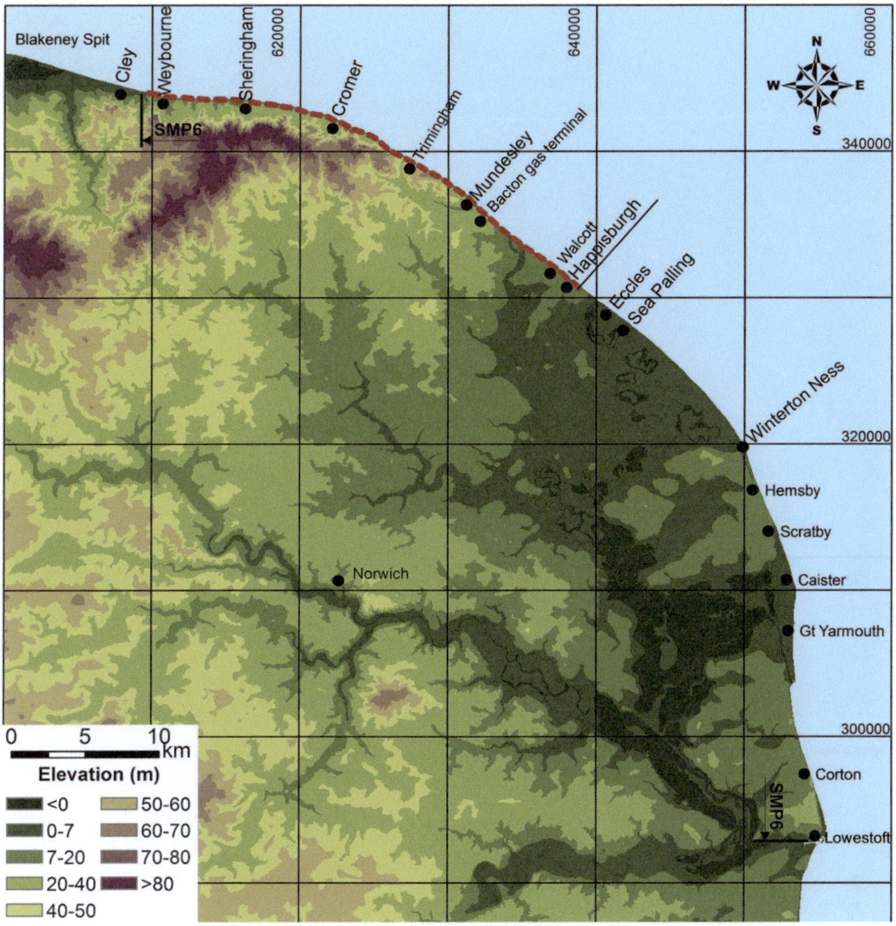

Fig. 1.5 Location of the study area (SMP6) in Eastern England, showing the location of the major settlements and land elevation. The area of major cliff erosion is indicated by the *dotted line*

1.4.2 Current Marine Climate

Tidal range for spring tides varies from mesotidal (3 m range) in the north and reduces to microtidal (1.6 m range) in the southern part of the domain. Large surges exceeding 2 m occur. This happened on 31 January/1 February 1953 in the infamous '1953 storm', and most recently, on 5/6 December 2013 causing some flooding in SMP6 and evacuation from coastal floodplains in Lowestoft and Great Yarmouth. However, due to the better defences and warning systems, impacts were lower in 2013 than in 1953. Mean relative sea levels at Lowestoft have risen at 2.24 ± 0.25 mm/year since 1957 (Haigh et al. 2009).

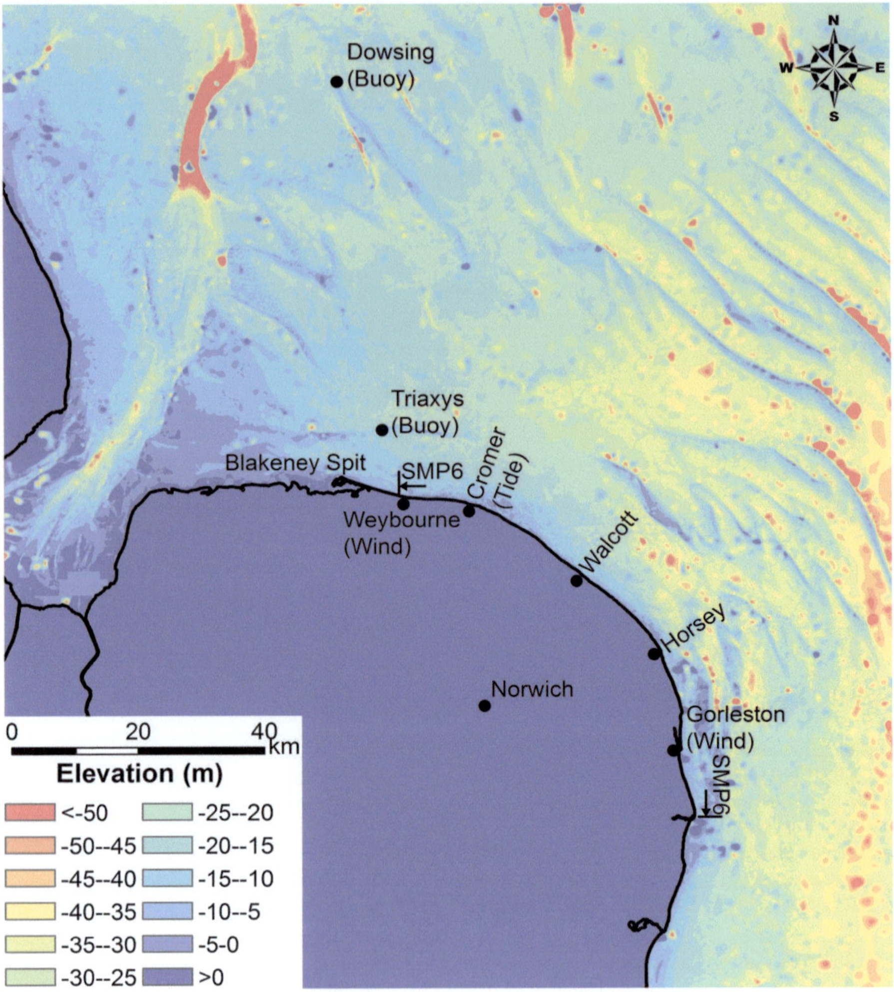

Fig. 1.6 Wave transformation modelling domain showing the bathymetry (metres below Ordnance Datum) off the East Anglian coastline

The coast is exposed to waves generated within the North Sea from directions between approximately 330°N and 120°N, but the largest waves arrive from the north through to the east where fetch lengths exceed 500 km. Monitoring data shows the waves are relatively small on average, the mean observed value being 1 m, with the most frequent waves coming from the northwest. The tidal sandbanks (Fig. 1.6) induce a loss of wave energy due to bottom friction and wave breaking (see Chap. 3). A spectral analysis of the measured significant wave height reveals a semidiurnal modulation with amplitude of 0.14 m. Previous numerical analysis showed that local wind speeds of less than 10 m/s do not influence the significant wave height (Kuang and Stansby 2004). In the case of more extreme winds (20 m/s), the significant wave height can be increased along the coast by about 0.6 m.

1.4.3 Coastal Geomorphology

The east coast of England has a dynamic history, as described in Box 1.5. In North Norfolk, between Weybourne and Cromer, there is a transition in cliff geology from predominantly chalk to predominantly till. Coincidentally, in this general area, there is also a divide in the direction of longshore sediment transport (Vincent 1979; Clayton 1989) and the beach sediments (Chang and Evans 1992). Beaches in the South of Cromer beaches are predominantly sandy, with a southerly littoral drift. Dunes are developed in front of the Norfolk Broads, and there is a large but poorly understood offshore movement of sand from the shore to the nearest bank at Winterton Ness. West of Cromer, the beaches have an increasing proportion of pebbles, and west of Weybourne, the beaches are composed largely of pebbles, ultimately forming the major shingle spit at Blakeney. Clayton (1989) estimated that the glacial till cliffs have retreated at rates averaging around 1 m/year over some 5,000 years (or roughly 5 km in total), whereas the chalk cliffs have retreated somewhat slower.

Box 1.5 The Ever-Changing East Coast of England

The coastline of the East of England has been dynamic for centuries. Hence, the threat of losing a settlement is not new in Norfolk, or elsewhere – although our capacity to interfere and engineer the coast has changed, as has our ability to understand these processes of change. Just off the coast at Cromer is the submerged village of Shipden which was washed away during the fourteenth century – in 1888, a tug reportedly struck the long since submerged church tower of St. Peter's. Further south lies the lost village of Eccles-on-Sea, which spreads over of 8 km^2 at its peak but is now a few hundred properties. Much of the village was washed away during a violent storm in 1604, and in 1895, the church tower slipped onto the beach and was gradually destroyed. Other lost villages on this coast include: Clare (near Mundesley), Keswick (near Bacton), Whimpwell (near Happisburgh) and Waxham Parva (near Horsey). Although erosion at Pakefield, immediately south of Lowestoft, has exposed evidence of the oldest known human settlement in Great Britain.

Just south of the map in Fig. 1.6, the town of Dunwich in Suffolk was one of England's most important settlements in the thirteenth century, with a population half the size of London. Its decline began in the fourteenth century when a huge storm in 1328 shifted the shingle of the seabed, changing the current and blocking the harbour, as well as destroying many houses, churches and windmills. Subsequent storms, loss of trade and steady recession of the coastline increased the rate of Dunwich's decline. Dunwich is now a small village with a population of 84 in the 2001 census.

Further north, the Holderness coast in Yorkshire has also seen many changes. Figure 1.7 shows the land and many settlements that have been lost since Roman times. More recently, land has been reclaimed within the estuary of the River Humber, but rapid erosion of the open coast continues.

(continued)

Box 1.5 (continued)

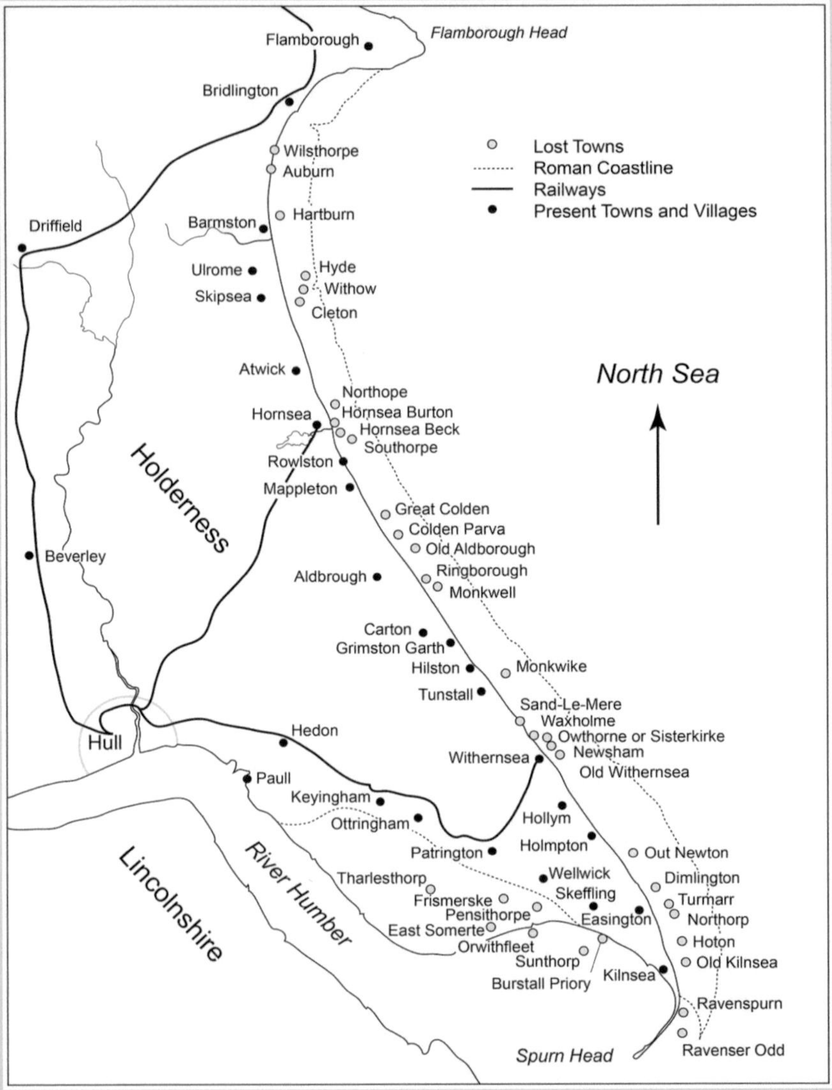

Fig. 1.7 Lost towns of the Holderness coast (Adapted from Sheppard 1912, title page)

1.4.4 Key Vulnerabilities

There are currently almost 20,000 properties at risk of coastal flooding between Weybourne and Lowestoft, over 3,000 of which are nonresidential (Fig. 1.8a). The majority of the people live in the Lowestoft and Great Yarmouth area. There are many beaches and tourist resorts along the Norfolk coast, with access to the shore being of major importance for the region's economy. Although still important, tourism is in relative decline, and retirement is a common reason to move to the coast (see Chap. 4). As with most coastal towns in the UK, deprivation is an issue (Humby 2013). Key infrastructure along the coast includes the Bacton Gas Terminal, one of

Fig. 1.8 SMP6. (**a**) Location of settlements, major transport infrastructure and the one in the 200-year-indicative floodplain (Provided by the Environment Agency of England and Wales). (**b**) Agricultural land quality

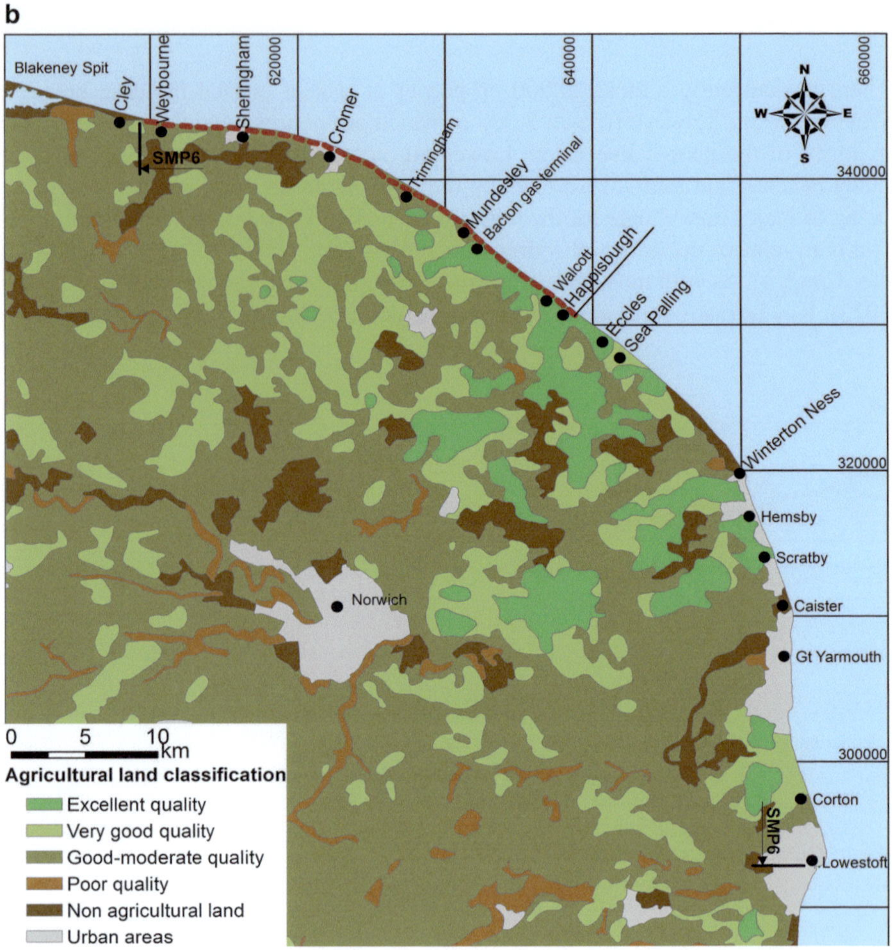

Fig. 1.8 (continued)

the three gas terminals in the UK receiving gas from the North Sea, and the ports at Great Yarmouth and Lowestoft.

Almost 50 % of the floodplain, which is almost 340 km^2, is premium agricultural land, with only 5 % classified as less productive agricultural land (Fig. 1.8b). Some 35 % is considered environmentally sensitive (with Sites of Special Scientific Interest, Ramsar sites and National Nature Reserve designations), including the Norfolk Broads – a National Park and the UK's largest nationally protected wetland, recognising its wildlife and distinct landscape character. There are sites of archaeological significance along the coastline, and recently, the oldest human footprints outside of Africa were discovered at Happisburgh. Along the 32 km stretch of cliffs, there are almost 1,400 properties within 100 m of the clifftop. Although narrow, over a third of this area is premium agricultural land.

1.4.5 Legacy of Erosion, Flooding and Coastal Engineering

A full timeline for engineering works along the cliffed coastline is shown in Fig. 7.2 (Chap. 7). The major settlements of Sheringham, Cromer and Mundesley have been protected by groynes and seawalls since the latter part of the nineteenth century. Overstrand was also protected, but did not have a seawall until the early twentieth century. South of Mundesley toward Happisburgh most major coastal engineering works are groynes and palisades that were installed between 1950 and 2000, many of which are now nearing the end of their effective life. However, it was deemed uneconomic to renew the defences at Happisburgh which fell into disrepair following storms in 1991 and 1996, and after becoming a threat to public safety, these were removed. The shoreline response to the loss of the defences was rapid with retreat rates of up to 10 m/year and the loss of 26 houses from 1992 to 2007 and evacuation of other property. However, these retreat rates are now significantly lower.

South of Happisburgh are large areas of low-lying land in the Norfolk Broads (from Eccles to Winterton, e.g. Fig. 1.4a) and smaller areas at Great Yarmouth and Lowestoft (Figs. 1.4 and 1.8a). Much of this low-lying hinterland is reclaimed marshland and is below normal spring high tides and therefore highly susceptible to flooding. Extensive flooding of the Broads area within SMP6 has occurred a number of times over the last 100 years, most notably in 1907, 1938, 1953 and 1978 (Mosby 1939; Steers 1953; Grieve 1959; Steers et al. 1979; Harland and Harland 1980): in the 1953 flood 17 people died in SMP6: seven in Sea Palling in the Norfolk Broads and ten people in Great Yarmouth (Kelman 2009). High sea levels on 5–6 December 2013 produced by Cyclone Xaver threatened a major flood in SMP6, but a combination of evacuation, improved defences and warning systems prevented large-scale impacts.

An extensive system of dunes has been reinforced by construction of seawalls and groynes (e.g. see Fig. 14.2a–c). In the 1990s, the progressive lowering of the beaches in front of these structures near Sea Palling, and consequent increase in flood risk, was addressed by the construction of nine detached breakwaters and extensive and ongoing beach nourishment (Thomalla and Vincent 2003) (see Fig. 1.4b). The dunes southeast of Happisburgh are protected by concrete seawalls with a combined height in excess of 10 m along most of this stretch of coastline. Further south at Winterton the beaches are wider with timber groynes, and there is a more extensive dune system, some of which is reinforced with a concrete wall. In the December 2013 storm, some chalets were lost along the unprotected section at Hemsby (Fig. 1.9). Further south again at Great Yarmouth, areas of land are sufficiently low to have been inundated during previous extreme events, such as the 1953 flood. Failure through undermining or wave overtopping of the structures protecting the promenade of Great Yarmouth is unlikely to lead to large-scale inundation due to the relative level of the immediate hinterland. However, wave overtopping and overflow of the harbour walls at Great Yarmouth are more likely, and extreme sea levels in the tidal flats of Breydon Water could result in inundation in areas of Great Yarmouth, such as Southtown (as occurred in 1953). Between Great Yarmouth and

Fig. 1.9 Chalets lost at Hemsby, an undefended stretch of the Norfolk coast in December 2013 following a tidal surge

Lowestoft, there are small protected areas at risk of flooding. As at Great Yarmouth, there is potential for quayside flooding in Lowestoft.

1.4.6 Drivers of Change for the North Norfolk Coastline

Climatic and socio-economic drivers are altering many of the processes described in Sects. 1.4.1, 1.4.2, 1.4.3, 1.4.4, 1.4.5 and 1.4.6. Our analysis considered changes due to:

- *Climate* – Changes to climatic conditions are likely to alter rates of erosion and sediment transport. Here, the UK Climate Projections were used to consider changes in sea level, including a possible acceleration of the current rate of sea-level rise. The UKCIP02 (Hulme et al. 2002) and UKCP09 (Fig. 1.10) climate projections (Jenkins et al. 2009) were applied in Phases 1 and 2, respectively. Sensitivity tests are used to explore possible changes to wind and wave conditions as well as possible changes to offshore bathymetry.
- *Demography and economy* – Changes to socio-economic conditions alter vulnerability to coastal risks. Alternative rates of socio-economic change associated with the Foresight Futures (DTI 2002) have been used to consider different patterns of residential and nonresidential development associated with a range of spatial planning policies.

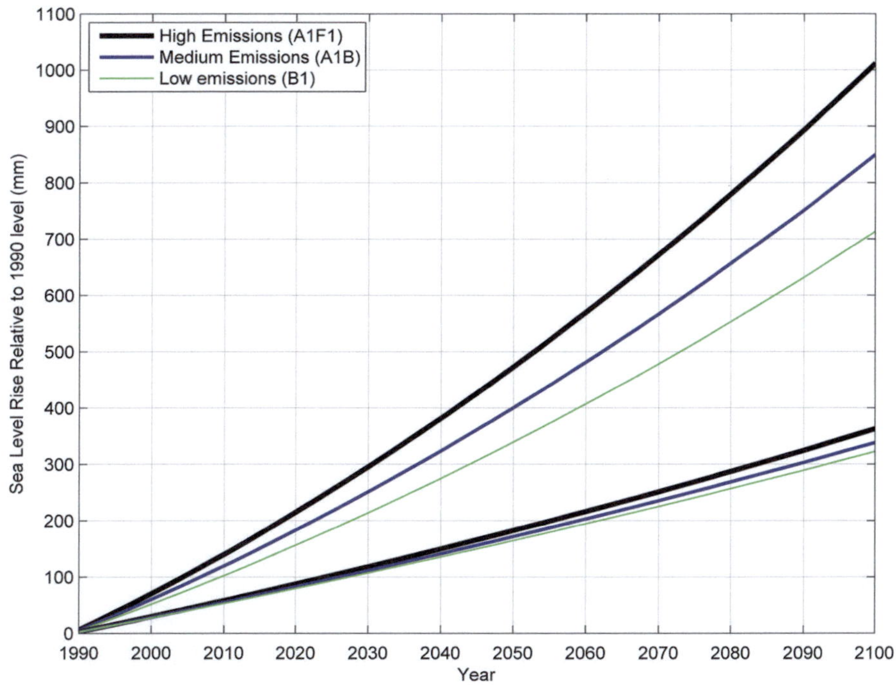

Fig. 1.10 The 5th and 95th percentiles change to mean sea-level rise associated with three global greenhouse gas emission scenarios (From Jenkins et al. 2009)

- *Shoreline management* – Changes to shoreline management policy have inevitable consequences for land owners and communities living in the coastal zone. In addition to considering current conditions, the implications of the SMP proposals and a range of options from the installation of a fully fortified coastline to a completely natural configuration with no structures were analysed, as well as beach nourishment options.

These drivers cannot be considered in isolation as they interact over wide spatial scales and extended timeframes. Further development and sea-level rise in particular will exacerbate existing pressures.

In the face of the uncertainties over the rate and magnitude of change, a number of possible climatic and socio-economic scenarios were tested to gain a broad appreciation of the scale of future flood and erosion risks that may need to be addressed (see Box 1.6). The effectiveness of different coastal management policies at managing these risks is tested in the context of these scenarios. The scenarios are fully described, and relevant variables are quantified in Appendix A.

The policy process of most importance was the implementation of Shoreline Management Plans. In 1996, the North Norfolk District Council adopted the first-generation SMP from Sheringham to Lowestoft, which covers the case study area.

Box 1.6 Using Scenarios to Explore the Future

The future is uncertain and hence analysis needs to explore this uncertainty. The adoption of 'scenarios' is a useful tool for dealing with long-term uncertainties in complex systems by exploring different pathways for how the future may unfold. The approach can take many forms, but the core principle of a scenario is a narrative storyline: this provides a qualitative description of the hypothetical, consistent future evolution of relevant social and environmental issues, sectors and drivers such as future greenhouse gas emissions (to drive climate models) (Nakićenović and Swart 2000), environmental change (Rounsevell and Metzger 2010), or coastal change (Nicholls et al. 2008b, 2012).

Storylines are intended as tools for forward planning (i.e. to help understand a 'what-if' question), rather than provide predictions of the future. These storylines provide both the basis for qualitative analysis of these tendencies as well as the development of more quantified realisations of a range of possible futures. Here, they have been used to develop consistent input parameters for quantitative models, as discussed in later chapters in this book.

Scenarios often capture processes of change operating at a broader scale than the analysis domain, yet the coastal management and climate change adaptation options are to some extent under the control of local decision makers. Example adaptation options considered in this book include hard-engineered structural works to act as flood defences and coastal protection to stop or reduce erosion, soft protection options such as beach nourishment and abandoning defences and its consequences, which might be living with erosion and flooding. To distinguish these adaptation options from scenarios, they are referred to as 'policy options' through the book. The separation of exogenous scenarios and local or regional scale policy options is to some extent artificial, as coastal policy develops in a national and international context, not in isolation. Furthermore, processes of long-term change at a national and global level are influenced by changes taking place on the coast. However, the approach followed is to test the effectiveness of coastal management policies in the context of a range of possible futures at a national and global level.

A second-generation SMP appeared in draft in 2004 to much public disquiet and concern because in many places the management policy had changed from a largely 'hold the existing defence line' approach to one of 'no active intervention'. Defences in these locations would no longer be replaced or maintained and erosion would be allowed to occur, putting at risk the homes and livelihoods of many clifftop communities. It was not the remit of the SMP process to consider how a transition in policy should be managed – but controversy surrounding such a shift in policy was inevitable (Milligan et al. 2009). Moreover, at the time, no other institutional arrangements were in place to deal with the expected consequences of such a change in policy. Following several consultations and document revisions, the new SMP6 was finally adopted in 2012 after 8 years of controversy and negotiation (East Anglian Coastal Group 2012). Chapter 12 examines the implications of the SMP for local adaptation to coastal change.

1.5 Conceptualising and Designing an Integrated Assessment for North Norfolk – The Tyndall Coastal Simulator

Over the two phases of the Tyndall Centre Coastal Research Programme, the design of the Tyndall Coastal Simulator evolved significantly.

While the technical focus in Phase 1 was around the integration of activities related to coastal erosion and flooding, it soon became apparent that integration needs to be a core activity from the start of the research programme. In particular, this involved investing significant time to identify, recruit and train researchers with the relevant expertise and capacity to work together across disciplines and contribute toward an ambitious integrated assessment. The IA was refined and expanded over a number of years into the framework shown in Fig. 1.11 and introduced in the following section.

1.5.1 Global and Regional Scale

At the global scale, the socio-economic futures drive emissions and hence climate change (e.g. Hall et al. 2005; Thorne et al. 2007). In line with most UK climate impact assessments, four global socio-economic futures are considered in the analysis to capture a range of possible changes in governance structures and social values. The implications of these socio-economic scenarios nationally for the whole UK are provided by the DTI (2002), with the most salient points for coastal management summarised in Appendix A.

For the coast of the Northwest European Region, results from the global scale Hadley Centre coupled Ocean–atmosphere General Circulation Model (OAGCM) have been downscaled to the regional scale (~25 km resolution) using a Regional Climate Model (RCM). Chapter 2 describes how this provides wind forcing for surge and wave models (Leake et al. 2008) and draws on the same model base as the UKCP09 marine simulations (Lowe et al. 2009). The storm surge modelling

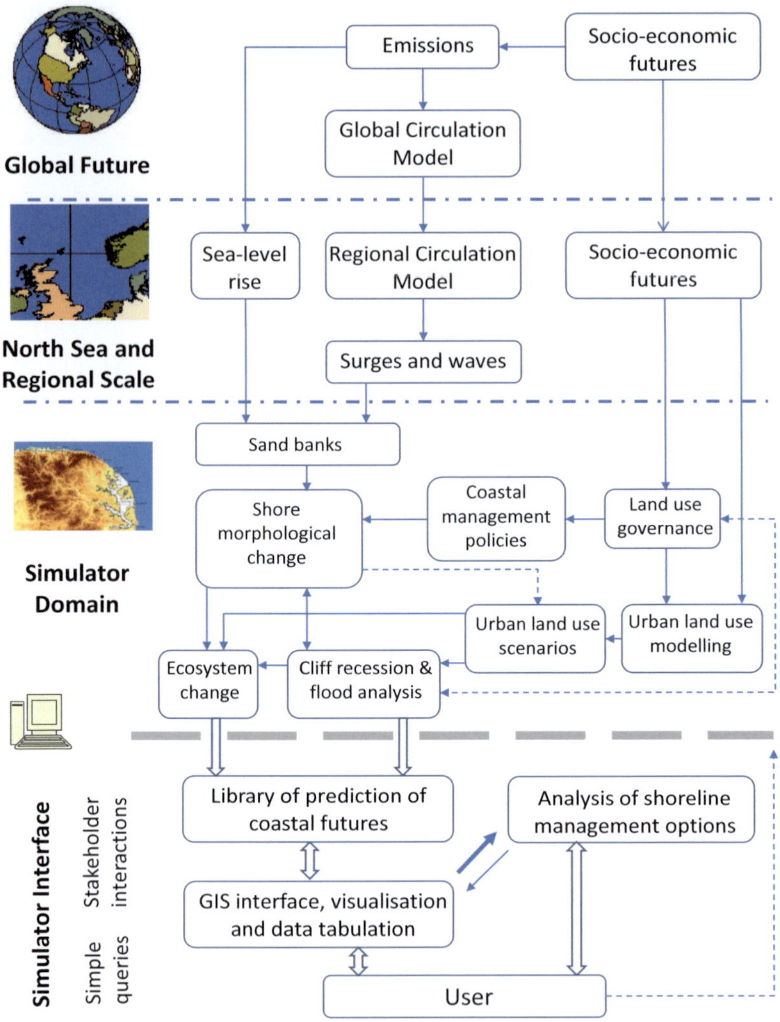

Fig. 1.11 The integrated framework for developing the Tyndall Coastal Simulator, distinguishing the different scales of modelling, analyses and the user interface (Adapted from Mokrech et al. 2011)

follows the work of Lowe and Gregory (2005), but with several improvements in methodology. The wave model uses a similar method of nested models to that of Wolf and Flather (2005) for a hindcast of the 1953 floods. For this IA, the whole Atlantic provides boundary conditions for a shelf-scale wave model with a grid of approximately 12 km resolution (Leake et al. 2008). For the Norfolk area, these new results suggest that wave and surge climate will remain stable. Hence in Phase

2, sea level is the only climate parameter which evolves with time. These model results provide boundary conditions to the Simulator domain and are described in Chap. 2.

1.5.2 Tyndall Coastal Simulator Domain (Management Scale)

At the scale of the Tyndall Coastal Simulator domain, the onshore propagation and transformation of waves is important. This is strongly influenced by the extensive offshore sandbanks, which themselves are potentially dynamic under changing climate. The effects of sandbanks and sea-level rise on sediment pathways and nearshore marine climate are considered for periods of several years up to a century using the TELEMAC modelling system as described in Chap. 3.

The resultant nearshore marine climate provides input to the SCAPE (Soft Cliff and Platform Erosion) cliff erosion model and the flood analysis. The process-based SCAPE model is used to simulate coastal evolution under different shoreline management options to provide information on sediment supply and beach size (Chap. 7). Information on cliff position from SCAPE feeds directly into an erosion risk model that is described in Chap. 9. Changes in beach volume, from SCAPE, influence the probability of flood defence failure. Where failures occur, a flood model for the adjacent coastal lowlands is used to calculate flood extent (Chap. 8) and subsequently changes to flood risk (Chap. 9). Probabilistic climate information that was available for Phase 2 of the Tyndall Coastal Simulator development (Lowe et al. 2009) was incorporated in the analysis through Monte Carlo simulations.

Process-based morphological models are often not available because of our limited understanding of broadscale geomorphic processes. Therefore, to identify broad patterns of shoreline change, such as narrowing of beaches and barriers, in response to changes in sea level and sediment supply, an outcome-driven approach which builds on results from the FutureCoast project (Burgess et al. 2002) and formalises expert knowledge and understanding has been developed (Chap. 5). This geomorphic change is also linked to habitat and ecosystem change which is examined by Chap. 6. Changes in habitats on the south coast of England (at Pennington) in SMP13 and in SMP5 are evaluated. These changes are a product of sea-level rise, storminess and the management approach. In general, significant intertidal creation is expected in the UK over the coming decades (cf. ABPMer 2013).

An important element of the Tyndall Coastal Simulator is the inclusion of social and economic change as this influences the magnitude of erosion and flood risk as well as its variability in time and space. While previous studies (e.g. Holman et al. 2005) have applied socio-economic scenarios based on informed expert judgement, the Tyndall Coastal Simulator has tested the development of scenarios of the built environment using algorithm-based methods, described in Chap. 4. This makes the underlying assumptions more transparent and allows new scenarios to be developed easily as new knowledge or questions emerge.

1.5.3 Stakeholder Interactions

Two other important elements, relating to stakeholder interaction, are addressed within the Simulator Framework. The first relates to stakeholder participation and participatory techniques using visualisation which is explored in Chap. 10. The interface that was constructed to enable exploration of results from the Tyndall Coastal Simulator is described separately in Chap. 11. The second element relates to governance and how it impacts on the formulation and delivery of coastal management responses and is considered in Chap. 12. The experience in North Norfolk shows that these two elements are essential for a successful SMP process. Stakeholder engagement was also important to communicate the Simulator results and discuss their implications.

1.6 Conclusions

In broad terms, the risks to human populations in coastal areas are growing due to climate and socio-economic changes, and these trends are likely to continue and accelerate during the twenty-first century. Due to their complexity, risks such as flooding and coastal erosion are usually considered separately. Furthermore, social and environmental impacts are often not considered alongside economic impacts associated with these risks. Yet frequently, these issues are interconnected – by mechanisms such as longshore exchange of sediments, rising sea levels, development and planning policies and the resulting broadscale coastal system evolution.

To understand these changing risks and the resulting choices and pathways to successful management and adaptation, broadscale integrated assessment is essential. Over a decade, the Tyndall Centre for Climate Change Research developed an ambitious integrated assessment of the 72 km shoreline of SMP6 on the East Anglian Coast of England describing its possible evolution over the twenty-first century. The study is unique in coastal management terms because of the large spatial scale and extended temporal scales over which the analysis is quantified and is also a site of significant controversy about how to manage flood and erosion risks in the coming century because of recent shifts in the Shoreline Management Plan.

This chapter has introduced the economic, social and environmental challenges in East Anglia, as well as the integrated assessment framework developed to analyse coastal futures. Each subsequent chapter introduces different components of the framework, with Chap. 13 providing an international outlook of other challenges for coastal integrated assessment, before concluding in Chap. 14 with reflections on transferability, the process and science of integrated assessment, and a manifesto for future research in this area.

Appendix: Integrated Analysis of Risks of Coastal Flooding and Cliff Erosion Under Scenarios of Long-Term Change

Richard J. Dawson, Mark Dickson, Robert J. Nicholls, Jim W. Hall, M.J.A. Walkden, Peter Stansby, Mustafa Mokrech, J. Richards, J. Zhou, J. Milligan, A. Jordan, S. Pearson, J. Rees, P.D. Bates, Sotiris Koukoulas, and Andrew R. Watkinson

Abstract The risks to human populations in coastal areas are changing due to climate and socio-economic changes, and these trends are predicted to accelerate during the twenty-first century. To understand these changing risks, and the resulting choices and pathways to successful management and adaptation, broadscale integrated assessment is essential. Due to their complexity the two risks of flooding and erosion are usually managed independently, yet frequently they are interconnected by longshore exchange of sediments and the resulting broadscale morphological system behaviour. In order to generate new insights into the effects of climate change and coastal management practices on coastal erosion and flood risk, we present an integrated assessment of 72 km of shoreline over the twenty-first century

Previously published in Climatic Change, 95, 249–288.

R.J. Dawson (✉)
School of Civil Engineering and Geosciences, Tyndall Centre for Climate Change Research, Newcastle University, Newcastle upon Tyne, NE1 7RU, UK
e-mail: richard.dawson@newcastle.ac.uk

M. Dickson
School of Environment, The University of Auckland, Private Bag, 92019 Auckland, New Zealand
e-mail: m.dickson@auckland.ac.nz

R.J. Nicholls
Tyndall Centre for Climate Change Research, Faculty of Engineering and the Environment, University of Southampton, Southampton SO17 1BJ, UK
e-mail: r.j.nicholls@soton.ac.uk

J.W. Hall
Environmental Change Institute, Oxford University Centre for the Environment, University of Oxford, South Parks Road, Oxford OX1 3QY, UK

M.J.A. Walkden
Tyndall Centre for Climate Change Research, School of Civil Engineering and Geosciences, Newcastle University, Newcastle upon Tyne, NE1 7RU, UK

P. Stansby
Tyndall Centre for Climate Change Research, School of Mechanical, Aerospace and Civil Engineering, University of Manchester, Manchester, M13 9PL, UK
e-mail: pr.k.stansby@manchester.ac.uk

on the East Anglian coast of England which is a site of significant controversy about how to manage coastal flood and erosion risks over the twenty-first century. A coupled system of hydrodynamic, morphological, reliability and socio-economic models has been developed for the analysis, implemented under scenarios of coastal management and climate and socio-economic change. The study is unique in coastal management terms because of the large spatial scale and extended temporal scale over which the analysis is quantified. This study for the first time quantifies what has for some years been argued qualitatively: the role of sediments released from cliff erosion in protecting neighbouring low-lying land from flooding. The losses and benefits are expressed using the common currency of economic risk. The analysis demonstrates that over the twenty-first century, flood risk in the study area is expected to be an order of magnitude greater than erosion risk. Climate and socio-economic change and coastal management policy have a significant influence on flood risk. This study demonstrates that the choices concerning coastal management are profound, and there are clear trade-offs between erosion and flood impacts.

M. Mokrech
Environmental Institute of Houston, School of Science and Computer Engineering, University of Houston Clear Lake, Houston, TX 77058, USA
e-mail: mokrech@uncl.edu

J. Richards
ABP Marine Environmental Research, Waterside House, Town Quay, Southampton S014 2AQ, UK

J. Zhou
Department of Civil Engineering, University of Liverpool, Brownlow Hill, Liverpool L69 3GH, UK

J. Milligan • A. Jordan
Tyndall Centre for Climate Change Research, School of Environmental Sciences, University of East Anglia, Norwich NR4 7TJ, UK

S. Pearson
British Geological Survey, Keyworth, Nottingham NG12 5GG, UK

J. Rees
Natural Environment Research Council, Polaris House, North Star Avenue, Swindon SN2 1EU, UK

P.D. Bates
School of Geographical Sciences, University of Bristol, University Road, Bristol BS8 1SS, UK

S. Koukoulas
Department of Geography, University of the Aegean, University Hill, Mytilene 881100, Greece
e-mail: skouk@geo.aegean.gr

A.R. Watkinson
School of Environmental Sciences, University of East Anglia, Norwich NR4 7TJ, UK
e-mail: a.watkinson@uea.ac.uk

A1 Introduction

Climate change with its associated rising sea level and possible increases in the frequency and/or the intensity of storms and changes in wave climate can be expected to increase the risks of coastal erosion and flooding in most coastal locations (Nicholls et al. 2007a). Changes in the pattern and extent of coastal erosion and flooding are all the more concerning because a significant proportion of the world's population reside in the coastal zone: in 1990, 1.2 billion people lived in the near-coastal zone (the area both within 100 km distance of the coast and 100 m elevation of sea level) at densities about three times the global mean (Small and Nicholls 2003). Urbanisation has resulted in a high concentration of the world's cities and settlements (Nicholls 1995; Small and Nicholls 2003) and a considerable portion of global GDP being produced in coastal zones (Turner et al. 1996; Nordhaus 2006). GDP and population density in the coastal zone are higher than the global average. These socio-economic and demographic trends often imply growth in flood and erosion risk areas, compounding the increase in risk due to climate change. Successful management of the coastal zone has consequently been identified as a major challenge for the twenty-first century (Sciberras 2002). However, integrated analysis of environmental and social change on the coast includes major uncertainties due to the many complex and interacting variables. In the face of these uncertainties, coupled scenarios of climate change projections and socio-economic change have been used to construct a number of alternative futures and to gain a broad appreciation of the scale of future flood and erosion risks that may need to be addressed (cf. Parry 2004; Holman et al. 2005; Thorne et al. 2007).

It is fundamental to recognise that there is a longshore connectivity between the various geomorphological features that comprise the natural coastal system, as recognised in the concept of littoral cells and subcells (Komar 1988; Leafe et al. 1998). Hence, evolution of one length of coastline influences the evolution in morphologically connected areas. This sediment movement along beaches and in shallow coastal waters provides a mechanism of interaction between erosion processes and flood risk. The behaviour of beaches is a key determinant of flood risk. Beach morphology has a critical role in dissipating incident wave energy. Moreover, on natural coasts, the beach is the 'defence', whilst on artificially managed coasts the reliability of structures usually depends on beach volume as the level of the adjacent beach influences the likelihood of toe scour, a critical failure mechanism for coastal structures. Sediments released from erosion may serve to increase beach volume and reduce the probability of structural failure on neighbouring coasts. Conversely, the existence of structures that disrupt sediment movement may starve downdrift beaches of much needed beach protection. Therefore, coastal managers should no longer consider erosion and flood hazards in isolation. Rather analysis of climate impacts and long-term coastal management must be implemented at an appropriate (broad) scale to account for this morphological interdependence.

Thus, there is a need for a systematic and integrative approach to analyse the impact of long-term change on both coastal flood risk and coastal erosion. Risk assessment provides a rational basis for supporting the development of coastal management policy, allocation of resources and monitoring the performance of coastal management activities on local, regional and national scales (Dawson and Hall 2006; Hall et al. 2003a). Estimates of risk, expressed in terms of expected annual damages, provide a common currency with which erosion impacts can be compared with flood impacts. Furthermore, in a situation where there are several organisations responsible for coastal management, as is the case in many countries, quantified risk analysis helps identify the organisation(s) most able to mitigate the risk.

Approximately 3 % of the land area of England and Wales is at risk from coastal flooding. Similar to global patterns, this relatively small area is highly developed, containing over one million residential and non-residential properties with expected annual damages to property of £1.0bn and lost agricultural production worth £5.9 m (Hall et al. 2006). In addition to the risk of flooding in coastal lowlands, of the 3,700 km coastline of England and Wales, 28 % is experiencing erosion greater than 10 cm a year (Burgess et al. 2007), despite the existence of coastal protection structures along more than 60 % of this coastline (Eurosion 2004).

In this paper, we describe the application of risk analysis under scenarios of future climate and socio-economic change to analyse alternative coastal management options on a 72 km stretch of UK coastline that has experienced flooding and cliff erosion for many centuries. This integrated assessment allows us to explore the effects of (1) a range of climate change factors, particularly (a) sea-level rise, (b) wave height and (c) wave direction, (2) different coastal defence management choices and (3) socio-economic development, on the evolution of flood and erosion risk. Whilst the concept of integrated assessment of coastal risks is not new (Holman et al. 2002; Hall et al. 2005), this analysis is notable in its deployment of physically based models over such a large spatial scale and extended timescale. The analysis has consequently enabled the quantified assessment of the feedbacks between different coastal hazards under a number of different scenarios of change. We begin with a description of the study site followed by the risk analysis framework and a description of each of the components in the coupled system of models used to conduct the assessment.

A2 Study Domain and Management Challenges

A2.1 The Coastline

The case study site is a 72 km length of coastline and hinterland between Weybourne and Lowestoft on the east coast of England (Fig. 1.12). It is designated as a ‘coastal subcell’ (number 3b) in UK coastal management planning, which means that it is reasonably self-contained in terms of sedimentary interactions with neighbouring coastlines (DEFRA 2006). The north of the site, between Weybourne and Happisburgh, is comprised of retreating chalk and till cliffs, whereas the southern area of the site between just south of Happisburgh and Great Yarmouth is generally low lying and at risk of coastal flooding, although there are some areas of raised ground and local cliffs between Winterton Ness and Great Yarmouth and south of Breydon Water and Lowestoft as shown in Fig. 1.12.

Between Weybourne and Cromer, there is a transition in cliff geology from predominantly chalk to predominantly till. Coincidentally, in this general area there is

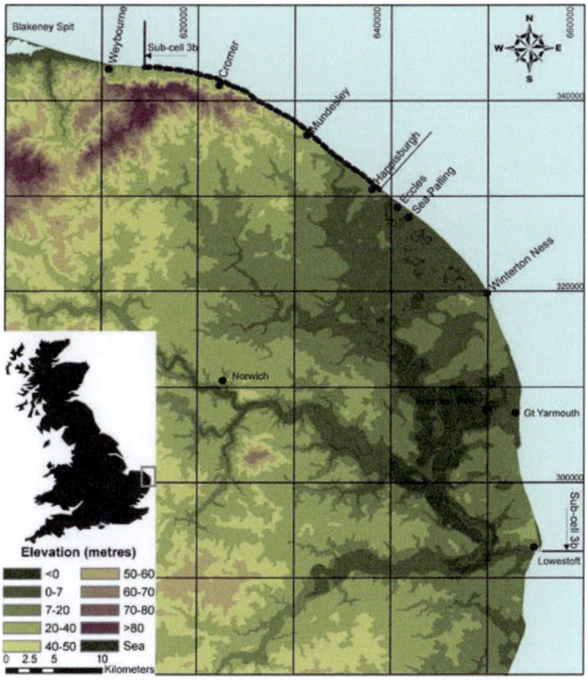

Fig. 1.12 Location of the study area in eastern England, showing the location of the major settlements and land elevation. The area of cliff erosion is indicated by the *dotted line*

also a divide in the direction of longshore sediment transport such that sediment is transported westwards, west of Cromer, and eastwards east of Cromer (Vincent 1979; Clayton 1989), as well as a change in the size of beach sediments (Chang and Evans 1992). Beaches south of Cromer are predominantly sandy with pebbles present in varying proportions, whereas beaches in front of the north-facing chalk cliffs have an increasingly high proportion of pebbles from Cromer through to Weybourne where beaches are composed largely of pebbles. Clayton (1989) estimated that the glacial-till cliffs have retreated at rates averaging around 1 m/year over some 5,000 years, whereas the chalk cliffs have retreated somewhat slower.

The major settlements of Sheringham, Cromer and Mundesley have been protected with groynes and seawalls since the latter part of the nineteenth century, with Overstrand only protected by groynes until early mid-twentieth century. South of Mundesley towards Happisburgh, most major coastal engineering works occurred between 1950 and 2000, with construction mainly comprising groynes and palisades many of which are now nearing the end of their effective life.

The coast southeast of Happisburgh includes large areas of low-lying land in the Norfolk Broads (from Eccles to Winterton) and smaller areas at Great Yarmouth and Lowestoft (Fig. 1.12). Much of this low-lying hinterland is reclaimed marshland and is below the level of mean high-water spring tides and therefore highly susceptible to flooding. Extensive flooding of the Broads area within subcell 3b has occurred a number of times over the last 100 years, most notably in 1907, 1938, 1953 and 1978

Fig. 1.13 Beach at Winterton with concrete seawall in front of sand dune on 31 October 2003 (the detached breakwaters are just visible on the horizon)

(Mosby 1938; Steers 1953; Grieve 1959; Harland and Harland 1980): in the 1953 flood, 17 people died in subcell 3b (Kelman 2003). An extensive system of dunes has been reinforced by construction of seawalls and groynes. In the 1990s the progressive lowering of the beaches in front of these structures near Sea Palling, and consequent increase in flood risk, was addressed by construction of nine detached breakwaters and extensive beach nourishment (Thomalla and Vincent 2003). The dunes southeast of Happisburgh are protected by concrete seawalls (Fig. 1.13) with a combined height of more than 10 m along most of this stretch of coastline.

Further south at Winterton the beaches are wider with timber groynes and there is a more extensive dune system. South again at Great Yarmouth, areas of land are sufficiently low to have been inundated during previous extreme events, such as the 1953 flood. Failure, through undermining or wave overtopping of the structures protecting the promenade of Great Yarmouth, is unlikely to lead to inundation due to the relative level of the hinterland. However, wave overtopping and overflow of the harbour walls at Great Yarmouth are more likely and extreme sea levels in the tidal flats of Breydon Water could result in inundation in areas of Great Yarmouth, such as Southtown (as occurred in 1953). South of Great Yarmouth and just north of Lowestoft, there are four sections of seawall, totalling 2.3 km in length, with a crest height of 5–5.5 m. The seawall is fronted by a sand/shingle foreshore for 1.7 km and rock armour for a length of 600 m. The defences protect a small area of low-lying land whose main economic assets are caravan parks.

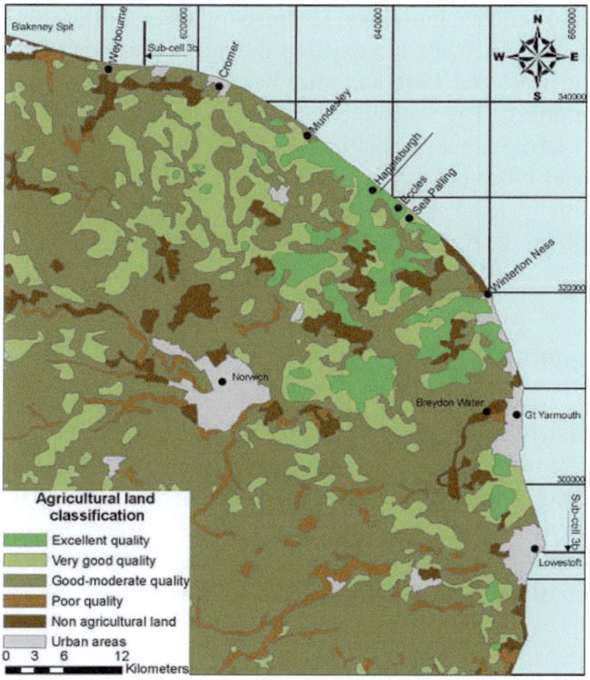

Fig. 1.14 Land use and location of settlements (agricultural classification corresponds to Table 1.3)

A2.2 Vulnerability to Coastal Risks

There are currently almost 20,000 properties at risk of coastal flooding in the model domain, over 3,000 of which are non-residential. Almost 50 % of the floodplain, which is almost 340 km^2, is premium agricultural land, with only 5 % less productive agricultural land (Fig. 1.14). Some 35 % is registered as being environmentally sensitive (with Sites of Special Scientific Interest, Ramsar, salt marsh, national nature reserve designations). Along the 32 km stretch of eroding coastline, there are almost 1,400 properties within 100 m of the clifftop. Of this area, 37 % is premium agricultural land and a further 50 % is less productive agricultural land.

The study area has a long history of erosion and flooding, but climate change exacerbates both these risks in terms of loss of buildings on the cliff coast and increased flood risk in the coastal lowlands. Erosion risk is exacerbated by trends for increased building in these areas, although planning regulations for clifftop areas do take specific account of risks associated with coastal erosion (NNDC 2006). At the national scale, a recent planning policy statement (DCLG 2006) provides guidance on floodplain development. A new shoreline management plan for subcell 3b (Halcrow et al. 2006), which has not yet been adopted, has proposed abandoning significant lengths of cliff defences over the next few decades, which would result in increased loss of clifftop buildings. Happisburgh exemplifies this strategy: it was deemed uneconomic to renew the existing defences which fell in to disrepair following storms in 1991 and 1996 and after becoming a threat to public safety were subsequently removed. The shoreline response to the collapse of the defences was rapid with short-term retreat rates of up to 10 m/year and the loss of 26 houses from 1992 to 2007 and other properties evacuated. Consequently, citizen groups have formed, who with the local council have started a fund for rocks for cliff protection, and Happisburgh has been the subject of national attention.

A2.3 Management Options

The UK government has been promoting a strategic approach to coastal management through the use of shoreline management plans (SMPs; Leafe et al. 1998; Cooper et al. 2002; DEFRA 2006) and more detailed local coastal strategy plans. These plans take a strategic view of coastal processes and associated risks over extended timescales (up to 100 years). A broad range of options, including managed retreat of the coastline as well as measures to prevent erosion and flooding, are considered.

Within this paper, we consider a wide range of contrasting cliff erosion management scenarios ranging from a highly artificial state in which the whole coast is permanently protected with a seawall (Scenario A), through to full abandonment for which all structures are removed (Scenario E). Intermediate management scenarios that correspond closely to suggested strategies that have been proposed in practice (cf. Halcrow et al. 2006) include those where some form of coastal defence (seawall, groynes or palisades) is maintained along about 71 % (the current state, Scenario B), 34 % (Scenario C) and 16 % (Scenario D) of the cliffed coast. Where removal of coastal structures is simulated, it is assumed to happen in a realistic manner, such that structures are removed first from areas with the lowest economic value and at the end of their expected lives, which is generally before 2030.

For this study, the flood defence structures are assumed to be maintained at their current crest level and condition. This enables the impact of strategic shoreline management decisions for the cliffed section of coast to be explored, as it enables changes to flood risk to be measured directly against a baseline scenario. Flood management options, which are not considered in this paper, include raising flood defences and adding sediment trapping structures (groynes, detached breakwaters, etc.) and beach recharge—many of these options are already ongoing. Although these are likely responses to a number of climate scenarios considered here, in the absence of more detailed analysis of the costs of such works, there are a few insights to be gained from analysis of scenarios of improving the flood defence system. More radical flood management options include a proposal to realign the flood defences such that approximately 6,500 ha of land is completely inundated or converted into intertidal habitat forming a new estuary (English Nature et al. 2003).

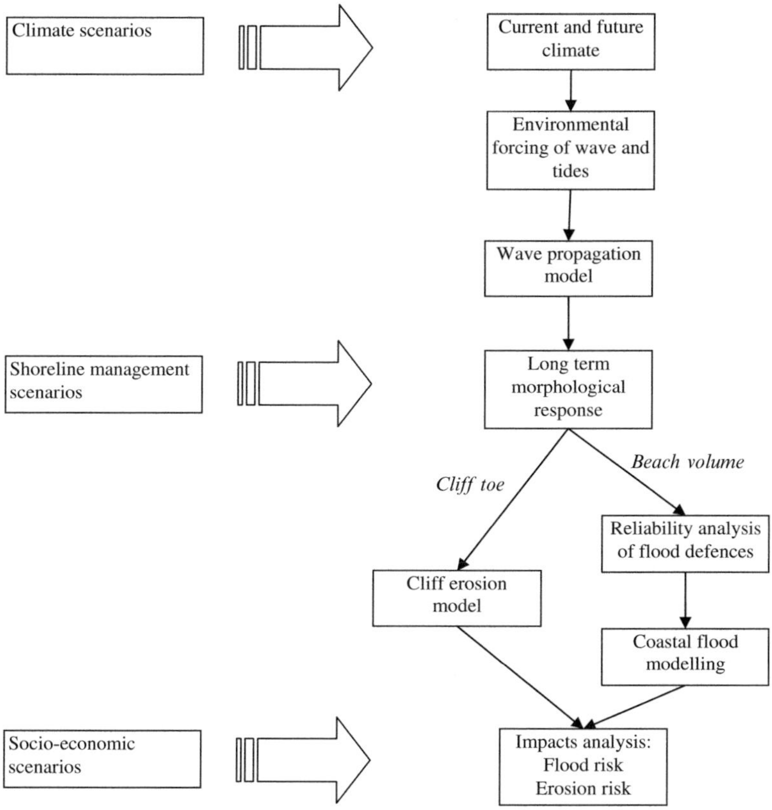

Fig. 1.15 Overview of the assessment methodology, including the role of the scenarios

A3 Assessment Methodology

The assessment has been conducted with a coupled system of hydrodynamic, morphological and impact models, as illustrated in Fig. 1.15. These models were used in a baseline assessment of the risk in 2003 and then modified to include scenarios of:

- Climate change
- Demographic and economic change
- Shoreline management options

implemented separately and/or in combinations. First we discuss the probabilistic framework for estimating the risks of coastal erosion and coastal flooding, before examining each of the main steps in the modelling process (Fig. 1.15) in more detail.

A3.1 Risk Analysis Framework

The occurrence of damaging erosion or flooding on the coast is determined by highly variable marine processes (surge tides, extreme waves) and the (uncertain) responses of the coastal system. It is therefore impossible to predict coastal flooding and erosion in deterministic terms. Instead, a risk-based approach is adopted in which coastal behaviour is predicted, through time, in probabilistic terms and combined with assessment of impacts in order to quantify the 'expected annual damage'. The primary metric for comparison of scenarios and management options here is economic risk, but this can be supplemented by indicators of potential social or environmental impact if available.

Spatial distribution of risk is defined relative to a baseline roughly parallel to the coast. Distances inland from the baseline are designated as the x direction and distances up and down the coast are designated the y direction. The probability of the damaging event, be it due to coastal erosion or flooding, is represented by a function, which will in general vary over space $x \times y$ and time t. The consequences of a damaging event will in general depend on its magnitude and will also vary over space and time. Thus, in general, the risk at a particular point in time can be written as the product of probability (obtained by integrating over the relevant probability density functions) and consequences. We consider two risks:

1. *Erosion risk*, which is calculated by combining probabilistic clifftop erosion predictions with maps of the location of coastal properties, modified as necessary to represent future economic and land-use changes
2. *Flood risk*, which is calculated by combining annual probabilities of flooding with maps of the location of properties in the coastal floodplain, modified as necessary to represent future economic and land-use changes

A3.1.1 Erosion Risk

Erosion of the material from the foot of a coastal cliff or bluff will steepen the cliff. Eventually, erosion will lead to a landslide collapse of the clifftop and a consequent reduction in the cliff slope. The recession proceeds at an irregular rate through occasional landslides. Generally recession involves small-scale failures, but occasionally large-scale failures do occur, as seen at Overstrand in 1990, 1992 and 1994 (Frew and Guest 1997).

The morphological modelling described below is used to predict shoreline (i.e. cliff toe) location, $x_s(y, t)$, at distances y along a shore-parallel baseline and time t years in the future. This prediction includes the effects of natural variability in the environmental forcing (waves, tides and storm surges) and is conditional upon a particular climate scenario. Given a particular shoreline location, x_s, we then generate a probability density function (p.d.f.), $f_c(x|x_s, y, t)$, of the clifftop location, x, at a distance y along the coast and in year t using the method described by Hall et al.

(2000). This p.d.f. expresses the uncertainty in the location of the clifftop, relative to a given shoreline. The expected annual damage due to erosion, r_c, is therefore given by:

$$r_c(t) = \iint f_c(x \mid x_s, y, t)\, c_c(x,y,t)\, dxdy \tag{1.1}$$

where $c_c(x,y,t)$ is a spatially variable damage function constructed from market values of property and agricultural land, which will depend upon the socio-economic scenario (cf. Thorne et al. 2007).

A3.1.2 Flood Risk

The failure of the flood defence system is dealt with via a simplified reliability analysis (see Dawson and Hall 2006). The flood defence system is considered to be comprised of n components each of which may be a 'failed' or 'non-failed' state. There are therefore 2^n combinations of failure and non-failure, which are labelled $S_i : i = 1, \ldots, 2^n$. The response of the flood defence system to environmental loading (i.e. wave heights, h, and water levels, w) is described in terms of a 'fragility function' (Dawson and Hall 2006), $P(S_i|H,W,x_s)$, which is the conditional probability of the system being in state S_i, given wave height H, water level W and shoreline location x_s. The flood risk is therefore calculated according to the following expression:

$$r_f(t) = \iint \sum_{i=1}^{2^n} P(S_i \mid H, W, x_s)\, f_f(H, W \mid t)\, c_f(H, W, S_i, t)\; dH\, dW \tag{1.2}$$

where $f_f(H, W \mid t)$ is the joint probability density function of wave height, H, and water level, W, at time t. Dependence upon t reflects the fact that the frequency of waves and extreme water levels (due to sea-level rise) is expected to change with climate change. $c_f(H,W,S_i,t)$ is a damage function, which:

- For a given combination, S_i, of flood defence failures, wave height H and water level W uses hydrodynamic inundation modelling to calculate flood depths in the floodplain
- Uses information on the location of properties, together with standard depth–damage functions (Penning-Rowsell et al. 2003b) to calculate the flood damage

Note how the causal sequence described above is reflected in Eq. 1.2:

1. Shoreline recession, x_s, is reflected in the probability of failure of the flood defence structures, $P(S_i|H,W,x_s)$.
2. The arrival of an extreme storm is described by the joint probability density function $f_f(H,W|t)$ and may cause failure of one or more of the defences structures, depending upon the conditional probability $P(S_i|H,W,x_s)$.

Table 1.1 Return periods for extreme water levels (mAOD) at Cromer and Great Yarmouth (Dixon and Tawn 1997)

Return period (year)	Cromer	Great Yarmouth
10	3.78	2.74
50	4.12	3.08
100	4.33	3.3
250	4.53	3.51
1,000	4.82	3.8

3. The severity of flood inundation depends upon the environmental loading conditions, *h* and *w*, and the number and locations of flood defence failures, which are reflected in $c_f(H,W,S_i,t)$.
4. The damage to properties located in the floodplain, for a given flood depth, is also reflected in $c_f(H,W,S_i,t)$. As with cliff erosion, the function $c_f(H,W,S_i,t)$ changes through time to reflect changing land use and socio-economic exposure in the floodplain.

Note that where *n* is large it is intractable to solve Eq. 1.2 precisely so an importance sampling approach, described in Dawson and Hall (2006), has been employed. The total risk *r*(*t*) at time *t* is simply the sum of the cliff erosion and flood risk:

$$r(t) = r_c(t) + r_f(t) \quad (1.3)$$

The various statistical and process-based models that generate predictions of the variables in Eqs. 1.1 and 1.2 are now described.

A3.2 The Environmental Forcing: Waves and Tides

The coast is exposed to waves generated within the North Sea from directions between approximately 330 and 120°N, but the largest waves arrive from the north through to the east where fetch lengths exceed 500 km. The most frequent waves are from the northwest. Estimates of extreme water levels corresponding to different return periods are shown in Table 1.1.

The morphological modelling described below is driven by water levels and wave heights at tidal time steps. Records of water levels were available for Cromer from 1988 to 2002. These were extended to generate a stationary synthetic time series of 1,000 years by resampling segments whilst preserving seasonality. Historic rates of mean sea-level rise were subsequently imposed upon this extended time series. Extreme values were resampled from a distribution constructed from estimates of extreme water levels provided by Dixon and Tawn (1997). Wave heights were hindcast from 23 years of wind data (from 1978 to 2001) and extended to 1,000 years (HR Wallingford 2002). This long record was constructed from month segments resampled from the original 23-year file, preserving seasonality. A distribution was fitted to extreme values and used to resample extreme events in the long synthetic series. This long time series is necessary to 'spin up' the morphological model (see Walkden and Hall 2005). Subsequently a time series representing changing climatic conditions over the timescale 2000–2100 was developed for predictive simulations as described below.

A3.3 Future Marine Climate

Three representative relative sea-level-rise scenarios were used which were designed to encompass the range of uncertainty in global, regional and local sea levels. This is mainly based on the Intergovernmental Panel on Climate Change (IPCC) Third Assessment Report (Church et al. 2001) and the related UK Climate Impacts Programme (UKCIP02, Hulme et al. 2002). The 'low' scenario of relative sea-level rise (RSLR) represented no anthropogenic influence, i.e. a continuation of the recent historic rate of 1.5 mm $year^{-1}$ (see also Miller and Douglas 2004). The 'medium' scenario follows the UKCIP02 medium-high scenario and results in a twenty-first century increase of 0.45 m by 2100. The 'high' scenario is based on the IPCC high limit plus an additional regional (North Atlantic scale) sensitivity of 50 % (following Hulme et al. 2002) to allow for spatial variability in thermal expansion (see Gregory et al. 2001). This scenario results in an increase in sea level of 1.2 m by 2100. All three sea-level-rise scenarios used include a regional subsidence rate based on long-term geological observations of 0.7 mm $year^{-1}$ (Shennan and Horton 2002).

Long-term predictions of changes in wind speed and direction from climate models, and hence wave height and direction, are considered to be very uncertain. However, climate model predictions are supported by observations in the North Sea and Atlantic Ocean, which suggest that extreme wave heights may be increasing (Bouws et al. 1996; Kushnir et al. 1997). Given the concern about increasing wave heights, the sensitivity of erosion and flood risk to a range of possible future conditions was explored rather than downscaling climate model outputs directly. Wave heights were unchanged in the 'low' scenario, whereas in the 'medium' and 'high' scenarios, offshore winter wave heights were increased up to a maximum of 7 % and 10 %, respectively, by 2100. These increases were applied linearly over the twenty-first century. In addition, the 'high' scenarios included +10° (clockwise) and −10° (anticlockwise) rotations of the offshore wave rose. Potential changes to wave period were not considered.

It was not feasible to test all the combinations of sea-level-rise, wave height and wave direction scenarios together with all the management options mentioned in Sect. A2.3 above. Therefore, there was a deliberate focus on the low and the high scenarios of relative sea-level rise, to provide bounds on the potential future outcomes. Table 1.2 summarises the combinations of scenarios and management options that were tested in the coupled analysis. A range of behaviour is noted in the different scenarios. Those we explore in this paper are highlighted in Table 1.2. Those that also consider socio-economic change use the same numbering scheme

Table 1.2 Scenario numbering scheme for the coastal management and climate change combinations considered in this study

Management scenario (% cliff protected) (%)	Wave (H_s) and sea-level-rise (SLR) climate scenario								
	Low SLR				Medium SLR	High SLR			
	H_s low	H_s high	H_s high +	H_s high –	H_s mid	H_s low	H_s high	H_s high +	H_s high –
A (100)	1	2	3	4	5	*6*	7	8	9
B (71)	*10*	11	12	*13*	*14*	*15*	16	17	18
C (34)	*19*	20	21	22	23	*24*	25	26	27
D (16)	28	29	30	31	32	*33*	34	35	36
E (0)	37	38	39	40	41	*42*	43	44	45

'+' and '–' indicate rotation of the wave rose by 10° clockwise and anticlockwise, respectively; percentages indicate proportions of cliff coast protected from erosion by structures. The *italicised* scenarios are those that are explored in more detail in this paper

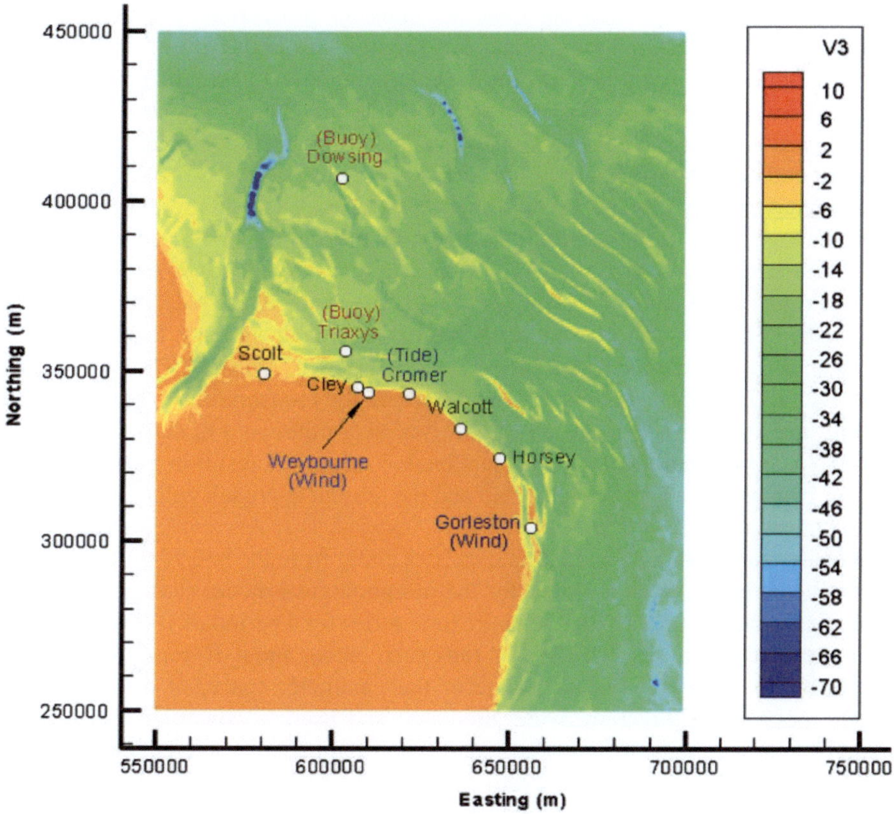

Fig. 1.16 Wave transformation modelling domain showing the bathymetry (metres below ordnance datum) off the East Anglian coastline

but are prefixed by WM, GS, LS and NE to correspond to the world market, global sustainability, local stewardship and national enterprise socio-economic scenarios, respectively (see Sect. A3.10).

A3.4 Wave Propagation Modelling

Propagation of waves from deep water off the coast of East Anglia over the complex bathymetry to the boundary of the morphological model was conducted with the TOMAWAC code (part of the Electricité de France TELEMAC suite; Kuang and Stansby 2004; Stansby et al. 2006) over an area of approximately 75 × 100 km (Fig. 1.16). The model operates on an unstructured (finite-element) grid which may be adapted to give high resolution where needed and accounts for shoaling, refraction, dissipation by white capping, bed friction and depth-limited breaking and wave–wave energy transfers but does not consider diffraction. The analysis for long-term predictions, involving domain calibration described below, is based upon the assumption that wave propagation inshore is not affected significantly by tidal (or wave-induced) currents, directionality (whether broad or narrow) and wave generation within the coastal domain. This has been previously justified for this area (Kuang and Stansby 2004), although very high wind speeds (greater than about 20 m/s) may modify the wave transformation somewhat. The offshore wave climate was obtained from onshore wind data assuming a long fetch with an unchanging wind direction, applying formulae for significant wave height and period from the Shore Protection Manual. This simplistic approach enabled historical predictions of cliff erosion from 1885 to be undertaken with reasonable success. Furthermore, it has been assumed that the bathymetry of the site remains constant. This is a significant assumption as the shallow sandbanks within the model domain (Fig. 1.16) are mobile and major changes have been observed over timescales of decades and centuries. Since both sea-level rise and sandbank erosion will result in greater depths, greater inshore wave heights will result and the present approach will underestimate shoreline erosion.

Long-term nearshore wave climate data resulting from the hourly time series for offshore conditions were required for the morphological model (introduced in the following section). Processing the entire time series for a period of 100 years would have involved about 10^6 TOMAWAC runs each taking about 30 min. Instead, 500 combinations, covering the full range of the parameter space, of offshore wave height, period and direction, were simulated and used to construct curves from which the remaining nearshore wave conditions could be interpolated without reusing TOMAWAC (Stansby et al. 2006) enabling nearshore wave climate time series to be generated in minimal computer time. Predictions of wave conditions inshore from the curves were compared with measurements made by the Environment Agency in 2002–2003, and nearshore wave height was predicted with little error, wave period was rather more scattered, and wave direction was not recorded by these measurements so could not be compared. The calibration process is described in full by Stansby et al. (2006).

A3.5 Long-Term Coastal Morphological Modelling

The effect of waves and tides on littoral sediment transport and erosion of the soft coastal cliffs and platform was simulated using the SCAPE model (Walkden and Hall 2005; Dickson et al. 2007; Walkden and Dickson 2008). SCAPE represents a number of processes and their interactions:

- Wave transformation from the nearshore points provided by the TOMAWAC model to the breaker point using linear wave theory.
- Sediment exchange between the beach and a nearshore bar using a simple parameterization of the COSMOS model (Nairn and Southgate 1993).

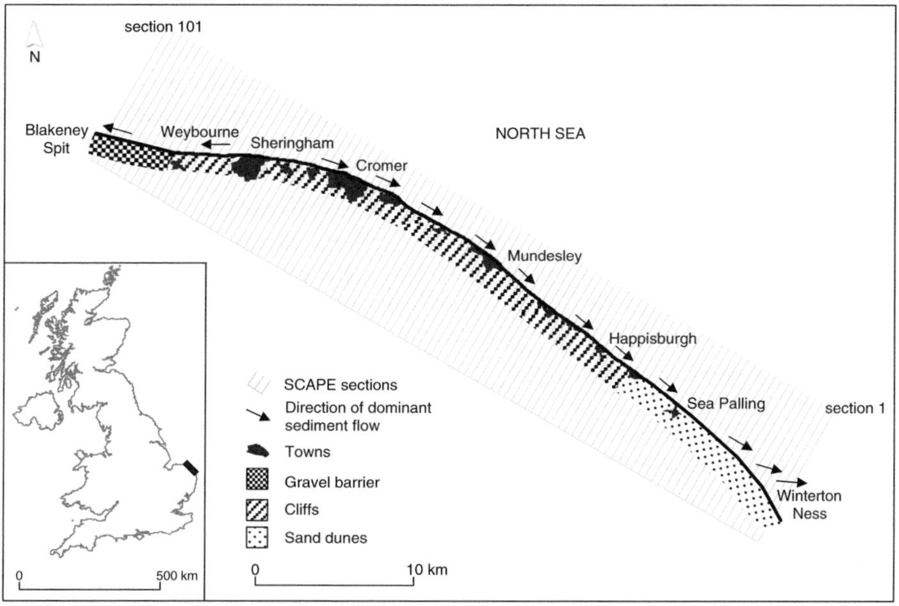

Fig. 1.17 Location of SCAPE model sections for simulating the long-term morphological response of the coast illustrating the major morphological features

- Longshore sediment transport using a one-line beach model (Pelnard-Considere 1956) of the form described in the Shore Protection Manual (CERC 1984).
- Erosion of the shore platform and cliff toe (Walkden and Hall 2005).
- Delivery of talus to the beach.
- The effect of shore-parallel coastal structures (seawalls and palisades) and groynes as follows: (1) seawalls prevent cliff toe retreat but do not stop lowering of the shore platform, (2) palisades reduce the heights of passing waves by 50 %, and (3) groynes reduce longshore sediment transport, except when beaches are wider than the structures.

This paper extends the SCAPE model to represent cliff failure and retreat of the clifftop (Hall et al. 2000). The shape of the coast, in both profile and plan view, emerges from the dynamic interaction between and within modules, which respond to the imposed loading (waves, tides and SLR) and coastal management interventions. After simulating the construction of coastal engineering structures between the later part of the 1800s and 2000, model behaviour was compared with historical recession data measured from Ordnance Survey maps. Once a satisfactory validation was achieved (described fully in Dickson et al. 2007), the model was run under combined climate change and management scenarios for the twenty-first century.

The southern part of the SCAPE model (Fig. 1.17) includes the low-lying coast between Happisburgh and Winterton, an area with no cliff, a low platform and a deep sandy beach. The absence of a cliff/platform means that all the beach sediment in this section arrives through easterly longshore transport. Consequently, the beach volume and berm levels here ultimately depend on patterns of sediment supply and longshore transport from the cliffed coast south of Sheringham. For this reason, shoreline management decisions to prevent cliff recession or to construct groynes cause the southern beaches to denude, whilst the removal of existing structures releases sediment that increases the beach volume.

The beach component in the SCAPE model has been extended south of the soft cliffs to the low-lying coast as far as Winterton Ness. South of the drift divide at Winterton Ness, the land rises above typical flood levels and is not at significant risk of flooding (Fig. 1.12). The main flood protection structures are in the Great Yarmouth and Lowestoft harbours and tidal river reaches. However, a small area of floodplain exists at Lowestoft Denes (just north of Lowestoft), and in parts of the town itself, although in all this section is significantly less important in terms of impacts. The influence of cliff management on this region is less significant than for the Happisburgh to Winterton section due to offshore losses of sand to nearshore banks and the presence of several barriers to longshore transport (e.g. the estuary and associated harbour breakwaters at Great Yarmouth; Futurecoast 2002). Moreover, the SCAPE model is most rationally deployed where there is significant interaction between the cliff and beach. Given the forgoing considerations a simplified approach based on the Bruun rule (described by Bruun 1962 and many others) was applied to the southern end of the site using past observations for calibration (Halcrow 1999; Futurecoast 2002), to provide an estimate of shoreline recession. This means that at the southern end of the site, the probability of failure due to toe erosion (and overtopping) is only influenced by changing sea level and wave conditions, with no longshore interaction.

A3.6 Cliff Failure Modelling

For economic appraisal of the impacts of coastal cliff recession, predictions are required of when individual clifftop assets will be lost due to coastal landsliding. Coastal landsliding is a consequence of a combination of cliff toe recession and geotechnical processes within the cliff slope. Landsliding on unprotected soft cliffs is preceded by a process of marine removal of material from the cliff toe, resulting in steepening of the coastal slope. Eventually the slope becomes unstable and a landslide occurs that reduces the coastal slope and delivers debris to the beach. The timing of the landslide is a function of the rate of removal of material from the cliff toe and other processes, primarily connected with pore pressure distributions within the cliff that influence cliff stability. The timing of a landslide cannot be predicted precisely. However, knowledge of the rate of shoreline retreat (from SCAPE) can be combined with an assessment of the geotechnical characteristics of the slope to generate an approximate probability distribution of the possible clifftop location following failure (Hall et al. 2000).

The approach adopted here is based on the notion of a Cliff Behavioural Unit (CBU) being a stretch of cliff line which behaves in broadly the same way. Within a

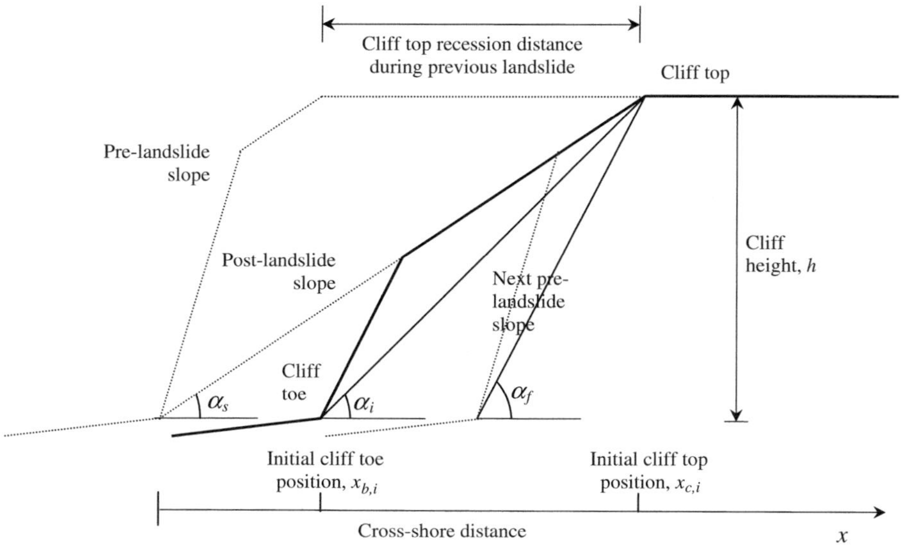

Fig. 1.18 Diagrammatic representation of the coastal landsliding model. The parameters are described in the text

CBU, the cliff can be expected to fail when it reaches an average angle α_f and will, after failure, adopt an angle α_s. Neither α_f nor α_s can be predicted precisely. They will vary because of temporal variations in pore pressure and local variations in cliff strength and composition. Even if all the required information were available, they could still not be predicted precisely because of uncertainties in our understanding of the processes of coastal landsliding. This uncertainty in α_f and α_s has been included in the analysis by representing both values as normally distributed random variables, with means and variances obtained from a geomorphological assessment of the CBU. The situation is illustrated diagrammatically in Fig. 1.18.

Further uncertainty is apparent in the initial cliff angle at the site. Within a CBU there will be a range of initial angles, whilst in this analysis (other than for very long CBUs), a prediction of clifftop recession has been generated for entire CBUs or, where appropriate, subsections thereof. The initial cliff angle has therefore also been represented as a normally distributed random variable, with mean and variance based on measurements of cliff angle within the CBU.

The clifftop position is calculated relative to the SCAPE prediction of cliff toe recession. From Fig. 1.18 we see that the distance the cliff toe has to retreat until the first landslide is $h(\cot\alpha_i - \cot\alpha_f)$ and the distance lost between each subsequent landslide to the next is $h(\cot\alpha_s - \cot\alpha_f)$. For year $t+1$, the pre-landslide slope is taken from the post-landslide slope of year t. The clifftop recession process can be calculated from:

- SCAPE predictions of cliff toe recession
- The cliff height (relative to the cliff base)

- The initial cliff angle $\alpha_I \sim \mathrm{N}(\mu_I, \sigma_I)$, i.e. it is a normally distributed random variable with mean μ_i and standard deviation σ_I
- Geomorphological assessments of the pre- and post-landslide angles $\alpha_F \sim N(\mu_F, \sigma_F), \alpha_s \sim N(\mu_s, \sigma_s)$

Estimates of the parameters μ_i, σ_i, μ_f, σ_f, μ_s and σ_s were obtained from limited geological data, site visits and analysis of aerial laser scanning and photogrammetric images of the cliffs. The probability of a cliff landslide occurring is estimated by sampling from α_i and α_f. The resultant clifftop position is then estimated by sampling from α_s. Large numbers of samples of these sequences of angles are used to generate a time series of distributions of potential clifftop locations at each year through the twenty-first century. It is assumed that the only climatic influence on cliff recession is via shoreline (i.e. cliff base) recession, and no changes to geotechnical factors (e.g. changing pore pressure from a different hydrological climate) are considered. The potential effects of hydrological changes on cliff stability were not considered.

A3.7 Reliability Analysis of Coastal and Flood Defences

The three primary failure modes for the coastal defence structures found on the East Anglian coast are:

1. *Overflow*: water level exceeds the flood defence crest level.
2. *Overtopping*: a combination of extreme waves and water levels results in waves running over the defence.
3. *Toe failure*: low beach levels and/or excessive scour at the base of the defence leads to lowering or collapse of the flood defence.

Overflow of the flood defences along the Norfolk Broads is unlikely, even for extreme water levels (Table 1.1), due to the relatively high crest level of the beach and defence structures.

Overtopping discharges are estimated according to the roughness, permeability and slope of the flood defence structure (HR Wallingford 1980, 1999). For example, the overtopping rate, Q, for an impermeable seawall is given by:

$$Q = gH_s T_m a \exp\left[\frac{-b(z_c - z_w)}{T_m \sqrt{g\ H_s}}\right] \tag{1.4}$$

where H_s and T_m are the significant wave height and the mean wave period at the toe of the structure, z_c is the crest level, z_w is the still water level, g is the acceleration due to gravity, and a and b are coefficients based on the slope and berm (HR Wallingford 1999). Overtopping volumes in the study area are generally very small due to the height of the flood defence structures (with higher discharges being estimated for the lower seawalls south of Great Yarmouth). These overtopping discharges were then input directly into the flood inundation model described in Sect. A3.8.

The dominant mechanism for breach of a coastal defence is toe scour (CIRIA 1986). This can lead to undermining and collapse of the coastal flood defence. This occurs when:

$$z_t - z_b < 0 \tag{1.5}$$

where z_t is the level of the beach at the toe and z_b is the critical beach level. The beach volume, taken from the SCAPE model, is assumed to be evenly distributed over the shore platform. The post-storm beach profile is calculated using the approach of Vellinga (1983) for each value of $f(H,W)$, and the beach level, z_t, at the toe of the flood defence is subsequently extracted. Toe erosion is currently insufficiently understood to establish z_b precisely (Reeve and Burgess 1993; HR Wallingford 2003). Therefore, a fragility function, which describes the probability of toe failure conditional on beach level, was generated on the basis of generic coastal defence design evidence (cf. Thomas and Hall 1992; CIRIA and CUR 1991), data from site investigations (Environment Agency 1996) and expert judgement. This provided estimates of the probability of one or more breaches, which was combined with information on breach width (Visser 1998; Wahl 1998; HR Wallingford 2004) to calculate flood discharges into the floodplain due to flood defence breaching.

A3.8 Coastal Flood Modelling

Simulation of inundation over low-gradient floodplains with dike structures requires at least a two-dimensional modelling approach with relatively high spatial resolution to represent the complex geometry of the floodplain. However, full two- or three-dimensional modelling remains computationally prohibitive if multiple scenarios are to be modelled. To reduce the computational burden of the hydrodynamic calculations for this study, a simple 2D raster-based inundation model called LISFLOOD-FP was selected. The model generates a spatial field of water levels from which the digital elevation map (DEM) can be subtracted to give a field of flood depths. Bates and De Roo (2000) describe the model in detail which has been successfully demonstrated at a number of coastal sites and has been shown to perform as well as full two-dimensional codes at predicting maximum flood inundation extents (Horritt and Bates 2001; Bates et al. 2005).

The model was set up using lidar (light detecting and ranging) surveys of the floodplain, with a spatial resolution of 2 m, with a reported rms error of 0.1 m. The model was implemented on a 250 m raster grid and validated against the 1938 flood event which provided a fitness measure of 91 % giving significant confidence. Calibration of this model is described in more detail by Bates et al. (2005), and more detailed consideration to calibration of the LISFLOOD-FP model is described in Horritt and Bates (2001) and Hall et al. (2005).

Table 1.3 Losses by area for different classes of agricultural land at 2003 prices (erosion losses of agricultural land are reduced by 45 % to account for government subsidies)

Land grade	Inundation losses (£/ha; Penning-Rowsell et al. 2003a)	Erosion value (£/ha; RICS 2003)
Excellent quality	1,160	5,683
Very good quality	770	5,683
Good quality	400	5,683
Moderate quality	160	5,683
Poor quality	50	4,571
Nonagricultural land	20	4,571

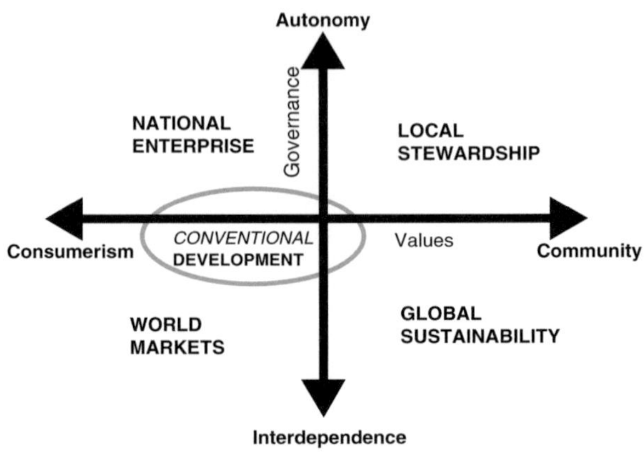

Fig. 1.19 The four socio-economic scenarios (DTI 2002) used to explore the impacts of socio-economic change on flood and erosion risk

The flood defence line was taken to be the open boundary of the model, with overtopping discharges and/or inflows through breaches simulated at appropriate points in the boundary. A dynamic tidal profile was used that corresponds in shape to the 1953 extreme event (Rossiter 1954). The maximum flood depth for each model grid cell was extracted from a total of 20,000 runs of the hydrodynamic model and used to evaluate flood damages in the baseline risk calculation.

A3.9 Impact Analysis

Two direct economic impacts are considered in the analysis:

1. Flood damage to property and agricultural production
2. Permanent erosion of clifftop properties and agricultural land

Flood damage was calculated using standard relationships between the depth of flooding and economic damage to different classes of property and agricultural land (Penning-Rowsell et al. 2003b). Erosion damage was calculated using the risk-adjusted market value of clifftop property and agricultural land. The regional average market value (VOA 2003; ODPM 2003) of £208,824 has been supplemented with more local information from the UK Land Registry to estimate an average market value of £150,000 per residential postal address. Typical flood losses and erosion value of agricultural land at 2003 prices are summarised in Table 1.3.

A3.10 Socio-economic Futures

The scenario framework shown in Fig. 1.19 was used to produce four socio-economic scenarios for the coasts and floodplain fringing subcell 3b. This framework examines alternative directions in which social, economic and technological change may evolve over the coming decades in the UK, with a broadly individual versus collective action on the x axis and a broadly localised versus globalised world on the *y* axis (DTI 2002). The scenarios are discussed extensively in terms of future erosion and flood risks in Hall et al. (2003b), whilst more complete details of the scenario methods used in this study are in Nicholls et al. (2005).

In contrast to earlier work that largely draws on expert judgement (e.g. Holman et al. 2005; Thorne et al. 2007), regional scenarios for changes to the built environment were developed and then spatially quantified across the case study domain using an algorithm-based methodology. The current baseline property distribution was kept fixed and then augmented with new properties based on the current regional housing policy forecasts until 2021 (e.g. GOEE 2005) and then to 2100 according to the four socio-economic scenarios (DTI 2002; Nicholls et al. 2005). The number of properties is projected to increase under all four socio-economic scenarios. The distribution of the new development is influenced by four attraction/repulsion factors for new build:

1. Existing settlements
2. The transport network
3. The coastline
4. The floodplain

These attraction factors reflect the urbanisation policies and the levels of attraction that areas and infrastructures are likely to generate under the different socio-economic scenarios. For example, under the local stewardship scenario all the factors are negative indicating a trend of moving development into the countryside, which contrasts to the urbanisation trend of the world market scenario. Socio-economic change would be expected to alter the value of agricultural land as well as the value of property and its contents. Increases to property and content values would be greatest under the world market and least (possibly even negative) under the local stewardship scenario, whilst increases to agricultural land values might only be expected under the national enterprise scenario (Thorne et al. 2007).

Table 1.4 shows the numbers of both residential and non-residential properties located within the floodplain, the strength of each attractor for the four socio-economic scenarios and the change in the number of non-residential properties relative to residential properties. The new properties were located according to a multi-criteria

Table 1.4 The strength of the attractors for development and the total number of new properties by 2100 in the model domain (see Fig. 1.12) and inside the 2003 floodplain under the four socio-economic scenarios

Scenario		Local stewardship	Global responsibility	National enterprise	World market
Ratio of non-residential to residential property		−10 %	0 %	0 %	10 %
Attractor/repulsor factor	Settlements	1/16	8	1	1
	Transport	1/8	1	1	8
	Floodplains	1/16	1/8	8	16
	Floodplains	1/16	1/8	8	16
Number of new properties in model domain shown in Fig. 1.13 by 2100	RP	62,607	98,158	136,053	171,790
	NRP	4,069	9,939	13,785	19,662
New properties in the floodplain	RP	1,343	10,882	19,550	19,719
	NRP	80	1,082	1,908	2,127
	NRP	17	44	52	85
	NRP	17	44	52	85

spatial weighting technique (Eastman et al. 1993). The attractors/repulsors were assigned the weights $w_i = [1/16, 1/8, 1, 8, 16]$ representing their relative importance under different socio-economic futures. The number of additional properties, p_j, added to flood damage cell j (each cell has an area of 250 × 250 m) is:

$$P_j = \frac{P}{C}\sum_{i=1}^{n} w_i \left(1 - \left[\frac{d_i - d_{\min}}{d_{\max} - d_{\min}}\right]\right) \tag{1.6}$$

where P is the total number of new properties to be added under a given socio-economic scenario, C is the number of damage cells in the model domain (not including inland water body cells and others constrained from development), n is the number of factors (note: $n > 4$, as there are multiple settlements, etc.), w_i is the weight of factor i and d_i the proximity of factor i to cell j, and $d_{\min}$ and $d_{\max}$ correspond to the minimum and maximum values of the proximity of factor i to cell j.

In 2003 the floodplain (defined as the area with probability of flooding greater than zero in Fig. 1.25a) contained 16,766 residential properties (RP) and 3,160 non-residential properties (NRP).

A4 Results

Results are presented in order of the modelling sequence (Fig. 1.15): (1) coastal morphology, (2) analysis of the probability of flooding and (3) risk analysis (validation of the wave propagation modelling is described by Stansby et al. 2006). Presentation in this order helps to explain the important and rather complex results obtained from the risk analyses, which are an aggregation of the preceding steps with the socio-economic scenarios. The baseline data used for all the results is 2003.

A4.1 Coastal Morphology

The predictions of coastal erosion under the different scenarios of climate change and coastal management varied alongshore and through time. Sea-level rise has the most significant impact on erosion rates. However, in some downdrift locations recession rates fell, because of a localised increase in beach volume caused by an increase in recession at the regional scale. This counter-intuitive result illustrates the necessity of accounting for longshore interactions when exploring coastal recession at the scale of shoreline management (i.e. a subcell). A fuller description and interpretation of these results is given by Dickson et al. (2007). The effects of increased offshore wave heights were mitigated by energy dissipation across the shallow bathymetry of the North Sea which was assumed to be constant (as described previously). Consequently this had a small impact on erosion rates. Changes in offshore wave direction by ±10° proved more important for erosion, with a clockwise rotation of the offshore wave rose generally increasing erosion rates and an anticlockwise rotation reducing them.

The importance of coastal management on recession rates can be seen in Fig. 1.20a, which shows the total recession predicted under Scenarios 15, 24 and 33. As can be seen in Table 1.2, these all assume high sea-level rise with unchanged waves and the three most plausible management policies. No cliff recession occurs where seawalls have been maintained. Scenario 24 assumes the loss of structures from Bacton and Trimingham, whilst under Scenario 33 the Mundesley and Overstrand structures are also allowed to fail. Structure loss results in high recession (e.g. at Overstrand under Scenario 33). However, longshore sediment transport has an important mitigating effect. For example, the removal of structures at Bacton has a much smaller effect under Scenario 33 than under Scenario 24 due to beach sediments released by the updrift recession of Mundesley and Overstrand. Figure 1.20b illustrates the high recession rates that initially follow structure failure as the coastline straightens and the shore platform at the cliff toe returns to a more natural form.

The release and southward transport of beach material also has an important influence on flood risk in the Norfolk Broads. Figure 1.21 illustrates the evolution of beach volumes under Scenario 33 over the twenty-first century. It can be seen that there is a general decrease in beach volume with increasing distance from Winterton. Beach volumes are particularly large around the shore-parallel reefs at Sea Palling. Diagonal bands of low/high (dark/light) beach volumes can be seen south of Mundesley, particularly before 2050, although also present elsewhere. These can be interpreted as the southward movement of sand waves along the coast at approximately 0.8 km/year. In the model the sand waves are promoted by the presence of engineering structures. This means that the spatiotemporal distribution of beach volume is more complex under scenarios with more engineering intervention. The relationship between the simulated and observed features has not been explored in

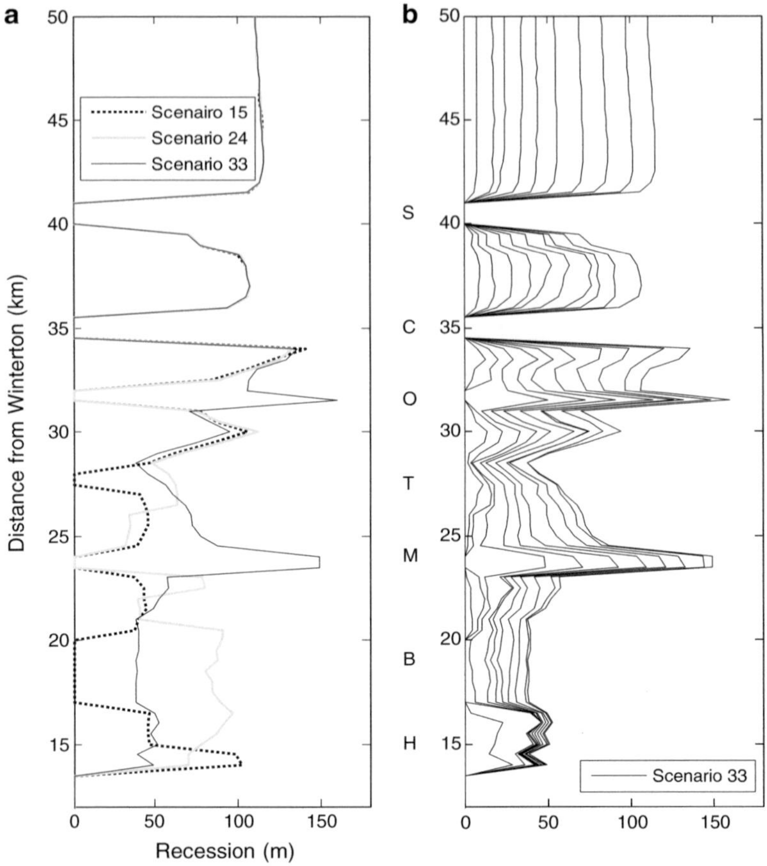

Fig. 1.20 (**a**) Total recession predicted along the cliffed frontage under Scenarios 15, 24 and 33 with 71 %, 34 % and 16 % coastal protection, respectively, and (**b**) recession distances under Scenario 33 in 10-year time steps. The letters *H*, *B*, *M*, *T*, *O*, *C* and *S* represent the locations of Happisburgh, Bacton, Mundesley, Trimingham, Overstrand and Sheringham, respectively

depth and so this model behaviour should be interpreted with caution. However, the model waves are coupled to patterns of erosion in a manner observed by Pringle (1985) at the Holderness Coast in the UK and there is anecdotal evidence of the existence of sand waves of similar scale and speed south of Mundesley (personal communication, Gary Watson, Environment Agency) which provides some confidence that these are not modelling artefacts. Figure 1.22 shows the frequency of occurrence of average beach volumes at Sea Palling throughout the twenty-first century under Scenarios 15, 24 and 33. These scenarios produce, respectively, increasingly stable beaches because they maintain decreasing levels of coast protection (i.e. there is less variance in beach level when a lower percentage of coastline is engineered).

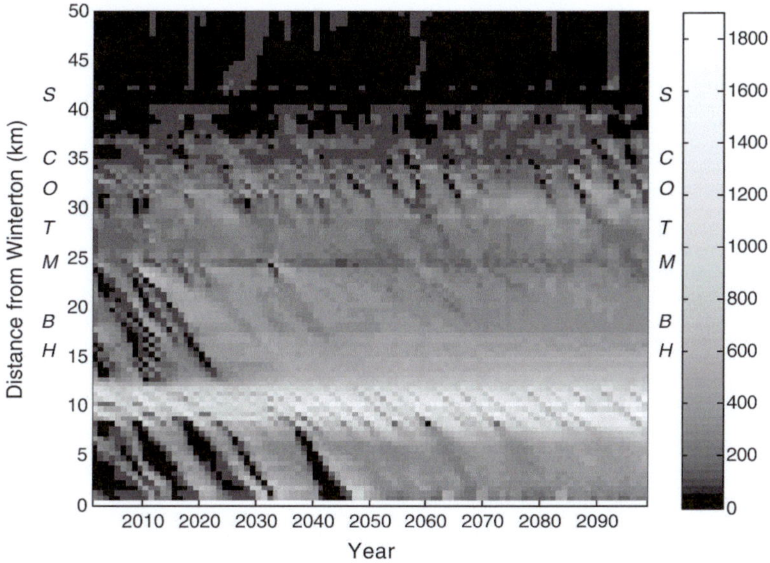

Fig. 1.21 Beach volumes (m^3/m) to 2100 predicted under Scenario 33 for the coastline north of Winterton. The letters on the *vertical axes* represent the locations of the settlements and the *right-hand bar* provides the scale

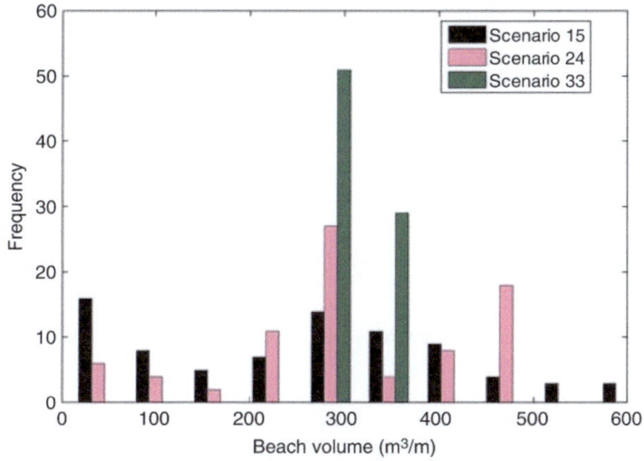

Fig. 1.22 Histogram of average annual beach volumes from 2020 to 2100 for Scenarios 15, 24 and 33 with 71 %, 34 % and 16 % coastal protection, respectively. There is increased variability in beach volume along the coast when more engineering structures are present

A4.2 Analysis of the Probability of Flooding

For all scenarios, the flood defence structures were assumed to be maintained to their current levels to explore the relative impacts of broadscale erosion management strategies on mitigating this flood risk.

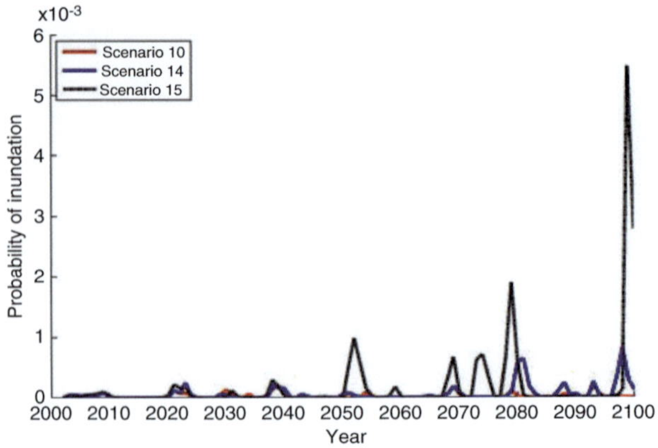

Fig. 1.23 Time series of annual flood probability (for a water depth greater than 0 m) for low, medium and high sea-level rise (Scenarios 10, 14 and 15) at E:645075, N:322775 just inland of the midpoint between Sea Palling and Winterton Ness

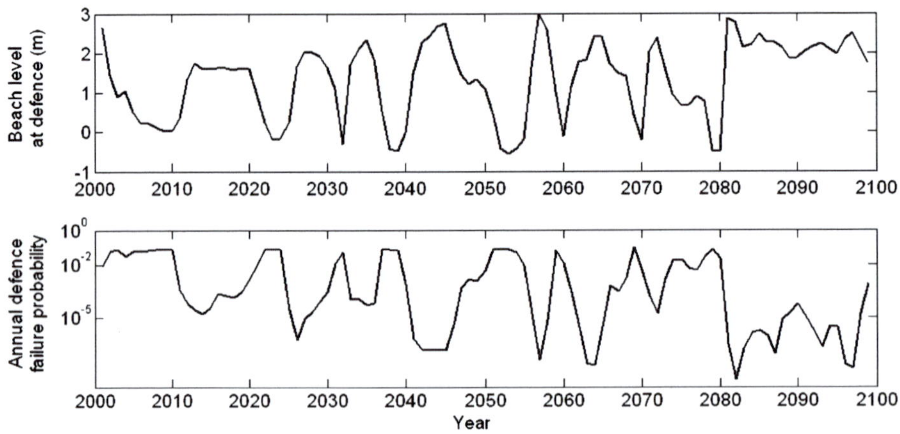

Fig. 1.24 The influence of beach morphology on defence failure probability for a flood defence in front of the Norfolk broads for Scenario 19

Figure 1.23 illustrates the changing annual probability of flooding in the Norfolk Broads. The plot shows considerable interannual, and longer, variability, but the general future trend suggests an increasing probability of flooding—particularly under the higher sea-level-rise scenario. Figure 1.24 illustrates how fluctuations in beach volume, and subsequently beach toe level, are amplified in the calculation of breach probabilities for the flood defences, which in this area are critically dependent on the beach level in front of them. This mechanism can be understood by consideration of the beach in Fig. 1.13, where it can be observed that in its present state the beach, seawall and dune together provide a reasonable level of protection, whilst beach lowering by several metres would greatly increase the probability of seawall overturning and rapid erosion of the dune. Figure 1.25 provides a broadscale

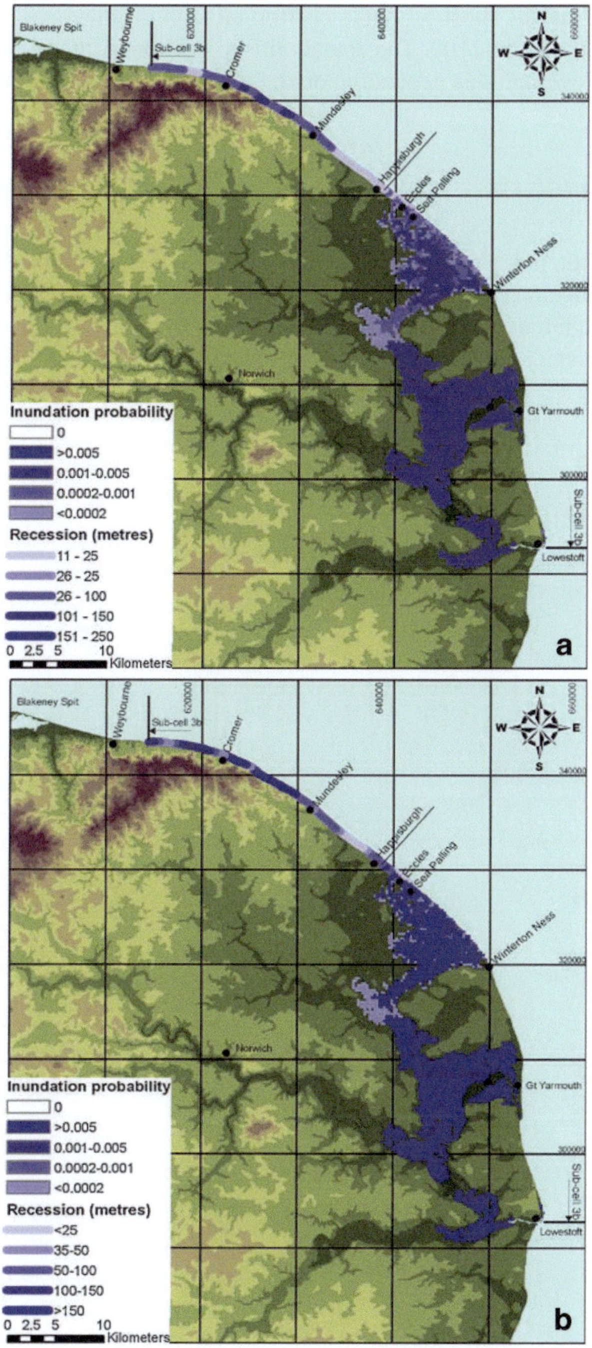

Fig. 1.25 Map of most probable (i.e. maxima of the p.d.f. $f_c(x|x_s,y,t)$ described in Eq. 1.1) clifftop erosion extent (metres) and the annual probability of flooding (to a depth of greater than 0 m) for (**a**) in 2003 and (**b**) in 2100 for Scenario 15 with high sea-level rise and 71 % of the coast protected. *Darker shades* imply greater probability of recession and flooding

overview showing the most probable clifftop erosion distance and the inundation probability in 2003 and 2100. The actual clifftop erosion distance may be substantially more or less than this depending on the cliff properties (Sect. A3.6).

A4.3 Risk Analysis

The results of the risk analysis are presented in 2003 prices. Figure 1.26 illustrates separately the erosion risk, r_c, and the flood risk, r_f. The flood risk at the site is at least an order of magnitude greater than the erosion risk in the base year and potentially increases at a greater rate. The flood risks grow exponentially during the latter half of the twenty-first century, whilst the erosion risk is predicted to remain relatively

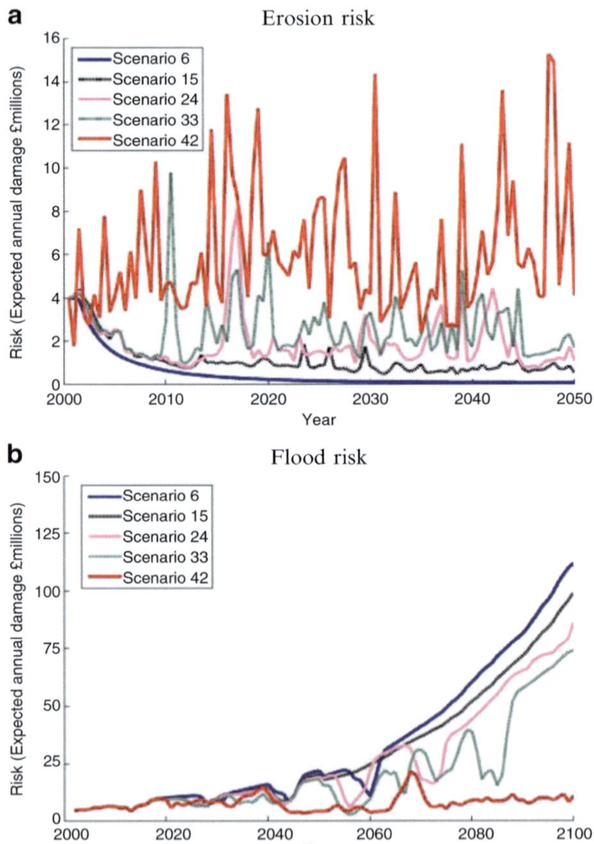

Fig. 1.26 Comparison of the evolution of (**a**) erosion and (**b**) flood risks over the twenty-first century under a high climate change scenario and assuming no change in socio-economic vulnerability: Scenarios 6,15, 24, 33 and 42 (Note: **a** and **b** are plotted to different scales) represent decreasing levels of cliff protection

constant through the twenty-first century, assuming the baseline socio-economic scenario. The difference between the two risks is further illustrated in Fig. 1.27, which shows how the erosion risk is concentrated in a narrow band near the coast, whereas the flood risk zone extends far in land. The differences between the risk functions associated with the two hazards arise because:

- Properties can be flooded on multiple occasions, whereas they can only be eroded from a clifftop once. Taking this to its most extreme: if an entire coastal settlement is lost to erosion over the course of the twenty-first century, the risk will decrease to near zero once the settlement has disappeared and all that remains is the agricultural hinterland.
- There are fewer properties at risk of erosion over the twenty-first century than at risk of flooding which reflects that the flood risk zone is larger and more populous than the erosion risk zone (Table 1.4).
- For a given return period, sea-level rise increases the area at risk of flooding and depth of inundation over the entire floodplain, whilst its influence on erosion is more complex as accelerated cliff recession generates additional sediment which can reduce recession rates elsewhere along the coastline (Dickson et al. 2007). In both cases, the relationship between sea-level rise and risk is not linear.

Figure 1.26 also illustrates the trade-off associated with different cliff management options. Whilst cliff protection reduces erosion risks (at a cost), it also reduces sediment supply to the flood-prone downdrift coast, leading to the potential for rapidly increasing flood risks over the twenty-first century (assuming no improvement in flood defence infrastructure). The erosion risk for the 100 % protection management scenario is never zero, because there is always a finite probability of a failure: despite the presence of cliff protection, landslides can still occur (cf. Frew and Guest 1997). However, over the twenty-first century the erosion risk associated with this scenario decreases due to the slope of the cliff tending towards an increasingly stable angle. Removal of cliff protection results in increasingly severe economic losses on the cliffed coastline according to the amount of protection removed. On the other hand, whilst the removal of cliff protection will result in significant economic losses on the cliffed coastline, the gains are usually of an order of magnitude greater, in terms of risk reduction, on the flood-prone coast—as shown in Fig. 1.26b by the time series plot of Scenario 42. Of particular note is that whilst the flood risk does increase from around £4 m to £11 m, the majority (60 %) of this increase comes from the floodplain south of Winterton Ness that does not benefit from the additional sediment influx from the eroding cliffs (i.e. where the main driver of changes to flood risk is from sea-level rise). The implications for coastal management of this insight are discussed further below.

Figure 1.28 illustrates the effect of different sea-level-rise scenarios on the projections of risk, assuming the baseline socio-economic scenario and 71 % coastal protection. The results reflect the predicted changes in the probability of flooding shown in Fig. 1.23. Current observed rates of sea-level rise are reflected in more or less steady risk levels over the twenty-first century. However, a tripling of the rate of sea-level rise from current rates to 4.5 mm/year over the twenty-first century results in

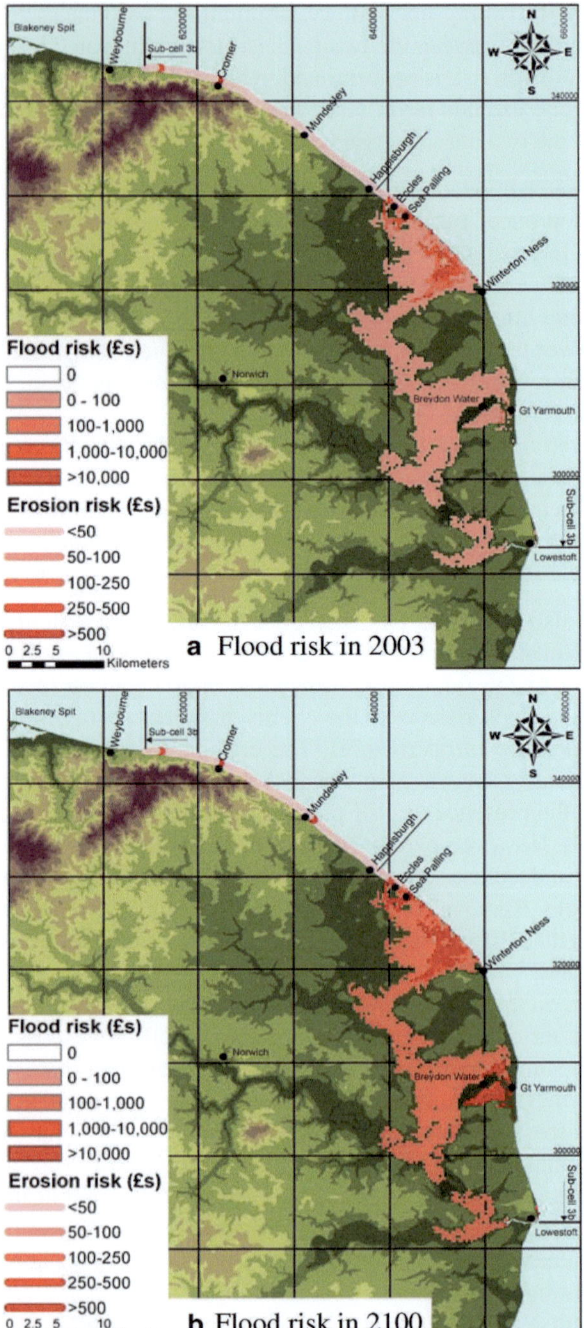

Fig. 1.27 Map of flood and erosion risks, expressed in terms of expected annual damages (**a**) at present and (**b**) in 2100 for Scenario 15 with constant socio-economic conditions. The cliff erosion risk has been scaled to the area of a flood cell (250 × 250 m). *Darker shades* imply a greater risk

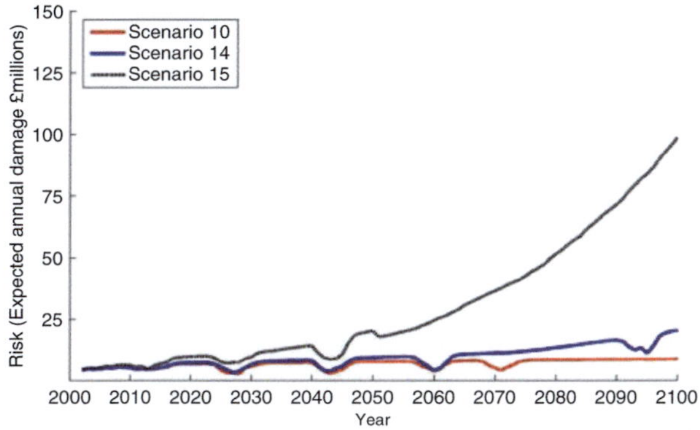

Fig. 1.28 The effect of sea-level rise on the evolution of flood risk over the twenty-first century for Scenarios 10, 14 and 15 (low, medium and high, respectively) assuming no socio-economic change

only an approximate doubling of the flood risk. The reduction is because rising sea levels will result in increasing release of sediments from the updrift cliffed coast (see Fig. 1.20). However, this feedback is insufficient to compensate for the effect of larger changes in climate: a 1.2 m sea-level rise over the twenty-first century results in greatly increased levels of flood risk. Somewhere in between the medium and high scenarios, a major threshold in coastal behaviour occurs, which could be thought of as constituting 'dangerous' climate change (according to the definition given by the UN 1994). The most marked differences become noticeable over the second half of the twenty-first century; before that time the difference between scenarios does not exceed the range of variability in the output from the coastal models.

The effects of increases in offshore wave heights had little impact on erosion rates and consequently flood risk. Whilst changes in offshore wave direction by ±10° proved more significant in terms of erosion, the consequent changes in available sediments were not sufficient to result in a noticeable impact on the flood risk.

All 45 scenarios representing different combinations of climate change, coastal management and socio-economic scenarios are summarised in Fig. 1.29 in terms of net present risks (i.e. total expected annual damages discounted to 2003 prices at a discount rate of 3.5 % for the first 30 years, 3 % for the subsequent 45 years and 2.5 % thereafter as recommended by HM Treasury (2003)). Figure 1.29 indicates that for high sea-level-rise scenarios, the lowest net present risk is to remove all cliff protection. However, for low and medium sea-level-rise scenarios, the increase in cliff erosion risk would not be offset by gains in flood risk reduction. In these cases the two interim management solutions that remove a portion of the existing protection would give the lowest net present economic risks. However, the use of a fixed discount rate over such an extended timescale is questionable, as is the use of a constant discount rate for different socio-economic scenarios which will have quite different associated rates of economic growth and attitudes towards the value of the future (Thorne et al. 2007). This is perhaps most evident when comparing low and medium sea-level-rise risks which, when discounted, are only marginally different

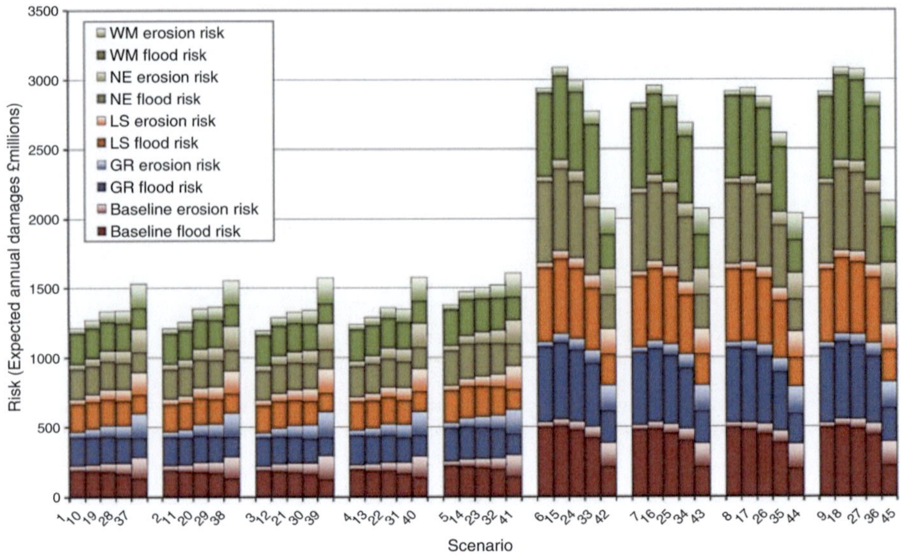

Fig. 1.29 Net present risks for each combination of management and socio-economic scenario clustered according to climate scenarios (see Table 1.2 for details)

as the two risk profiles only diverge later in the twenty-first century. Our aim in presenting discounted risks is not to provide a basis for cost–benefit analysis, but merely as a convenient means of summarising a large number of results integrated over time. Where a higher discount rate is to be used, the expected damages from flooding towards the end of the twenty-first century, would be weighted less, thereby reducing the net present benefits of removing cliff protection and vice versa.

As might be intuitively expected, socio-economic growth in the region acts to increase flood risk as analysed in terms of changes to the built environment. In all cases the damages are increased, with average increases in direct damages (at 2003 property and agricultural prices) of 6 %, 9 %, 16 % and 19 % under the local stewardship, global sustainability, national enterprise and world market futures, respectively. Thus, the effects of this growth on flood risk are, given the assumptions made in this study, of a similar order of magnitude when compared to a medium sea-level-rise scenario and relatively insignificant when compared to the highest potential sea-level-rise scenarios. However, these damages assume a constant depth–damage relationship through time, whilst recent experience has observed these functions to be dynamic due to the rising value of goods in properties (Penning-Rowsell et al. 2003a). A key insight from this work is that doubling of the number of properties (as occurs in the world market scenarios) does not imply a doubling of flood risk. This is because the location of development is crucial in determining the change in risk: development in floodplain locations with lower inundation probability contributes less towards flood risk when compared to development in areas of the floodplain that experience more frequent inundation. Likewise, development on higher ground within the floodplain will lead to lower damages for a given flood event because the depth of flooding is likely to be less. The sensitivity of risk estimates to these and other socio-economic changes should be explored in future studies.

A5 Implications for Coastal Management

The results from this study have quantified the connectivity between the various geomorphological features that comprise the natural coastal system and shown, quantitatively for the first time, that the vulnerabilities of the different parts of a coastal subcell are not independent. Clearly it is inappropriate to consider erosion and flood hazards in isolation. Rather, analysis of climate impacts and long-term coastal management must be implemented at a broader scale that accounts for this morphological interdependence. Currently, this is generally not the case. Coastal erosion and flooding are typically assessed independently and the governance structures do not always match the scale at which management is required. In this paper we have presented a method for integrated analysis of flood and erosion risks over extended timescales that enables management scenarios to be appraised under a range of plausible future climatic and socio-economic conditions using risk as a common currency to illustrate trade-offs and support decision-making.

Whilst the research presented here is considered to be a substantial contribution towards providing decision-makers with some of the necessary information to support integrated coastal zone management, key challenges remain, including the reduction of the vulnerability of coastal populations to climate impacts and the implementation of appropriate governance processes for the management of a transition towards a more sustainable configuration. Within the UK there are, however, moves to develop a more holistic approach to flood and erosion management that is risk based (DEFRA 2005), and recently there have been a range of strategies, consultations and planning documents produced that impact on the management of the coast (Milligan et al. 2006). Of particular importance are shoreline management plans which have defined subcells and cells, providing a framework for the methods described here (MAFF et al. 1995; DEFRA 2001, 2006). Other important initiatives include Making Space for Water (DEFRA 2004), planning guidance through Planning Policy Statement 25 (DCLG 2006) and Local Development Frameworks (ODPM 2004) and a number of coastal and maritime strategies (Milligan et al. 2006, Environment Agency 2005, National Trust 2005). Integrated coastal zone management is also being developed within Europe (European Commission 2000) and more widely (cf. RIKZ et al. 2005). Whilst it is clear that the various authorities involved in the management of the coastline are increasingly working together (Wright 2004), there are a range of issues, both in the UK (Milligan et al. 2006) and elsewhere (Moser 2005), that remain to be addressed around the extent to which integrated coastal zone management should be developed, funding and institutional arrangements, planning issues and public perceptions and expectations.

The progressive transformation of policy in the UK is proving controversial due to commonplace expectation amongst the general public that a ‘hold the line’ policy will be maintained. As identified in the review of management options, this is becoming increasingly challenged by policymakers and shoreline planners. Nevertheless, strengthening community and stakeholder involvement is a key principle underlying the new planning system in England and Wales (ODPM 2004). Inclusion remains a difficult task because of the dynamic complexity of coastal systems and the difficulty of communicating uncertainty to a wide range of stakeholders (Brown et al. 2006). These difficulties become especially apparent when one considers the need to 'involve stakeholders at all levels of risk management' and achieve a better balance 'between the three pillars of sustainable development (economic, social and environmental) in … risk management activities' (DEFRA 2005). However, as demonstrated in this paper, economic arguments do not necessarily conflict with the environmental reasons for allowing some of the coastline to return to a more natural configuration. However, whilst this work goes some way to supporting decision-makers, this type of analysis is incomplete in itself as environmental and equity issues must also be considered—we do not, for example, underestimate the serious social implications of allowing coastal settlements to be lost.

With the increase in emphasis on improving participation and consultation, there is a need to consider new techniques that can aid the communication of coastal information to the public (Jude et al. 2006). GIS visualisation techniques provide one such method to motivate and engage stakeholders (Jude et al. 2005, 2006; Nicholson-Cole 2005). Outputs from the erosion model presented here have been linked with a GIS to develop a visualisation of the evolving coastal environment (Fig. 1.30); the credibility of the resulting virtual landscapes are enhanced by their derivation from scientific data provided by the simulation model (Brown et al. 2006).

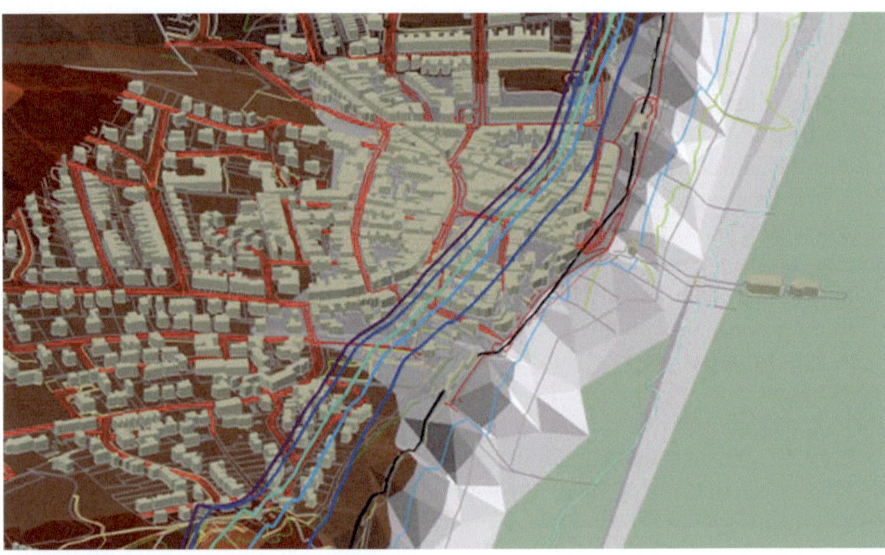

Fig. 1.30 Virtual reality visualisation of coasts (each *line* represents a 10-year central estimate of the cliff toe position)

The major implications for coastal management are, therefore, that:

1. Management of the coastline and the governance structures on which that management depends need to reflect the connectivity and consequential trade-offs in management policy, between the various geomorphological features that comprise the natural coastal system. This paper demonstrates that the tools are now available to conduct the analyses required to support such management, at least for open-coast settings. However, the principles shown here could be applied in other environments such as estuaries and deltas (which have been identified as very vulnerable to climate change by Nicholls et al. 2007b) using appropriate models for the analysis.
2. The economic analysis strengthens the argument for allowing some of the coastline to return to a more natural configuration and for negotiating compensation to landowners and householders to facilitate a change back to a more natural and dynamic shoreline. Such decisions should be based on multiple criteria encompassing economic, environmental and social dimensions.
3. Given the uncertainties in the future sea-level rise, and the extent to which this influences the effectiveness of different management strategies, an adaptive strategy that is continuously reviewed in the light of new information is most appropriate.
4. Further work is needed to better understand the socio-economic issues associated with coastal risk management. However, visualisation of the results from the simulation models provides a potentially powerful tool for motivating and engaging policymakers and stakeholders.

A6 Conclusions

A systematic appraisal of the effects of climate and socio-economic change and coastal management policy on risks associated with coastal erosion and flooding has been described. This has been achieved through coupled modelling of water levels, waves, coastal morphology and risk analysis over the whole of a reasonably self-contained coastal system (UK subcell 3b). The outcomes of coastal management and climate change scenarios for two natural hazards have, for the first time, been presented quantitatively in terms of changes in expected annual damage thereby enabling direct economic evaluation of local effects of modifying coastal cliff erosion compared with potential broader-scale effects on flood risk. The analysis indicates that the main drivers for flood risk over the twenty-first century in northern East Anglia are a consequence of a combination of:

- Rising sea levels
- Local and broadscale natural and anthropogenically driven morphological change that may lower or raise beach levels
- Changes in exposure to flooding and erosion due to socio-economic changes

Increases in wave height are found to be less significant than these factors owing to attenuation across the nearshore and are only felt on the coast if accompanied by high sea-level rise. Possible changes in wave direction have a more noticeable influence on coastal morphology, but this does not translate into significant changes in economic risk. Although the risk profiles will be different for other locations, the key drivers are likely to be the same in similar settings around the world's coasts.

A rapidly increasing sensitivity to sea level is predicted for rates of mean sea-level rise greater than about 4.5 mm/year. This corresponds to an increase in global average surface air temperature of 3.3 °C and total atmospheric CO_2 of 715 ppm, whilst the high sea-level-rise scenario corresponds to 970 ppm and 4.5 °C (Church et al. 2001; Hulme et al. 2002), although there is uncertainty in this relationship about the mean values presented here. Risk functions will differ between regions (e.g. Nicholls 2004), but the possibility of such a high degree of non-linearity for potential impacts will have significant implications for policymakers concerned with global emission mitigation. Further investigation of these thresholds for the scale of subcells up to regions and even globally would be useful.

The results demonstrate that over the twenty-first century, significant benefits in terms of mitigating flood risk can be obtained by allowing previously defended cliffs to erode naturally. These benefits are greatest under high sea-level-rise scenarios. However, the economic and social implications for clifftop communities of such a policy are significant. The changes in risk are predicted to be quite steady over the first half of the twenty-first century, allowing time for planned adaptation. However, given the slow turnover of housing and infrastructure, plans for managed retreat from parts of the coast need to be implemented now if excessive adaptation costs are not to be incurred in the latter half of the twenty-first century.

Because of the uncertainty in future sea-level rise, and the extent to which this rise influences the effectiveness of the removal of cliff protection on flood risk reduction, an adaptive strategy that can be modified according to the actual sea-level rise is appropriate. Management of the coastline and the governance structures on which that management depends need to reflect the connectivity between the various geomorphological features that comprise the natural coastal system. This paper shows that the tools are now available to conduct the analyses required to support such management. Given the benefits in terms of risk reduction that this study has demonstrated, a cogent case can be made for compensation of clifftop communities who have to make way to allow coastal erosion to proceed.

The uncertainties in generating predictions of coastal erosion and flood risk over extended timescales are considerable. We have grounds for some confidence in the results because of the validation and track record of the TOMAWAC and LISFLOOD models and critically (for simulation of long-term change) the validation and stability of the morphological model (Dickson et al. 2007). However, major uncertainties, associated with climate change and socio-economic change, have merely been dealt with in terms of scenarios. Extension to include probabilistic climate scenarios and dynamical downscaling of waves and surges from regional climate models is planned. Whilst we have sought to give a reasonable coverage of combinations of climate change, socio-economic change and coastal management, not all possible options have been tested and that other combinations of greenhouse gas emissions, coastal management and socio-economic scenarios could be realised.

The analysis has integrated knowledge on complex natural, engineered and social systems: regional climate predictions have driven hydrodynamic and morphological models and captured the dynamic effects of climate change and its interaction with the natural, engineered and social systems. Such a broad analysis can never be complete, nor can we expect to deal with individual process within this interacting system in utmost detail. However, the rationale is that that by representing the dominant interactions that determine the long-term behaviour of the system the response to changes, be they due to climatic change or coastal management, can be simulated with some degree of confidence. By quantifying the impacts of these changes in terms of the common currency of risk, better and more consistent information is available to decision-makers and the management policies concerning different hazards and involving multiple stakeholders can be more effectively understood and communicated.

Acknowledgements This research was funded by the *Tyndall Centre for Climate Change Research* through the ongoing coastal programme (http://www.tyndall.ac.uk/research/programme5/). The *Environment Agency of England and Wales* provided lidar data, surveys of flood protection structures and the National Property Database. Tide data were supplied by the *British Oceanographic Data Centre* as a part of the function of the National Tidal and Sea Level Facility. The wave time series was supplied by *HR Wallingford Ltd.*, who also determined the storm conditions under which material moved to and from the offshore. The *British Geological Survey* provided geomorphic information for model calibration. Information on the history of construction along the cliffed coastline was provided by Peter Lawton (*St. La Haye Consulting Engineers*) and *North Norfolk District Council*. Previous development of SCAPE has been funded through the *UK Engineering and Physical Sciences Research Council*, as well as the *North Norfolk District Council* and *Halcrow*. Previous development of the flood risk analysis methodology has been funded through the joint research and development programme of the *Department for Environment, Food and Rural Affairs* and *Environment Agency of England and Wales* as the RASP (Risk Assessment for Strategic Planning) project. Mark Dickson was supported by a fellowship provided by the *Foundation for Research, Science and Technology* (New Zealand) through part of this research.

References of Appendix

Bates, P. D., & De Roo, A. P. J. (2000). A simple raster-based model for floodplain inundation. *Journal of Hydrology, 236*, 54–77.

Bates, P. D., Dawson, R. J., Hall, J. W., Horritt, M. S., Nicholls, R. J., Wicks, J., et al. (2005). Simplified two-dimensional modelling of coastal flooding for risk assessment and planning. *Coastal Engineering, 52*(9), 793–810.

Bouws, E., Jannink, D., & Komen, G. J. (1996). The increasing wave height in the North Atlantic Ocean. *Bulletin of the American Meteorological Society, 77*(10), 2275–2277.

Brown, I., Jude, S., Koukoulas, S., Nicholls, R., Dickson, M., & Walkden, M. (2006). Dynamic simulation and visualisation of coastal erosion. *Computers, Environment and Urban Systems, 30*, 840–860.

Bruun, P. (1962). Sea-level rise as a cause of shore erosion. *Proceedings of the ASCE Journal of Waterways Harbour Division, 88*, 117–130.

Burgess, K., Jay, H., Nicholls, R. J., Green, C., & Penning-Rowsell, E. C. (2007). Assessment of future coastal erosion risk. In C. Thorne, E. Evans, & E. C. Penning-Rowsell (Eds.), *Future flooding and coastal erosion risks*. London: Thomas Telford.

CERC. (1984). *Shore protection manual*. Vicksburg: US Army Engineer Waterways Experiment Station Coastal Engineering Research Centre.

Chang, S.-C., & Evans, G. (1992). Source of sediment and sediment transport on the East Coast of England: Significant or coincidental phenomena? *Marine Geology, 107*, 283–288.

Church, J. A., Gregory, J. M., Huybrechts, P., Kuhn, M., Lambeck, K., Nhuan, M. T., et al. (2001). *Climate change 2001: The scientific basis. Intergovernmental Panel on Climate Change Third Assessment Report*. Cambridge: Cambridge University Press.

CIRIA. (1986). *Performance of sea walls: Technical note 126*. London: CIRIA.

CIRIA and CUR. (1991). *Manual on the use of rock in coastal and shoreline engineering* (CIRIA Special Publication 83/CUR Report 154). London: CIRIA.

Clayton, K. M. (1989). Sediment input from the Norfolk Cliffs, Eastern England—A century of coast protection and its effect. *Journal of Coastal Research, 5*, 433–442.

Cooper, N., Bray, M., Carter, D., & Barber, P. (2002). Shoreline management plans: A national review and engineering perspective. *Proceedings of the Institution of Civil Engineers Water Maritime and Energy, 154*(3), 221–228.

Dawson, R. J., Dickson, M., Nicholls, R. J., Hall, J., Walkden, M. J. A., Stansby, P. K., Mokrech, M., Richards, J., Zhou, J., Milligan, J., Jordan, A., Pearson, S., Rees, J., Bates, P. D., Koukoulas, S., & Watkinson, A. (2009). Integrated analysis of risks of coastal flooding and cliff erosion under scenarios of long-term change. *Climatic Change, 95*, 249–288. (see Appendix).

Dawson, R. J., & Hall, J. W. (2006). Adaptive importance sampling for risk analysis of complex infrastructure systems. *Proceedings of the Royal Society of London A, 462*(2075), 3343–3362.

DCLG. (2006). *Planning policy statement 25: Development and flood risk*. London: TSO.

DEFRA. (2001, June). Shoreline management plans: A guide for coastal defence authorities.

DEFRA. (2004, July). Making space for water: Developing a new government strategy for flood and coastal erosion risk management in England. A consultation exercise. http://www.defra.gov.uk/environ/fcd/policy/strategy.htm

DEFRA. (2005). Taking forward a new government strategy for flood and coastal erosion risk management in England. First government response to the autumn 2004 making space for water consultation exercise. http://www.defra.gov.uk/environ/fcd/policy/strategy/1stres.pdf

DEFRA. (2006, March). Shoreline management plan guidance—Volume 1: Aims and requirements. Volume 2: Procedures. http://www.defra.gov.uk/environ/fcd/pubs/smp/revisedsmpguidancefinal.pdf

Dickson, M. E., Walkden, M. J. A., & Hall, J. W. (2007). Systemic impacts of climate change on an eroding coastal region over the twenty-first century. *Climatic Change, 84*(2), 141–166.

Dixon, M. J., & Tawn, J. A. (1997). *Estimates of extreme sea conditions* (Proudman Oceanographic Laboratory Report 112), p 217.

DTI. (2002). *Foresight futures 2002: Revised scenarios and guidance*. London: UK Department of Trade and Industry.

Eastman, J. R., Kyem, P. A. K., Toledano, J., & Jin, W. (1993). GIS and decision making. In *Explorations in Geographic Information Systems technology* (Vol. 4). Geneva: United Nations Institute for Training and Research (UNITAR).

English Nature, DEFRA, Environment Agency, Natural Environment Research Council and LIFE-EU. (2003, April). *Winterton Dunes, Coastal habitat management Plan*.

Environment Agency. (1996). *East Anglian sea defence survey*. Peterborough: Environment Agency.

Environment Agency. (2005). *Cleaner coasts, healthier seas*. Working for a better marine environment: Our strategy for 2005–2011. http://www.environment-agency.gov.uk/subjects/conservation/208776/1209976

European Commission. (2000). *Communication from the commission to the council and the European parliament on integrated coastal zone management: A strategy for Europe*. Commission of the European Communities, COM 2000, 547 final, Brussels.

Eurosion. (2004). Living with coastal erosion in Europe: Sediment and space for sustainability: Major findings and policy recommendations of the EUROSION project. European Commission, Directorate General Environment, Service contract B4-3301/2001/329175/mAr/B3.

Frew, P., & Guest, S. (1997). *Overstrand coast protection scheme*. In Proceedings of MAFF conference of river and coastal engineers. London: MAFF.

Futurecoast. (2002). *CD-ROM, produced for Department for Environment* (Food & Rural Affairs). Swindon: Halcrow.

GOEE. (2005). *Regional housing strategy for the east of England*. Government Office for the East of England.

Gregory, J. M., Church, J. A., Boer, G. J., Dixon, K. W., Flato, G. M., Jackett, D. R., et al. (2001). Comparison of results from several AOGCMs for global and regional sea-level change 1900–2100. *Climate Dynamics, 18*, 225–240.

Grieve, H. (1959). *The Great Tide: The story of the 1953 flood disaster in Essex* (833 pp). Chelmsford: Essex County Council.

Halcrow. (1999). *Gorleston to Lowestoft coastal strategy study*. Halcrow Group Ltd.

Halcrow, English Nature, Environment Agency, North Norfolk District Council, Great Yarmouth Borough Council, Waveney District Council and Great Yarmouth Port Authority. (2006). *Kelling to Lowestoft Ness shoreline management plan review*. Anglian Coastal Authorities Group.

Hall, J. W., Sayers, P. B., & Panzeri, M. (1861). Impacts of climate change on the coast of England and wales: 2030–2100. *Philosophical Transactions of the Royal Society A, 364*, 1027–1049.

Hall, J. W., Lee, E. M., Meadowcroft, I. C. (2000, September). Risk-based benefit assessment of coastal cliff protection. *Proceedings of the Institution of Civil Engineers Water Maritime and Energy, 142*, 127–139.

Hall, J. W., Dawson, R. J., Sayers, P. B., Rosu, C., Chatterton, J. B., & Deakin, R. (2003a). A methodology for national-scale flood risk assessment. *Water Maritime and Energy ICE, 156*(3), 235–247.

Hall, J. W., Evans, E. P., Penning-Rowsell, E. C., Sayers, P. B., Thorne, C. R., & Saul, A. J. (2003b). Quantified scenarios analysis of drivers and impacts of changing flood risk in England and Wales: 20302100. *Global Environmental Change 5*(3–4), 51–65.

Hall, J. W., Tarantola, S., Bates, P. D., & Horritt, M. S. (2005). Distributed sensitivity analysis of flood inundation model calibration. *Journal of Hydraulic Engineering ASCE, 131*(2), 117–126.

Harland, M. G., & Harland, H. J. (1980). *The flooding of Eastern England*. Peterborough: Minimax Books.

Holman, I. P., Loveland, P. J., Nicholls, R. J., Shackley, S., Berry, P. M., Rounsevell, M. D. A., et al. (2002). *REGIS-Regional climate change impact and response studies in East Anglia and North West England*. Oxford: UKCIP and DEFRA.

Holman, I. P., Rounsevell, M. D. A, Shackley, S., Harrison, P. A., Nicholls, R. J., Berry, P. M., & Audsley, E. (2005). A regional, multi-sectoral and integrated assessment of the impacts of climate and socio-economic change in the UK: I methodology. *Climatic Change, 71*(1–2), 9–41.

Horritt, M. S., & Bates, P. D. (2001). Predicting floodplain inundation: Raster-based modelling versus the finite element approach. *Hydrological Processes, 15*, 825–842.

HM Treasury. (2003). *The green book: Appraisal and evaluation in central government*. London: HM Stationery Office.

HR Wallingford. (1980). *Design of seawalls for wave overtopping*. Report EX924, HR Wallingford.

HR Wallingford. (1999). *Overtopping of seawalls: Design and assessment manual*. (R&D Technical Report W178).

HR Wallingford. (2002). *Overstrand to mundesley strategy study: Hydrodynamics*. (HR Wallingford Ltd. Report).

HR Wallingford. (2003). *Beach lowering in front of coastal structures* (DEFRA/EA R&D Scoping Study Report FDW16/TR) London: DEFRA.

HR Wallingford. (2004). *Investigation of extreme flood processes & uncertainty* (IMPACT, EC Research Project no. EVG1-CT2001-00037). www.impact-project.net

Hulme, M., Jenkins, G. J., Lu, X., Turnpenny, J. R., Mitchell, T. D., Jones, R. G., Lowe, J., Murphy, J. M., Hassell, D., Boorman, P., McDonald, R., & Hill, S. (2002). Climate change scenarios for the United Kingdom: The UKCIP02 scientific report. In *Tyndall centre for climate change research, school of environmental sciences* (120 pp). Norwich: University of East Anglia.

Jude, S., Brown, I., Koukoulas, S., Dickson, M., Walkden, M., Jones, A., et al. (2005). Exploring future coastlines: Initial steps towards the development of an integrated coastal simulation and visualisation system. In E. Buhmann, P. Paar, I. J. Bishop, & E. Lange (Eds.), *Trends in real-time landscape visualization and participation* (pp. 69–76). Heidelberg: Herbert Wichmann Verlag.

Jude, S., Jones, A. P., Andrews, J. E., & Bateman, I. J. (2006). Visualisation for participatory coastal zone management: A case study of the Norfolk coast, England. *Journal of Coastal Research, 22*, 1527–1538.
Kelman, I. (2003, November 10). 1953 storm surge deaths: U.K. version 4. Downloaded from http://www.ilankelman.org/disasterdeaths/1953DeathsUK.doc
Komar, P. D. (1988). *Beach processes and sedimentation* (2nd ed.). Upper Saddle River: Prentice Hall.
Kuang, C. P., & Stansby, P. S. (2004). Modelling directional random wave propagation inshore. *Proceedings of the Institution of Civil Engineers Water Maritime and Energy, 157*, 123–131.
Kushnir, Y., Cardone, V. J., Greenwood, J. G., & Cane, M. A. (1997). The recent increase in North Atlantic wave heights. *Journal of Climate, 10*(8), 2107–2113.
Leafe, R., Pethick, J., & Townend, I. (1998). Realizing the benefits of shoreline management. *The Geographical Journal, 164*(3), 282–290.
Miller, L., & Douglas, B. C. (2004). Mass and volume contributions to twentieth-century global sea level rise. *Nature, 428*, 406–409.
Milligan, J., O'Riordan, T., & Watkinson, A. (2006). *Designing coastlines fit for the future* (English Nature Research Reports, No 702). Peterborough: English Nature.
Ministry for Agriculture, Fisheries-and-Food, Welsh Office, Association of District Councils, English Nature and National Rivers Authority. (1995). *Shoreline management plans: A guide for coastal defence authorities*. London: MAFF.
Mosby, J. E. G. (1938). The Horsey Flood, 1938: An example of storm effect on a low coast. *The Geographical Journal, 93*(5), 413–418.
Moser, S. C. (2005). Impacts assessments and policy responses to sea-level rise in three US states: An exploration of human dimension uncertainties. *Global Environmental Change, 15*, 353–369.
Nairn, R. B., & Southgate, H. N. (1993). Deterministic profile modelling of nearshore processes. Part 2, sediment transport and beach profile development. *Coastal Engineering, 19*, 57–96.
National Trust. (2005). Shifting shores: Living with coastal change. http://www.nationaltrust.org.uk/main/w-shifting_shores.pdf
Nicholls, R. J. (1995). Coastal megacities and climate change. *Geojournal, 37*(3), 369–379.
Nicholls, R. J. (2004). Coastal flooding and wetland loss in the 21st century: Changes under the SRES climate and socio-economic scenarios. *Global Environmental Change, 14*, 69–86.
Nicholls, R. J., Mokrech, M., Richards, J., Bates, P., Dawson, R., Hall, J., Walkden, M., Dickson, M., Jordan, A., & Milligan, J. (2005, October). *Assessing coastal flood risk at specific sites and regional scales: Regional assessment of coastal flood risk* (Tyndall Centre Technical Report No. 45). http://www.tyndall.ac.uk/research/theme4/final_reports/t2_46.pdf
Nicholls, R. J., Tol, R. S. J., & Hall, J. W. (2007a). Assessing impacts and responses to global-mean sea-level rise. In M. E. Schlesinger, H. S. Kheshgi, J. Smith, F. C. de la Chesnaye, J. M. Reilly, T. Wilson, & C. Kolstad (Eds.), *Human-induced climate change* (pp. 119–134). Cambridge: Cambridge University Press.
Nicholls, R. J., Wong, P. P., Burkett, V. R., Codignotto, J. O., Hay, J. E., McLean, R. F., et al. (2007b). Coastal systems and low-lying areas. Climate Change 2007: Impacts, adaptation and vulnerability. In M. L. Parry, O. F. Canziani, J. P. Palutikof, P. J. van der Linden, & C. E. Hanson (Eds.), *Contribution of Working Group II to the fourth assessment report of the Intergovernmental Panel on Climate Change* (pp. 315–356). Cambridge: Cambridge University Press.
Nicholson-Cole, S. (2005). Representing climate change futures: A critique on the uses of images for visual communication. *Computers, Environment and Urban Systems, 29*, 255–273.
NNDC. (2006). North Norfolk Local Plan: B35 Chapter 7—Environmental safety. North Norfolk District Council, updated March 31st 2006. www.north-norfolk.gov.uk/planning/
Nordhaus, W. (2006). Geography and macroeconomics: New data and new findings. *Proceedings of the National Academy of Sciences, 103*, 3510–3517.
ODPM. (2003). *Housing market: Simple average house prices*. London: Office of the Deputy Prime Minister.
ODPM. (2004). *Planning Policy Statement 12: Local development frameworks*. London: Office of the Deputy Prime Minister.

Parry, M. L. (2004). An assessment of the global effects of climate change under SRES emissions and socio-economic scenarios. *Global Environmental Change, 14*, 1–99.

Pelnard-Considere, R. (1956). Essai de theorie de l'evolution des forms de ravage en plage de sable et de galets. In *4th Journees de l'Hydraulique, Les Energies de la Mer. Question III* (pp. 792–808).

Penning-Rowsell, E. C., Johnson, C., Tunstall, S. M., Tapsell, S. M., Morris, J., Chatterton, J. B., et al. (2003a). *The benefits of flood and coastal defence: Techniques and data for 2003*. London: Flood Hazard Research Centre/Middlesex University.

Penning-Rowsell, E. C., Johnson, C., Tunstall, S. M., Tapsell, S. M., Morris, J., Chatterton, J. B., et al. (2003b). *The benefits of flood and coastal risk management*. London: Flood Hazard Research Centre/Middlesex University.

Pringle, A. W. (1985). Holderness coast erosion and the significance of ords. *Earth Surface Processes and Landforms, 10*, 107–124.

Reeve, D. E., & Burgess, K. A. (1993). A method for the assessment of coastal flood risk. *IMA Journal of Mathematics Applied in Business and Industry, 5*, 197–209.

RICS. (2003). *Rural land market survey Great Britain: Second Quarter 2003*. London: Royal Institution of Chartered Surveyors.

RIKZ, UNESCO-IHE, CRC. (2005). *ICZM in practice with examples from different countries*. RIKZ Work document nr. RIKZ/KW/2005.102w.

Rossiter, J. R. (1954). The North sea storm surge of 31st January and 1st February 1953. *Philosophical Transactions of the Royal Society Series A, A246*(915), 317–400.

Sciberras, A. M. (2002). Integrated coastal zone management—A challenge for the EU in the 21st century. *EIPASCOPE, 2002*(1), 1–5.

Shennan, I., & Horton, B. (2002). Holocene land- and sea-level changes in Great Britain. *Journal of Quaternary Science, 17*, 511–526.

Small, C., & Nicholls, R. J. (2003). A global analysis of human settlement in coastal zones. *Journal of Coastal Research, 19*(3), 584–599.

Stansby, P. K., Zhou, J. G., Kuang, C. P., Walkden, M. J. A., Hall, J. W., & Dickson, M. E. (2006). Long-term prediction of nearshore wave climate with an application to cliff erosion. In *Proceedings of the 30th international conferences on coastal engineering*, San Diego.

Steers, J. A. (1953). The east coast floods. *The Geographical Journal, 119*(3), 280–295.

Thomalla, F., & Vincent, C. E. (2003). Beach response to shore-parallel breakwaters at Sea Palling, Norfolk, UK. *Estuarine, Coastal and Shelf Science, 56*(2), 203–212.

Thomas, R. S., & Hall, B. (1992). *Seawall design*. London: CIRIA.

Thorne, C., Evans, E., & Penning-Rowsell, E. (2007). *Future flooding and coastal erosion risks*. London: Thomas Telford.

Turner, R. K., Subak, S., & Adger, W. N. (1996). Pressures, trends, and impacts in coastal zones: Interactions between socioeconomic and natural systems. *Environmental Management, 20*(2), 159–173.

UN. (1994). *United Nations framework convention on climate change*. United Nations.

Vellinga, P. (1983). *Verification of predictive computational model for beach and dune erosion during storm surges*. Delft: Delft Hydraulics Laboratory.

Vincent, C. E. (1979). Longshore sand transport rates—A simple model for the East Anglian coastline. *Coastal Engineering, 3*, 113–136.

Visser, P. J. (1998). *Breach growth in sand defences. Communication on hydraulic and geotechnical engineering* (TU Delft Report No. 98-91).

VOA. (2003, Spring). *Property market report*. Valuation Office Agency.

Wahl, T. L. (1998). *Prediction of embankment dam breach parameters: A literature review and needs assessment*. Dams Safety Office: DSO-98-004. www.usbr.gov/pmts/hydraulics_lab/twahl/

Walkden, M., & Dickson, M. (2008). Equilibrium erosion of soft rock shores with a shallow or absent beach under increased sea level rise. *Marine Geology, 251*(1–2),75–84.

Walkden, M. J. A., & Hall, J. W. (2005). A predictive mesoscale model of the erosion and profile development of soft rock shores. *Coastal Engineering, 52*, 535–563.

Wright, M. (2004). Coastal defence—Current UK practice. *Proceedings of the ICE Munich Engineering, 157*, 147–155.

References

ABPMer. (2013). *The online managed realignment guide*. http://www.abpmer.net/omreg/. Cited December 1, 2013.

Adger, W. N., Agrawala, S., Mirza, M. M. Q., Conde, C., O'Brien, K. L., Pulhin, J., Pulwarty, R., Smit, B., & Takahashi, K. (2007). Assessment of adaptation practices, options, constraints and capacity. In M. L. Parry, O. F. Canziani, J. P. Palutikof, C. E. Hanson, & P. J. van der Linden (Eds.), *Climate change 2007: Impacts, adaptation and vulnerability* (Contribution of Working Group II to the Fourth Assessment Report of the Intergovernmental Panel on Climate Change). Cambridge: Cambridge University Press.

Adger, W. N., Lorenzoni, I., & O'Brien, K. L. (2009). *Adapting to climate change – Thresholds, values, governance*. Cambridge: Cambridge University Press.

Agardy, T., Alder, J., Dayton, P., Curran, S., Kitchingman, A., Wilson, M., Catenazzi, A., Restrepo, J., Birkeland, C., Blaber, S., Saifullah, S., Branch, G., Boersma, D., Nixon, S., Dugan, P., Davidson, N., & Vörösmarty, C. (2005). Coastal systems. In *Millenium ecosystem assessment: Ecosystems & human well-being* (Current State and Trends, Vol. 1). Washington, DC: Island Press.

Burgess, K., Orford, J., Dyer, K., Townend, I., & Balson, P. (2002). FUTURECOAST – The integration of knowledge to assess future coastal evolution at a national scale. *Proceedings of the 28th International Conference on Coastal Engineering, World Scientific, 3,* 3221–3233.

Chang, S.-C., & Evans, G. (1992). Source of sediment and sediment transport on the east coast of England: Significant or coincidental phenomena? *Marine Geology, 107*, 283–288.

Clayton, K. M. (1989). Sediment input from the Norfolk cliffs, Eastern England – A century of coast protection and its effects. *Journal of Coastal Research, 5*, 433–442.

Collins, M. B., Shimwell, S. J., Gao, S., Powell, H., Hewitson, C., & Taylor, J. A. (1995). Water and sediment movement in the vicinity of linear banks: The Norfolk Banks, southern North Sea. *Marine Geology, 123*, 125–142.

Crossland, C. J., Kremer, H. H., Lindeboom, H. J., Marshall Crossland, J. I., & Le Tissier, M. D. A. (2005). Coastal fluxes in the anthropocene. In *The land-ocean interactions in the coastal zone project of the International Geosphere-Biosphere Programme* (Global change – The IGBP series). Berlin: Springer.

Dasgupta, S., Laplante, B., Meisner, C., Wheeler, D., & Yan, J. (2009). The impact of sea-level rise on developing countries: A comparative analysis. *Climatic Change, 93*, 379–388.

Dawson, R. J. (2007). Re-engineering cities: A framework for adaptation to global change. *Philosophical Transactions of the Royal Society A, 365*(1861), 3085–3098.

Dawson, R. J., Dickson, M., Nicholls, R. J., Hall, J., Walkden, M. J. A., Stansby, P. K., Mokrech, M., Richards, J., Zhou, J., Milligan, J., Jordan, A., Pearson, S., Rees, J., Bates, P. D., Koukoulas, S., & Watkinson, A. (2009). Integrated analysis of risks of coastal flooding and cliff erosion under scenarios of long-term change. *Climatic Change, 95*, 249–288.

DEFRA. (2010). *Adapting to coastal change: Developing a policy framework*. London: DEFRA. http://archive.defra.gov.uk/environment/flooding/documents/manage/coastalchange-policyframework.pdf. Cited December 19, 2013.

DEFRA. (2011). *Shoreline management plans: Guidance*. London: Department for Environment, Food and Rural Affairs. https://www.gov.uk/government/publications/shoreline-management-plans-guidance. Cited December 19, 2013.

DTI. (2002). *Foresight futures 2020: Revised scenarios and guidance*. London: Department of Trade and Industry.

East Anglian Coastal Group. (2012). *Kelling to Lowestoft Ness Shoreline Management Plan* (SMP6). East Anglian Coastal Group. http://www.eacg.org.uk/smp6.asp. Cited December 19, 2013.

Ford, J. D., Berrang-Ford, L., Lesnikowski, A., Barrera, M., & Heymann, S. J. (2013). How to track adaptation to climate change: A typology of approaches for national-level application. *Ecology and Society, 18*(3), 40. http://dx.doi.org/10.5751/ES-05732-180340.

Foresight. (2011). *Migration and global environmental change* (Final project report). London: The Government Office for Science.

Grieve, H. (1959). *The great tide*. Chelmsford: County Council of Essex.

Haigh, I., Nicholls, R. J., & Wells, N. (2009). Mean sea level trends around the English Channel over the 20th century and their wider context. *Continental Shelf Research, 29*(17), 2083–2098.

Hall, J. W., Dawson, R. J., Sayers, P. B., Rosu, C., Chatterton, J. B., & Deakin, R. (2003). A methodology for national-scale flood risk assessment. *Proceedings of the ICE – Water and Maritime Engineering, 156*(3), 235–247.

Hall, J. W., Sayers, P. B., & Dawson, R. J. (2005). National-scale assessment of current and future flood risk in England and Wales. *Natural Hazards, 36*, 147–164. doi:10.1007/s11069-004-4546-7.

Hall, J. W., Dawson, R. J., Walsh, C. L., Barker, T., Barr, S. L., Batty, M., Bristow, A. L., Burton, A., Carney, S., Dagoumas, A., Evans, S., Ford, A. C., Glenis, V., Goodess, C. G., Harpham, C., Harwatt, H., Kilsby, C. G., Kohler, J., Jones, P., Manning, L., McCarthy, M., Sanderson, M., Tight, M. R., Timms, P. M., & Zanni, A. (2009). *Engineering cities: How can cities grow whilst reducing emissions and vulnerability?* Norwich: The Tyndall Centre for Climate Change Research.

Harland, M. G., & Harland, H. J. (1980). *The flooding of Eastern England*. Peterborough: Minimax Books.

Hay, J. E. (2009). *Institutional and policy analysis of disaster risk reduction and climate change adaptation in Pacific Island Countries*. Suva: United Nations International System for Disaster Reduction (UNISDR) and the United Nations Development Programme (UNDP). http://www.unisdr.org/we/inform/publications/18869. Cited December 19, 2013.

Holman, I. P., Rounsevell, M. D. A., Shackley, S., Harrison, P. A., Nicholls, R. J., Berry, P. M., & Audsley, E. (2005). A regional, multi-sectoral and integrated assessment of the impacts of climate and socio-economic change in the UK: I Methodology. *Climatic Change, 71*, 9–41.

Horrillo-Caraballo, J. M., & Reeve, D. E. (2008). Morphodynamic behaviour of a nearshore sandbank system: The Great Yarmouth Sandbanks, U.K. *Marine Geology, 254*(1–2), 91–106.

Hulme, M., Jenkins, G. J., Lu, X., Turnpenny, J. R., Mitchell, T. D., Jones, R. G., Lowe, J., Murphy, J. M., Hassell, D., Boorman, P., McDonald, R., & Hill, S. (2002). *Climate change scenarios for the United Kingdom: The UKCIP02 scientific report*. Norwich: Tyndall Centre for Climate Change Research.

Humby, P. (2013). *A profile of deprivation in larger English seaside destinations, 2007 and 2010*. London: Office for National Statistics.

Huthnance, J. M. (1982). On one mechanism forming linear sandbanks. *Estuarine, Coastal and Shelf Science, 14*, 79–99.

Jenkins, G. J., Murphy, J. M., Sexton, D. S., Lowe, J. A., Jones, P., & Kilsby, C. G. (2009). *UK Climate Projections: Briefing report*. Exeter: Met Office Hadley Centre.

Kates, R. W., Travis, W. R., & Wilbanks, T. J. (2012). Transformational adaptation when incremental adaptations to climate change are insufficient. *Proceedings of the National Academy of Sciences, 109*, 7156–7161.

Kelly, R. A., Jakeman, A. J., Barreteau, O., Borsuk, M. E., ElSawah, S., Hamilton, S. H., Henriksen, H. J., Kuikka, S., Maier, H. R., Rizzoli, A. E., van Delden, H., & Voinov, A. A. (2013). Selecting among five common modelling approaches for integrated environmental assessment and management. *Environmental Modelling & Software, 47*, 159–181.

Kelman, I. (2009). *1953 storm surge deaths: U.K. Version 5*. http://www.ilankelman.org/disasterdeaths/1953DeathsUK.doc. Cited December 19, 2013.

Klein, R. J. T., Nicholls, R. J., Ragoonaden, S., Capobianco, M., Aston, J., & Buckley, E. N. (2001). Technological options for adaptation to climate change in coastal zones. *Journal of Coastal Research, 17*(3), 531–543.

Kuang, C. P., & Stansby, P. K. (2004). Efficient modelling for directional random wave propagation inshore. *Proceedings of the ICE – Maritime Engineering, 157*(3), 123–131. doi:10.1680/maen.2004.157.3.123.

Leafe, R., Pethick, J., & Townend, I. (1998). Realising the benefits of shoreline management. *The Geographical Journal, 164*, 282–290.

Leake, J., Wolf, J., Lowe, J., Hall, J., & Nicholls, R. (2008). *Response of marine climate to future climate change: Application to coastal regions*. Proceedings of ICCE 2008, Hamburg.

Lichter, M., Vafeidis, A. T., Nicholls, R. J., & Kaiser, G. (2011). Exploring data-related uncertainties in the analyses of land area and population in the "Low-Elevation Coastal Zone" (LECZ). *Journal of Coastal Research, 27*(4), 757–768.

Linham, M. M., & Nicholls, R. J. (2010). *Technologies for climate change adaptation: Coastal erosion and flooding* (TNA guidebook series). Roskilde: UNEP Risø Centre on Energy, Climate and Sustainable Development.

Lowe, J. A., & Gregory, J. M. (2005). The effects of climate change on storm surges around the United Kingdom. *Philosophical Transactions of the Royal Society A – Mathematical Physical and Engineering Sciences, 363*(1831), 1313–1328.

Lowe, J. A., Howard, T., Pardaens, A., Tinker, J., Jenkins, G., Ridley, J., Leake, J., Holt, J., Wakelin, S., Wolf, J., Horsburgh, K., Reeder, T., Milne, G., Bradley, S., & Dye, S. (2009). *UK Climate Projections science report: Marine and coastal projections*. London: DEFRA.

Lumbroso, D. M., & Vinet, F. (2011). A comparison of the causes, effects and aftermaths of the coastal flooding of England in 1953 and France in 2010. *Natural Hazards and Earth System Sciences, 11*, 2321–2333.

McRobie, A., Spencer, T., & Gerritsen, H. (2005). The big flood: North Sea storm surge. *Philosophical Transactions of the Royal Society A, 363*, 1261–1491.

Milligan, J., O'Riordan, T., Nicholson-Cole, S. A., & Watkinson, A. R. (2009). Nature conservation for sustainable shorelines: Lessons from seeking to involve the public. *Land Use Policy, 26*(2), 203–213.

Misdorp, R. (2011). *Climate of coastal cooperation*. Leiden: European Coastal and Marine Union.

Mokrech, M., Hanson, S., Nicholls, R. J., Wolf, J., Walkden, M., Fontaine, C. M., Nicholson-Cole, S., Jude, S. R., Leake, J., Stansby, P., Watkinson, A. R., Rounsevell, M. D. A., Lowe, J. A., & Hall, J. W. (2011). The Tyndall coastal simulator. *Journal of Coastal Conservation, 15*(3), 325–355. doi:10.1007/s11852-009-0083-6.

Mosby, J. E. G. (1939). The Horsey flood, 1938: An example of storm effect on a low coast. *The Geographical Journal, 93*, 413–418.

Moser, S. C., Williams, S. J., & Boesch, D. F. (2012). Wicked challenges at land's end: Managing coastal vulnerability under climate change. *Annual Review of Environment and Resources, 37*, 51–78. doi:10.1146/annurev-environ-021611-135158.

Motyka, J. M., & Brampton, A. H. (1993). *Coastal management: Mapping of littoral cells* (HR Wallingford report SR328). Wallingford: HR Wallingford.

Nakićenović, N., & Swart, R. (2000). *Emissions scenarios* (Special report of the Intergovernmental Panel on Climate Change). Cambridge: Cambridge University Press.

Nicholls, R. J., Wong, P. P., Burkett, V. R., Codignotto, J. O., Hay, J. E., McLean, R. F., Ragoonaden, S., & Woodroffe, C. D. (2007). Coastal systems and low-lying areas. In M. L. Parry, O. F. Canziani, J. P. Palutikof, P. J. van der Linden, & C. E. Hanson (Eds.), *Climate change 2007: Impacts, adaptation and vulnerability* (Contribution of Working Group II to the Fourth Assessment Report of the Intergovernmental Panel on Climate Change, pp. 315–356). Cambridge: Cambridge University Press.

Nicholls, R. J., Hanson, S., Herweijer, C., Patmore, N., Hallegatte, S., Corfee-Morlot, J., Château, J., & Muir-Wood, R. (2008a). *Ranking port cities with high exposure and vulnerability to climate extremes: Exposure estimates* (Environment Directorate Working Papers, (ENV/WKP(2007)1), 62 pp). Paris: Organisation for Economic and Co-operative development (OECD).

Nicholls, R. J., Wong, P. P., Burkett, V. R., Woodroffe, C. D., & Hay, J. E. (2008b). Climate change and coastal vulnerability assessment: Scenarios for integrated assessment. *Sustainability Science, 3*(1), 89–102.

Nicholls, R. J., Woodroffe, C. D., Burkett, V., Hay, J., Wong, P. P., & Nurse, L. (2012). Scenarios for coastal vulnerability assessment. In E. Wolanski & D. S. McLusky (Eds.), *Treatise on estuarine and coastal science* (Vol. 12, pp. 289–303). Waltham: Academic Press.

Nicholls, R. J., Townend, I. H., Bradbury, A., Ramsbottom, D., & Day, S. (2013). Planning for long-term coastal change: Experiences from England and Wales. *Ocean Engineering, 71*, 3–16. doi:10.1016/j.oceaneng.2013.01.025.

Park, H. B., & Vincent, C. E. (2007). Evolution of Scroby Sands in the East Anglian coast, UK. *Journal of Coastal Research, SI50*, 868–873.

Penning-Rowsell, E., Haigh, N., Lavery, S., & McFadden, L. (2013). A threatened world city: The benefits of protecting London from the sea. *Natural Hazards, 66*, 1383–1404.

Ranger, N., Reeder, T., & Lowe, J. A. (2013). Addressing 'deep' uncertainty over long-term climate in major infrastructure projects: Four innovations of the Thames Estuary 2100 project. *EURO Journal on Decision Processes, 1*, 233–262.

Reeve, D. E., Li, B., & Thurston, N. (2001). Eigenfunction analysis of decadal fluctuations in sandbank morphology. *Journal of Coastal Research, 17*(2), 371–382.

Rounsevell, M. D. A., & Metzger, M. J. (2010). Developing qualitative scenario storylines for environmental change assessment. *Wiley Interdisciplinary Reviews: Climate Change, 1*, 606–619.

Sheppard, T. (1912). *The lost towns of the Yorkshire coast*. London: A. Brown & Sons.

Steers, J. A. (1953). The east coast floods. *The Geographical Journal, 119*, 280–298.

Steers, J. A., Bayliss-Smith, T. P., Stoddart, D. R., Spencer, T., & Durbridge, P. M. (1979). The storm surge of 11 January 1978 on the east coast of England. *The Geographical Journal, 145*(2), 192–205.

Thomalla, F., & Vincent, C. E. (2003). Beach response to shore-parallel breakwaters at Sea Palling, Norfolk, UK. *Estuarine, Coastal and Shelf Science, 56*(2), 203–212.

Thorne, C. R., Evans, E. P., & Penning-Rowsell, E. (2007). *Future flooding and coastal erosion risks*. London: Thomas Telford.

UN Habitat. (2008). *State of the world's cities 2008/2009: Harmonious cities*. London: Earthscan.

Valiela, I. (2006). *Global coastal change*. Oxford: Blackwell.

Van Koningsveld, J. P., Mulder, M., Stive, M. J. F., van der Valk, L., & van der Weck, A. W. (2008). Living with sea-level rise and climate change: A case study of the Netherlands. *Journal of Coastal Research, 24*(2), 367–379.

Vincent, C. E. (1979). Longshore sand transport rates – A simple model for the East Anglian coastline. *Coastal Engineering, 3*, 113–136.

Walsh, C. L., Dawson, R. J., Hall, J. W., Barr, S. L., Batty, M., Bristow, A. L., Carney, S., Dagoumas, A., Ford, A., Tight, M. R., Watters, H., & Zanni, A. M. (2011). *Assessment of climate change mitigation and adaptation in cities*. Proceedings of the ICE – Urban design and planning, Special issue on Urban Development and Sustainability, *164*(DP2), 75–84. doi:10.1680/udap.2011.164.2.75.

Warren, R., de la Nava, S. S., Arnell, N. W., Bane, M., Barker, T., Barton, C., Ford, R., Füssel, H. M., Hankin, R. K. S., Klein, R., Linstead, C., Kohler, J., Mitchell, T. D., Osborn, T. J., Pan, H., Raper, S. C. B., Riley, G., Schellnhüber, H. J., Winne, S., & Anderson, D. (2008). Development and illustrative outputs of the Community Integrated Assessment System (CIAS), a multi-institutional modular integrated assessment approach for modelling climate change. *Environmental Modelling & Software, 23*, 592–610.

Westphal, M. L., Hughes, G. A., & Brömmelhörster, J. (Eds.). (2013). *Economics of climate change in East Asia*. Mandaluyong City: Asian Development Bank.

Wolf, J., & Flather, R. A. (2005). Modelling waves and surges during the 1953 storm. *Philosophical Transactions of the Royal Society of London A, 363*(1831), 1359–1375.

Wong, P. P., Losada, I. J., Gattuso, J., Hinkel, J., Khattabi, A., McInnes, K. L., Saito, A., & Sallenger, A. (2014). Coastal systems and low-lying areas. In C. B. Field, V. R. Barros, D. J. Dokken, K. J. Mach, M. D. Mastrandrea, T. E. Bilir, M. Chatterjee, K. L. Ebi, Y. O. Estrada, R. C. Genova, B. Girma, E. S. Kissel, A. N. Levy, S. MacCracken, P. R. Mastrandrea, & L. L. White (Eds.), *Climate change 2014: Impacts, adaptation, and vulnerability* (Part A: Global and Sectoral Aspects. Contribution of Working Group II to the Fifth Assessment Report of the Intergovernmental Panel on Climate Change, pp. 361–409). Cambridge/New York: Cambridge University Press.

World Bank. (2010). *Climate risks and adaptation in Asian coastal megacities: A synthesis report.* Washington, DC: World Bank Group.

Chapter 2
Climate Downscaling: Local Mean Sea Level, Surge and Wave Modelling

Judith Wolf, Jason Lowe, and Tom Howard

Abstract Coastal areas are at increased risk of flooding and erosion in the future, due to a rise in mean sea level and potential changes in tides, surges and waves. The investigation of future climate impacts at the coast thus requires sufficiently detailed projections for the offshore waves and sea levels in both the present day and under future climate scenarios, to provide boundary conditions for detailed coastal impact models. The process of model downscaling, in a climate modelling sense, is generally taken to refer to the generation of locally relevant data from the output of global circulation models of the atmosphere and ocean. The aim is to connect global-scale projections and regional dynamics to generate regionally specific projections relevant to coastal management decision-making.

We discuss research conducted as part of Phase 2 of the Tyndall Coastal Simulator research project on the future changes in surge and wave climate forced by winds and sea-level atmospheric pressures from a version of the Met Office Hadley Centre Climate Model (HadCM3), assuming various greenhouse gas emission scenarios and for various climate model parameter choices. The local spatial variation in mean sea level is also taken into account, incorporating deviations from global mean sea-level caused by regional variations in ocean density and circulation. The UK is still subject to glacial isostatic adjustment after the last ice age, counteracting recent sea-level rise. Dynamical downscaling from a global coupled atmosphere–ocean model, using a regional climate model, has been used to provide more realistic and detailed simulations of wind and pressure over the NW European continental shelf. These were then used to drive storm surge and wave models.

J. Wolf (✉)
Marine Systems Modelling Group, National Oceanography Centre,
Joseph Proudman Building, 6 Brownlow Street, Liverpool L3 5DA, UK
e-mail: jaw@noc.ac.uk

J. Lowe
Department of Meteorology, Met Office, University of Reading,
Earley Gate, PO Box 243, Reading RG6 6BB, UK
e-mail: jason.lowe@metoffice.gov.uk

T. Howard
Storm Surges, Hadley Centre, Met Office, Fitzroy Road, Exeter EX1 3 PB, UK
e-mail: tom.howard@metoffice.gov.uk

R.J. Nicholls et al. (eds.), *Broad Scale Coastal Simulation*, Advances in Global Change Research 49, DOI 10.1007/978-94-007-5258-0_2

This study was used in the Tyndall Coastal Simulator to understand the effects of different uncertainties in the climate system, in terms of marine parameters for the north Norfolk coast. Hence, the dominant drivers could be identified for analysis in later chapters. The changes in wave and surge conditions were less than suggested by earlier analyses, and the wave and surge conditions were assumed to be unchanged compared to today (Appendix A). These results were also used directly in UKCP09 (Lowe et al, UK Climate Projections science report: marine and coastal projections. Met Office Hadley Centre, Exeter, 2009). In the future, larger ensembles of wave simulations would provide a better description of future conditions.

Keywords Storm surge • Waves • Sea level • Modelling • Downscaling

2.1 Introduction

Most work on climate projections has been done on atmospheric climate projections at a global to regional scale. However, broadscale simulation of coastal change for impact and adaptation assessment, on a specific stretch of coastline, requires a range of input data, including estimates of the changes in sea level on various timescales (mean sea level, tides and storm surges, wave magnitude and direction) at a local scale. Surges and waves contribute to the extremes of coastal sea level through direct wind set-up and also wave set-up. Waves on top of high sea level can overtop and damage sea defences and cause coastal erosion and flooding (Wolf 2008, 2009). In order to produce meaningful local projections, we thus have to perform an exercise of downscaling from the global and regional models to local scales. Here, we distinguish between the three different processes (mean sea level, surges and waves) and treat them separately; although there are interactions between these processes, initially they can be treated independently.

For mean sea level (MSL), there are large regional and local deviations from the global mean sea level which is calculated in global climate models: absolute sea level is rising globally, but at certain locations, sea level may fall relative to the land, due to local land uplift, changes in ocean circulation and large-scale gravitational perturbations. Another issue to be considered is the change in rate of sea-level rise (SLR). There is recent evidence of acceleration in the rise in MSL, which may enhance future projections.

Surges and waves are generated by storm events which are localised in space and time, and so we need to know the intensity and frequency of extreme events and how these will change in a future climate. The importance of having long time series data for assessing the statistics of extremes has been identified, e.g. Weisse et al. (2009). These time series data can be generated using proxy data from time series of other variables, e.g. sea-level pressure (SLP) data as a proxy for storminess, or can be generated from long hindcasts of dynamical models. Projections of future impacts require assumptions that the relationships between proxy variables will remain the same in a future climate or may be made by running dynamic models into the future. Local impact models are highly dependent on the accuracy of projections of storminess in global climate models (where storminess may be defined as a measure of the frequency and intensity of storms). Changes in the occurrence of storms are affected by changes in regional weather patterns.

Storm surges are generated by SLP (the inverse barometer effect). Surges and waves are also driven by wind stress, which (approximately) increases with the cube of the wind speed. This means that underestimates of extreme winds can significantly bias the impact model results. Storms are poorly resolved in global models, due to their relatively small spatial extent, and in general, this leads to underestimation of extreme winds and waves (Caires and Sterl 2005). In the UK, we are mainly concerned with changes in the location and intensity of the storm tracks of midlatitude depressions in the North Atlantic.

Issues for impact models such as wave and surge models (leading to models of coastal impacts such as flooding and erosion) are (1) quantification of model accuracy for past events and (2) understanding the uncertainty for future projections. This uncertainty consists of (1) uncertainty in greenhouse gas (GHG) emission (climate scenarios), (2) uncertainty in climate model projections of sea level and storms and (3) uncertainty in the surge and wave models. As models improve, the model uncertainty may be reduced, but there remain uncertainty in the emissions and some of the model physics.

Downscaling from global to regional climate model projections is thus vital for the study of meaningful local impacts. Downscaling, in a climate modelling sense, is generally taken to refer to the generation of locally relevant data from the output of global circulation models (GCMs). The aim is to use global-scale projections, using accepted greenhouse gas emission scenarios to generate regionally specific and useful forecasts, with increased spatial and temporal resolution, and including processes which are not resolved in a coarser model, if possible. Downscaling can be done in several ways: (1) using process models, (2) using empirical/statistical relationships and (3) using hybrid methods, e.g. using pattern recognition. Nesting a regional climate model (RCM) into an existing GCM is an example of the first method, termed dynamical downscaling, and is the approach developed here. An RCM is a dynamic model, like a GCM, but it can give higher-resolution results. At the large scale, it is essentially driven by the GCM, but it uses its own physics-based equations to resolve local effects. Downscaling can also be done using statistical regression, at various levels of complexity. This aims to capture the essential relationships between the global model and local variables. Another strategy is to use stochastic weather generators, identifying weather patterns and their changing statistics of occurrence; although these have in the past mainly focussed on rainfall patterns for hydrological impact studies (e.g. Boé et al. 2006; Jones et al. 2009), recent work has shown how more difficult, skewed variables such as winds can also be included.

In the Tyndall Coastal Simulator project, the winds and SLP from the Hadley Centre GCM (HadCM3) and RCM (HadRM3) were used to provide forcing for surge and wave models on the NW European continental shelf, on a grid of approximately 12 km resolution. These in turn were used to force local area models for the north Norfolk coast (see Chap. 3). Note that future storms and their impacts in Liverpool Bay have also been investigated in a separate study using the same GCM/RCM data to model waves and surges (Brown et al. 2012). The GCM/RCM runs selected were for various emission scenarios, A2, B2 and A1B, which represent medium-high, medium-low and medium emissions, respectively (see Chap. 1). Changes in sea level were not applied to the surge and wave models, which were not run in coupled mode in this study.

The results presented here are also discussed in the UK Climate Projections report (Lowe et al. 2009), hereafter referred to as UKCP09. In the present chapter, we focus on the issues related to downscaling and identify future directions and references to recent work. The following sections provide detail on the MSL projections and downscaling (Sect. 2.2), present and future storm climate (Sect. 2.3), model forcing from HadCM3 and HadRM3 models (Sect. 2.4), surge and wave modelling methodology and downscaling (Sect. 2.5), results (Sect. 2.6) and conclusions (Sect. 2.7).

2.2 Regional Scale Variability in Sea Level

First we discuss how global trends in climate and sea level translate into regional changes in MSL, reviewing the present state of knowledge of global sea-level rise and its spatial variability. Much of the work of collating this research has been done by the Intergovernmental Panel on Climate Change (IPCC), in the most recent Fourth Assessment Report, referred to as AR4. There are still uncertainties in projecting sea-level rise into the future with recent work on ice-sheet melting and SLR acceleration.

2.2.1 Global Sea Level in a Warming Climate

The evidence for past global sea-level rise in the last century is summarised in Church et al. (2010). The largest contributions since the mid-twentieth century come from ocean thermal expansion (0.8 mm $year^{-1}$) and the melting of glaciers and ice caps (0.7 mm $year^{-1}$), with Greenland and Antarctica contributing about 0.4 mm $year^{-1}$.

2.2.2 Regional Spatial Deviations from the Global Mean

Regional patterns in sea-level change depart from the global average, due to the changing mass of ice sheets and glaciers and the exchanges of water (both naturally and anthropogenically) between the continents and the ocean. These changes in the surface loads on the planet cause the solid Earth to deform locally, resulting in perturbations in the gravitational field which can be measured from space (Tamisiea and Mitrovica 2011). These can lead to both near-field (i.e. near the loads) and far-field effects. Regional MSL changes also occur due to variations in the ocean circulation and due to changes in the mean atmospheric circulation. The large differences in regional variations of proposed, future sea-level change as simulated in different global climate models (GCMs) are due to many factors, but largely to the way that the different models treat the ocean circulation.

The relative sea-level rise, which is observed at a given location, depends partly on the local vertical land movement (Bradley et al. 2011). In order to interpret the cause of sea-level rise, corrections to sea-level observations must be made for these movements, which may be due to, for example, glacial isostatic adjustment (GIA), tectonics (e.g. earthquakes) or subsidence of land due to groundwater extraction. Geodynamic models exist at present only for GIA. Contributions to vertical land movement, of whatever origin, are now being measured at many sites using advanced geodetic techniques (e.g. Ostanciaux et al. 2011).

2.2.3 Present and Projected Changes in Sea Level Around the UK

Global sea level is believed to have risen at an average rate of about 1.7 mm/year during the twentieth century (Bindoff et al. 2007), with some evidence for acceleration in the rate between the nineteenth and twentieth centuries (Woodworth et al. 2011). For the UK, the twentieth-century SLR was a little lower than the global average, at around 1.4 mm/year (Woodworth et al. 2009).

In order to make a best estimate of future MSL change around the UK, the most plausible approach is to take the global average estimates from the climate models, assume that these also apply to the UK and then allow for regional variations in relative sea level due to vertical land movement. The latter can be derived from geological and geodetic measurements. For example, Shennan et al. (2011) discuss the variability of sea level around the UK. For the British Isles, present-day relative uplift shows an ellipse of (relative) sea-level fall ~1.2 mm/year, broadly centred on the deglaciated mountains of Scotland. Relative land-level change (the negative of relative sea-level change) is not the same as vertical land motion, because of changes in the geoid. There is a spatial pattern in the difference between relative land-level change and vertical land motion, with differences at present of approximately −0.1 to −0.3 mm/year around the British Isles. A new solution for relative sea-level rise around the British Isles is given in Bradley et al. (2011); see Fig. 2.1. Panel (a) shows a present-day relative sea-level fall in NW Scotland with a rise of up to 0.4 mm/year in SE England. In the area of interest, the north Norfolk coast, a present-day estimate is about 0.4 mm/year.

The urgent question we need to answer is 'how much will sea-level rise in the next 100 years?' The IPCC AR4 projected a global sea-level rise of 18–59 cm from

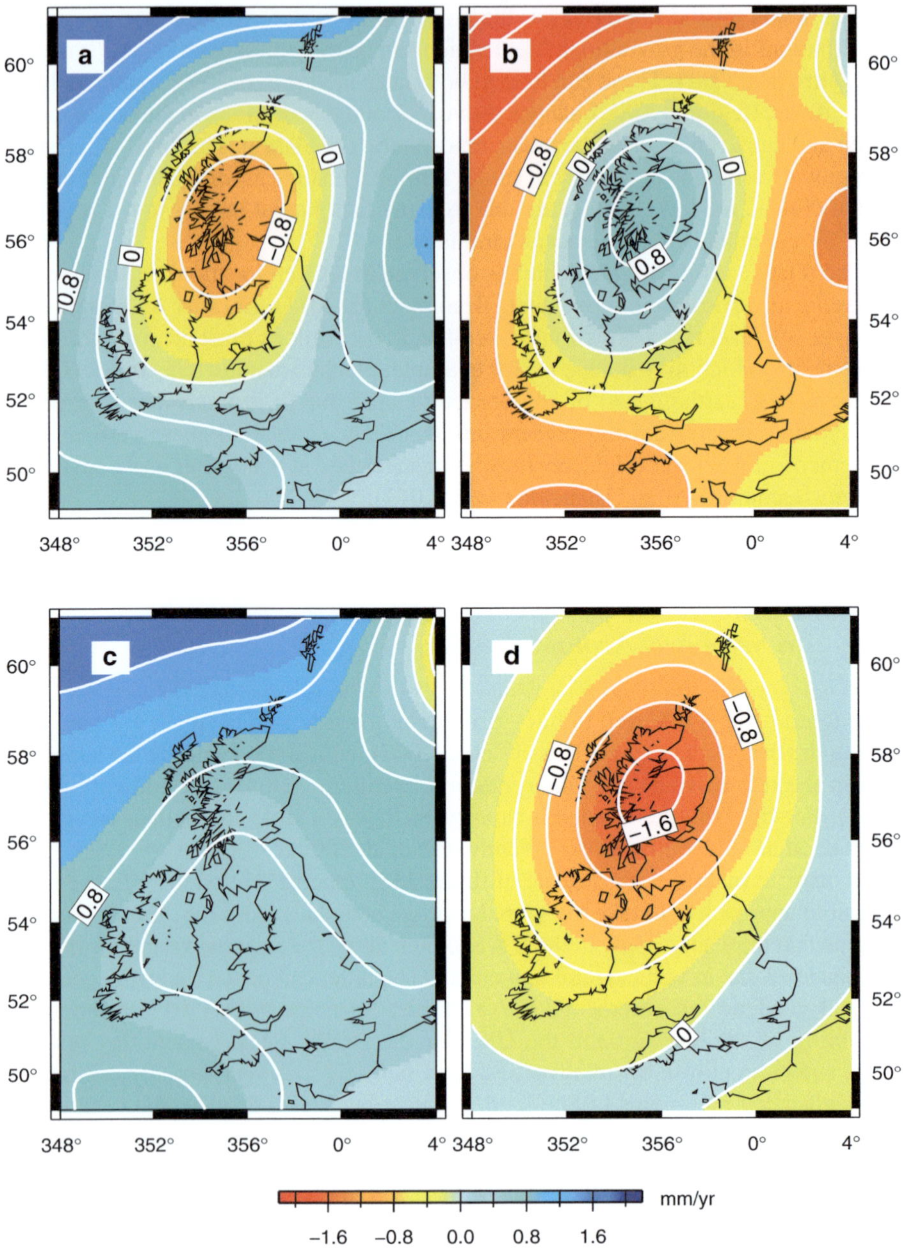

Fig. 2.1 Predictions of present-day sea-level change (**a**) and vertical land uplift (**b**) for the new ice model and its optimum Earth model. The lower frames show the contribution to and (**a**) from all ice sheets except the British–Irish Ice Sheet (*BIIS*) (**c**) and from the BIIS only (**d**) (From Bradley et al. 2011)

1990 to 2090, plus some unspecified amount that could come from changes in the large ice sheets covering Greenland and Antarctica. Recently, questions have been raised about the possibility of accelerating SLR in the twenty-first century based on the idea that the rate of SLR is proportional to the amount of global warming (the warmer it gets, the faster ice melts). Jevrejeva et al. (2010) have used past sea-level and temperature data to quantify this effect and find for different emission scenarios the global sea-level rise could vary from 0.6 to 1.6 m, with confidence limits of 0.59 and 1.8 m. This uncertainty is being addressed by the next IPCC Assessment Report (AR5, in preparation at the time of writing). Although the projections continue to improve as more understanding of the present-day processes is achieved, there will still be uncertainty of the magnitude. The issues are summarised in Church et al. (2011). Some of the major opportunities for improving projections of sea-level rise include more and better observations and the continuing improvement of models, including better understanding of ice-sheet dynamics.

2.3 Storms in the Present and Future Climate

Midlatitude winter storms, originating in the North Atlantic, typically track across the UK from SW to NE (Brown et al. 2010). Studies of storm surges over many years (e.g. Heaps 1983) have shown how surges on the east coast and west coast of the UK are related to somewhat different types of storm events, their track, translation speed and intensity. Wolf and Woolf (2006) investigated the effects of varying storm parameters on waves, compared to strengthening of the prevailing westerly winds. Brown et al. (2010) studied the effect of storms on surges and waves in the Irish Sea, finding that on average one to two storms per year could generate a storm surge >1.5 m at Liverpool, corresponding to wind speeds of Beaufort force 8 (gale force) or larger. Similarly, there are likely to be a few storms per year where there is a significant storm surge in the North Sea.

There has been much recent research aimed at understanding past and future changes in storminess over the UK, due to the changes in midlatitude storms over the NE Atlantic. Trends in severe wind storms around the UK are difficult to identify, due to low numbers of such storms, their decadal variability and the unreliability of direct wind speed observations (Wang et al. 2009). Their analysis shows that storminess conditions in this region have undergone profound decadal or longer timescale fluctuations, with considerable seasonal and regional differences. The most notable differences are seen between winter and summer and between the North Sea area and other parts of the region. Over the last century, the number of winter storms has decreased and then increased again. The observational evidence indicates that the strength of midlatitude SLP gradients and associated westerly circulation has increased in the northern hemisphere, especially during winter, since at least the late 1970s. Behaviour of the North Atlantic storm tracks seems to be a key to understanding present and future changes in storminess. Future climate model projections have a particularly large spread between models and a low signal to noise ratio over Europe compared to other midlatitude regions (Hawkins and Sutton 2009). Woollings (2010) identifies future European climate as particularly uncertain because (1) the spread between the predictions of current climate models is still considerable and (2) Europe is particularly strongly affected by several processes which are known to be poorly represented in current models such as the small-scale structure of storms.

Some of this variability seems to be related to the large-scale atmospheric patterns such as the North Atlantic Oscillation (NAO), which is related to the SLP difference between Iceland and the Azores. Low-frequency SLP variability, and with it the variability of the NAO, is essentially caused by the distribution of the synoptic systems. The NAO index is mostly given as a monthly or seasonal mean.

2.4 Model Forcing from Hadley Centre Climate Model

Dynamical downscaling is carried out using an RCM (HadRM3) at 25 km resolution, set-up over the region of interest (the European coastal region, Fig. 2.2), to simulate climate over Europe in more detail (Murphy et al. 2007). The current generation of RCMs tends to be atmosphere only, using the same atmospheric model formulation as the parent global climate model and being driven at its boundaries by appropriate fluxes of heat, moisture and momentum from the global model. Sea surface temperatures are also provided as a boundary condition from the global

Fig. 2.2 Domain of the storm surge model POLCS3 (*shaded*) on a background of the regional climate model (*outer square*) domain (From UKCP09)

model. The atmosphere-only nature of the model avoids the need for long, multi-year spin up and the cold start problem. The coupled version of the Met Office Hadley Centre (MOHC) global climate model (using a perturbed parameter ensemble, PPE) is used to provide atmospheric boundary conditions. The coupled PPE consists of 11 global model variants, of which one is the standard, or unperturbed, model. Each of the 11 variants drives a corresponding version of the regional model, which has equivalent parameter perturbations. These changes are equivalent to, rather than always identical to, the global model parameter changes because some parameter schemes are scale dependent, and this must be accounted for. The RCM can then be considered a downscaled version of the PPE global projection, which is suitable for driving the surge and wave models described below.

2.4.1 North-East Atlantic Storm Tracks

One feature of particular interest here is the position and intensity of the North Atlantic storm track. Natural variability is an important source of uncertainty for the next few decades and is particularly large over Europe (Hawkins and Sutton 2009). Lowe et al. (2009) define two indices: the magnitude of the storm track and its latitudinal position, and then consider future changes in these storm indices (Fig. 2.3). This figure includes a comparison of a multi-model ensemble (MME) compared to the NOHC PPE. See Sect. 2.4.2 for more details of the model ensembles. The PPE

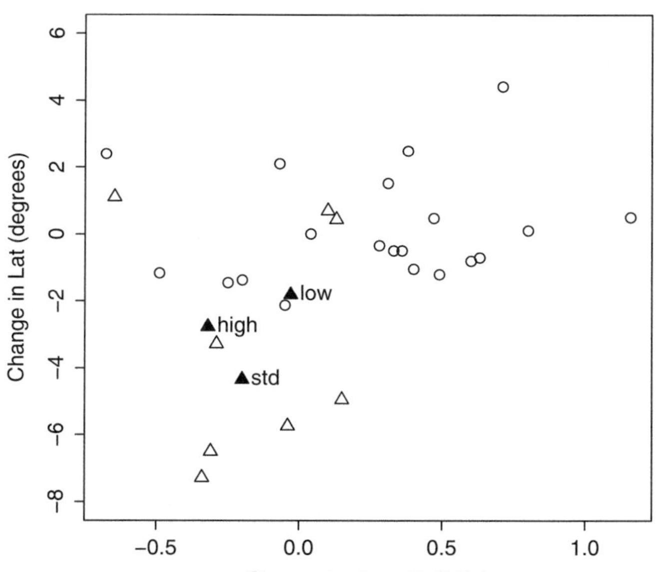

Fig. 2.3 Simulated twenty-first-century change in storm track intensity and location at 4° west using the indices described by Lowe et al. (2009). *Open circles* show the MME and *triangles* show the MOHC PPE; the *filled triangles* show the subset used for wave model forcing

changes can be seen to mainly occur in the movement of the storm track, typically in a southerly direction. A small weakening in the measure of storm intensity is also evident. The MME shows bigger changes in storm intensity, with some models having large positive changes not matched by the PPE. It would be desirable to look at changes in storm surge and waves resulting from a regional model driven by both the global PPE and MME, but regional model simulations are available only for the PPE.

The PPE ensemble typically shows a slight weakening of the winter storm track accompanied by a southward movement. The MME ensemble has a mix of strengthening and weakening storm tracks, with some members showing northward movement and some a southward movement. Of the members of the MME that moved south, the amount of southward displacement is less than that seen in the majority of the PPE ensemble.

A comparison of each model's present-day simulation with ERA-40 reanalysis data suggests that many (although not all) of the MME ensemble members show a greater error when predicting present-day conditions than the PPE ensemble, increasing confidence in the PPE.

2.4.2 Climate Models

Our best tool for forecasting century-scale changes in atmospheric storminess is a global coupled atmosphere–ocean general circulation model (AO-GCM) such as the Met Office Hadley Centre Climate Model, HadCM3 (Gordon et al. 2000; Pope et al. 2000; Collins et al. 2001). GCMs provide a physically based representation of how the atmosphere and oceans behave. HadCM3 is composed of two components: the atmospheric model HadAM3 and the ocean model (which includes a sea ice model). Equations of motion on a rotating sphere and conservation equations for heat and moisture in the atmosphere and heat and salinity in the ocean are solved numerically on a grid. The HadCM3 model uses a grid of 2.5° latitude by 3.75° longitude and has 19 vertical levels. The ocean component has a resolution of 1.25° latitude and 1.25° longitude on 20 vertical levels. The effect of processes on scales below the model's grid scale is parameterised using relationships based on theoretical understanding and statistical fits to observational data.

Different climate models can simulate credible reproductions of many features of near present-day climatology, including surface temperatures, rainfall and mean SLP. However, they typically simulate different amounts of future change for the same future greenhouse gas scenario (this is referred to as the climate model sensitivity). Our primary interest here is a model's ability to simulate global mean sea-level rise, regional spatial variations of the time-averaged absolute sea-level change and atmospheric storms, which drive storm surges and waves. Unlike earlier atmosphere-only GCMs, HadCM3 does not need flux adjustment (additional 'artificial' heat and freshwater fluxes at the ocean surface) to produce a good simulation. The higher ocean resolution of HadCM3 is a major factor in this; other factors include a good match between the atmospheric and oceanic components and an improved ocean mixing scheme. HadCM3 has been run to produce simulations for periods of over a thousand years, showing little drift in its surface climate. For the generation of climate models considered here, input is provided in the form of atmospheric concentrations of greenhouse gases and in some cases either the radiative forcing from sulphate aerosols or the emissions of sulphate aerosol precursors.

Uncertainty in climate model predictions can be partitioned among three sources (Hawkins and Sutton 2009): internal uncertainty of the climate system (natural variability), model uncertainty and scenario uncertainty. The first is most apparent on shorter time and space scales, e.g. in regional models of times less than a couple of decades. The second is the response of the system to a certain level of GHG emissions, and the latter is uncertainty in the emission scenarios.

The precise climatology simulated for present-day greenhouse gas concentrations varies between models, and its agreement with observations is often used as a measure of quality of a given model. The HadCM3 coupled model has been shown to have a remarkably stable climate when forced with preindustrial greenhouse gas concentrations (Megann et al. 2010). The global mean ocean surface temperature error remains within 0.28 °C of climatological estimates over more than four centuries, despite the absence of any flux correction. The ocean circulation is also broadly in agreement with reality, despite having a relatively coarse horizontal resolution. There remain a few significant errors in the ocean component of HadCM3, including a cold error of up to 4 °C over a large area of the North Pacific, progressive surface freshening, slow drifts in the temperature and salinity below 1,000 m depth and changes in the meridional overturning circulation over centennial timescales, with the southward flow of North Atlantic Bottom Water occurring at increasingly shallower depths.

Within a climate model, there are a number of parameter values that are not precisely defined, because of limited understanding or observation, creating the possibility of having multiple climate model versions with the same structure that can each be considered physically plausible. This collection of versions is often referred to as a perturbed parameter ensemble (PPE). At modelling centres around the world, alternative, structurally different global climate models have been developed and together can be analysed as a multi-model ensemble (MME). Many such models exist (see IPCC AR4 for the multi-model ensemble referred to here), and their projections of changes to the magnitude and location of the North Atlantic storm tracks vary significantly (Fig. 2.3).

Recently, new developments in climate modelling have suggested that changes in storminess may be affected by processes not included or well simulated in the IPCC fourth generation of models. These include the role of sea ice distribution, solar forcing and stratospheric processes. However, long simulations and multi-model or perturbed parameter ensembles of these models are not yet available, so it is not yet possible to reliably assess the effects on storm surges and waves.

2.5 Surge and Wave Models and the Downscaling Methodology

For the surge simulations, the RCMs supply wind speed components (at 10 m above the surface) and SLP to the storm surge model. The shelf sea hydrography model requires the same quantities, plus heat and water fluxes, and river outflow (from a separate, RCM-driven model). For the shelf sea wave simulations, the GCMs and RCMs only need to provide the winds.

The projected changes in shelf sea hydrography and circulation were driven from a single ensemble member of the PPE (see UKCP09): the unperturbed model (i.e. the model run with the default parameters). The changes in offshore waves used three model variants of the PPE: the unperturbed version and the low- and high-sensitivity members (Fig. 2.3). In this case, the climate model sensitivity is defined as an average from the global model. The storm surge projections were driven by atmospheric climate changes from all 11 PPE members.

The downscaling methodology for the climate model forcing is a complex procedure, requiring checks to ensure the resulting atmospheric circulation is compatible and dynamically consistent with the larger-scale forcing. In general, the surge and wave models, especially nearshore, are much more dissipative and less affected by initial conditions. Thus, the downscaling for these models mainly consists of a nesting procedure whereby the high-resolution model is driven by boundary conditions from a coarser model, with local forcing at high resolution. Surges in the deep ocean are mainly generated by the inverse barometer effect (Flather 2000), whereas in shallow water, wind stress becomes more dominant. Thus, the surge boundary condition beyond the shelf edge can be specified by the atmospheric pressure, and no further model nesting is required. On the other hand, waves can travel thousands of kilometres across the deep ocean, and swell input from the Atlantic Ocean is important, mainly on the western part of the continental shelf; thus, a coarse-grid Atlantic wave model was run to provide wave boundary forcing.

For a given choice of GCM forcing data, there are several methods whereby this can be used to drive a regional hydrodynamic-ecosystem model. The latter model is particularly concerned with changes in hydrography (temperature and salinity of the ocean) over time. The downscaling methods can be grouped into five different approaches:

(i) Direct forcing
(ii) Delta change approach
(iii) Bias correction (linear) or statistical downscaling (nonlinear)
(iv) Dynamic downscaling of atmosphere only
(v) Dynamic downscaling using an atmosphere–ocean model

Problems with different methods relate to model bias and the loss of dynamical consistency where a fully coupled atmosphere–ocean model is not yet available. Various levels of sophistication may be attempted for bias correction. Spatial corrections (e.g. to correct the location of the storm track) are significantly more involved. An assumption may be made regarding the future bias being similar to that of the present day (which is not necessarily correct). Method (iv) is used here, but will ultimately lead to a loss of dynamical consistency between the oceanic and regional solution over time (see Holt 2011).

2.5.1 *Surge Modelling*

The tide–surge model was based on the 2D operational flood forecasting model which is run twice daily at the UK Met Office, on a continental shelf model grid of 1/6° longitude by 1/9° latitude (here termed POLCS3). POLCS3 (Flather 2000) is a barotropic (depth-integrated) tide and surge model of ~12 km resolution covering the whole of the northwest European coastal shelf (Fig. 2.2). It is used operationally to forecast individual storm tides, typically at lead times of 1 or 2 days. In this operational model, POLCS3 is driven at its lateral boundaries by the 26 largest astronomical tidal constituents and at the water surface by wind and pressure fields from the Met Office forecast models, and output is validated against observations on a monthly basis.

In order to forecast changes in the storm surge climate, a different approach is required. In this case, we are not so much interested in the characteristics of individual events, but rather the changes in the statistics of the extremes of the distribution of high water levels, which in turn depend on changes in local MSL and changes in the atmospheric storminess. Several authors (e.g. Lowe et al. 2001; Howard et al. 2010) agree that these two effects can be added linearly to first order for coastal locations around the North Sea. So typically, an experimental design for forecasting century-scale changes in the North Sea surge climate will include an assessment of regional relative MSL change and an independent assessment of the impact of changes in storminess.

The atmospheric winds and pressures from the RCM ensemble are then used to drive the surge model. This approach gives a distribution of simulated extreme water levels which are much more nearly comparable with observations than those produced by driving the surge model directly with GCM wind and pressure fields. Owing to the computational expense, only a single emission scenario (the A1B scenario) was considered for the PPE.

2.5.1.1 Potential Changes Under a Strengthened Storm Track

The emergent range of North Atlantic storm track responses of the HadCM3 PPE does not span the full range of the corresponding responses of the many different valid atmosphere–ocean coupled models considered in AR4. At least one of these models predicts a significant strengthening of the North Atlantic storm track, a response which is not seen in the HadCM3 ensemble. So, in addition to the HadCM3 ensemble results, a surge model simulation was produced based on the IPCC AR4 model with the largest increase in North Atlantic storminess. Since regional climate model data for this model is not available, a crude scaling of wind and pressure fields from one of the regional atmospheric ensemble members was developed, based on the band-passed filtered SLP variability (which is a well-used large-scale measure of storminess) to give a representative simulation of a potential future storm surge climate under a strengthened storm track.

2.5.1.2 Statistical Approach

Many different approaches can be envisaged for tackling the problem of generating sufficient model for extreme water level data to seek statistically significant trends in the extremes. In earlier work (http://www.ukcip.org.uk/ukcp09/ukcip02/, hereafter UKCIP02), 30-year time slices representing present-day and future climatology were compared. By the time of the UKCP09 study, increases in computing power facilitated analysis of the whole of a long simulation extending from present day into the future (1960–2100). Three possible measures of extreme water suggest themselves:

1. Extreme water levels of a simulation which does not include the astronomical tide
2. Extreme water levels of a simulation which does include the astronomical tide
3. Some measure of the meteorological component in isolation, for example, the surge residual (tide plus surge run minus tide alone)

Each measure has advantages and disadvantages. By omitting the astronomical tide, we ensure that each intense atmospheric event is represented as a high-water event, whereas in a simulation including the astronomical tide surges which come at low tide or which coincide with a high neap tide may be overlooked if the extreme high water is used as a measure, meaning that fewer of the meteorological extremes contribute to the statistical analysis. On the other hand, it is well recognised that nonlinear interaction between the tide and surge modifies the surge (Prandle and Wolf 1978) and this interaction is lost in a simulation which does not include the astronomical tide. A compromise is to attempt to isolate the meteorological component of a simulation which does include the astronomical tide. The most impact-relevant measure here is the skew surge (de Vries et al. 1995), which is simply the difference between the astronomical high tide and the nearest actual (or modelled) high water level. A further compromise is involved in choosing the number of extreme events to include in the analysis. Following many previous workers, we use the five largest events per year, subject to a separation of 60 h between events to avoid double counting.

2.5.2 *Wave Modelling*

The wave model which was used is based on the well-tested 3rd-generation spectral model, WAM (Komen et al. 1994), implemented on two grids: a coarse $1° \times 1°$ grid for the whole Atlantic to provide wave boundary conditions just outside the continental shelf, using the deepwater wave approximation, and the same 12 km grid for the NW European continental shelf as used for the surge model (here termed WAMCS3). The version of the WAM model used here includes modifications for regions of shallow water and fine resolution (Monbaliu et al. 2000). The grid nesting is shown in Fig. 2.4, showing the full Atlantic model on the left with the nested WAMCS3 model.

The earlier part of the study (UKCIP02) used two 30-year time slices, as discussed above, comparing the present day (represented by 1960–1990) with a future climate (2070–2100). However, it can be demonstrated that the variability between different 30-year periods can dominate (see Fig. 2.10). In the second phase (UKCP09), the model was run using three sets of atmospheric model wind forcing covering the whole period 1960–2100. The cases comprised a simulation with medium climate sensitivity (unperturbed case), and high and low members of the PPE run for a medium emission scenario (SRES A1B), as indicated in Fig. 2.3. Only the unperturbed case was run for the whole 140-year period, due to limitations in computational resources, as the wave model takes much longer to run than the surge model. It may be seen from Fig. 2.3 that the two perturbed cases, selected from the global model results, do not span the whole range of storm track variability in the North Atlantic; thus, it would be desirable to run the rest of the 11 PPE cases for the whole 140 years to explore more fully the variability in the wave projections. A limited view of the uncertainty is gained from the results presented here.

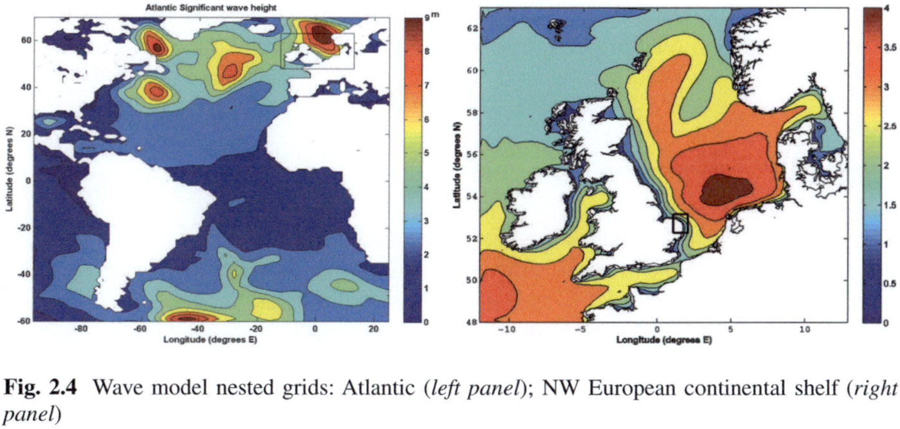

Fig. 2.4 Wave model nested grids: Atlantic (*left panel*); NW European continental shelf (*right panel*)

2.6 Model Results for Future Climate Scenarios

Here, we summarise the results of the downscaling of waves, surges and sea level for the whole continental shelf and especially for the East Anglian coast which is drawn upon for Chap. 3.

2.6.1 *Surge Model Results*

Increasing storminess would be expected to increase surge heights, but more direct effects on extreme water levels can be attributed directly to increased MSL. Results show that large increases in MSL (even up to 5 m) have very little effect on the dynamics of extreme surge events, the primary effect being on the speed of propagation of tide and surge (Howard et al. 2010) which produces changes in phase, usually preventing the peak surge from occurring at tidal high water.

In all locations around the UK coast, the changes in the skew surge extremes (not including changes in regional MSL) diagnosed from the PPE as a whole are physically small (Fig. 2.5, left-hand panel), for example, a projected change of 8 cm over the twenty-first century could be compared with observed global MSL rise during the period 1961–2003 of around 1.8 mm/year (Church et al. 2010), and in most locations they are not statistically significant (in other words, in most locations they are consistent with the size of random changes that we would expect as a result of the natural short-term variability).

Changes, diagnosed from a scaling of wind and pressure fields to give a representative simulation of a potential future storm surge climate under a strengthened storm track, were found to be much larger all around the UK coastline with increases of the order of 1 m around the Solway Firth and East Anglia (Fig. 2.5, right-hand panel).

The contrast between these two results illustrates the impact of the large uncertainties in current model projections of changes in the North Atlantic storm track.

2.6.2 *Wave Model Results*

Results for the A2 and B2 scenarios were discussed in Leake et al. (2008). The main pattern is consistent, with the future wave climate showing an increase in wave height to SW of the UK and a decrease to the north of Scotland and experiencing little change in the southern North Sea. The A2 scenario generates larger changes than B2, which is consistent with larger increases in CO_2 and temperature. This pattern corresponds to a southerly shift of the North Atlantic storm track over the UK. The main results shown here relate to significant wave height (SWH) for the A1B scenario. Figure 2.6 shows the mean winter SWH for the present day

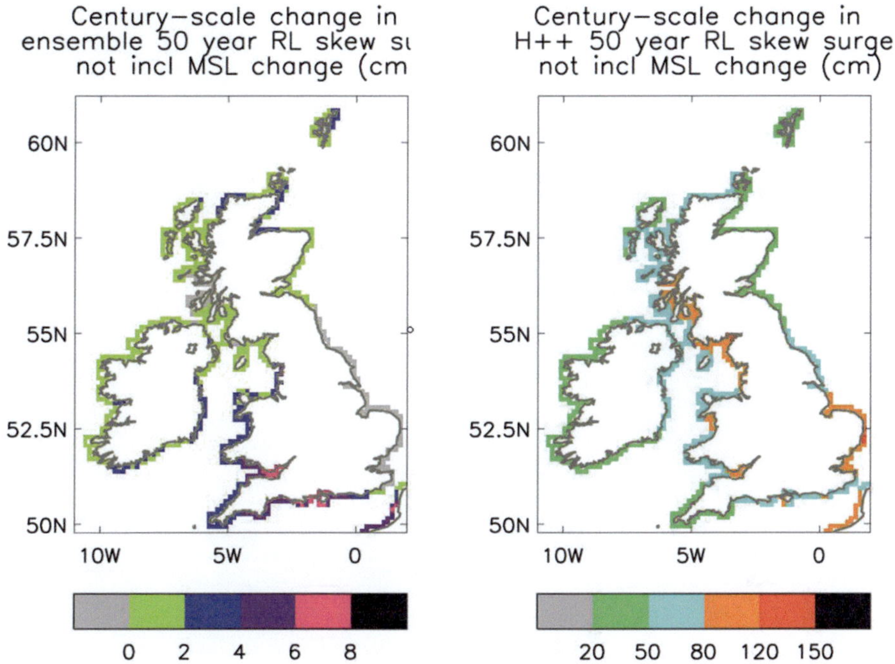

Fig. 2.5 Projected century-scale change in a 50-year level of skew surge from changes in atmospheric storminess only (centimetres). *Left-hand panel*: mean projection of the perturbed physics ensemble of HadCM3. *Right-hand panel*: crude estimate of potential changes under a strengthened storm track. Both *panels* show projected changes from 1990 to 2090

(1960–1990), showing a reduction from west to east across the UK, as the east coast is more sheltered from the North Atlantic storms.

Figure 2.7 shows the differences in SWH from the present day, for the three PPE ensemble members (i.e. future minus present day). The pattern is similar in each case with an increase in SWH in the south and a decrease to the north, although there are also distinct differences. Changes in wave period and direction are also of interest for changes in overtopping and erosion of coastal defences (not shown here; see Chap. 3).

Figure 2.8 shows changes in the winter maximum wave height. This has large error bars especially with the projected extreme values, as in this case a generalised extreme value distribution has been fitted and the confidence limits are used to identify where the differences are significant or not. Dark-shaded areas correspond to areas where the differences are not statistically significant at the 95 % confidence level. Projections of longer return period waves reflect the same pattern but with larger error bars.

Figure 2.9 (right panel) shows results for the trend in mean winter wave height for four locations, which correspond to wave buoys (left panel). The only significant trend is for the K7 buoy location to the north of Scotland with a steady reduction in

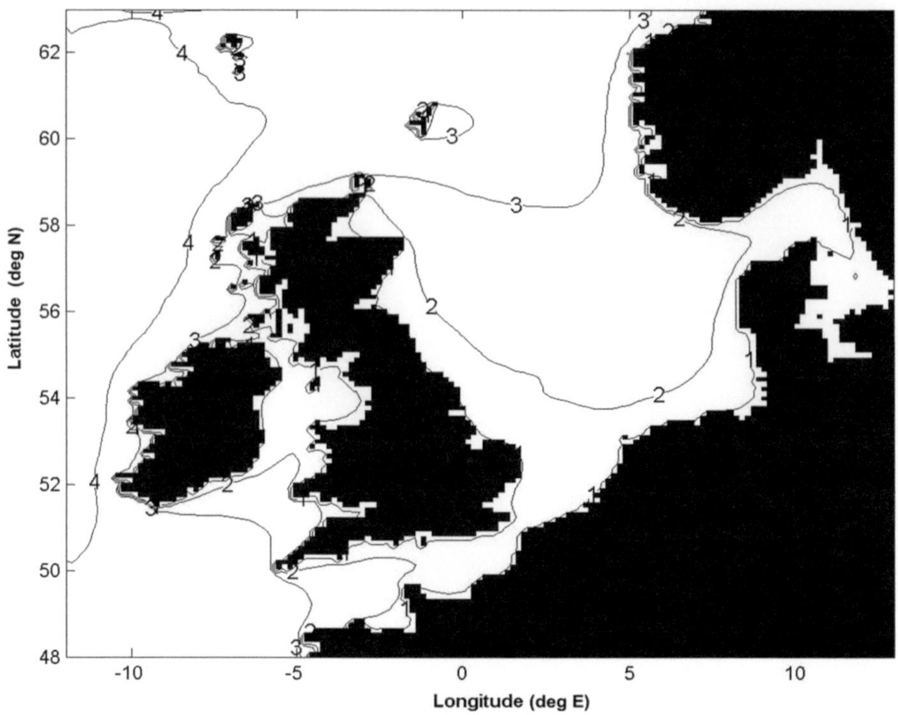

Fig. 2.6 Mean winter SWH for present day (1960–1990)

wave height over 140 years, although there is a large amount of variability around this trend line.

2.7 Conclusions and Next Steps

There is large uncertainty in projected changes in storminess for the NE Atlantic region, with different climate models providing conflicting results for the future. The behaviour of the North Atlantic storm tracks over NW Europe is rather variable, leading to most uncertainty in longer-term projections in this area. Some models project a poleward (northerly) shift in the North Atlantic storm track, although the HadCM3 model projects a southerly movement, under increased GHG scenarios. Increasing storminess is expected to increase surge heights, but more direct effects can be attributed to rising mean sea level. Results from this study show that even large increases in MSL have very little effect on the occurrence of extreme surge events, the primary effect being on the speed of propagation of tide and surge and on the total water level, which in terms of impacts is the most significant parameter. From the wave model results, seasonal mean and annual maximum wave heights are

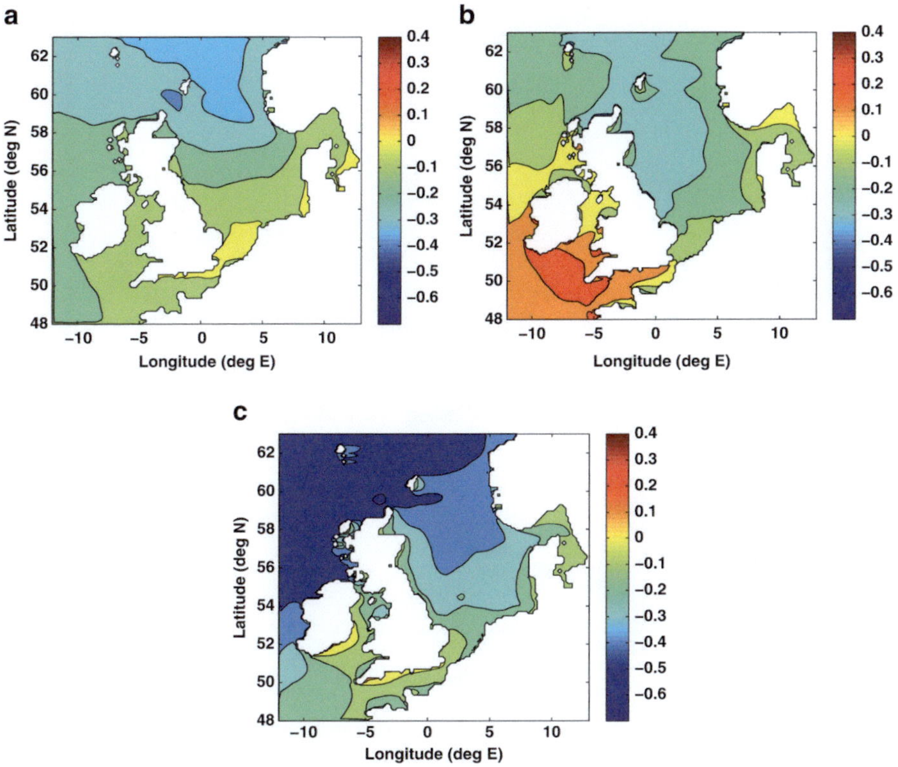

Fig. 2.7 Differences in mean winter wave height from present day for selected PPE members – standard, low and high sensitivity, respectively (*left to right*). *Colour scale* is wave height in metres

generally expected to increase to the SW of the UK, reduce to the north of the UK and experience little change in the southern North Sea. This pattern of change is consistent with a southerly shift of the storm track over the UK. However, there are large uncertainties, especially in the projected extreme values, due to large, decadal-scale fluctuations in the wind regime.

Decadal variability is illustrated by Fig. 2.10, which shows the 30-year running mean for a point offshore of the Norfolk coast. The large difference between successive 30-year time slices can be seen, as well as the difficulty of choosing the correct 30-year time slices to compare. The amount of variability means that although a long-term upward trend in wave height is predicted, it is not necessarily statistically significant. This illustrates the limitations of the time-slice approach.

Changes in storm tracks in climate models are the key to understanding the storm surge and wave projections. The HadCM3 model shows a southerly movement of storm tracks and little change in intensity over the UK, whereas other models show changes in intensity but little latitudinal shift. New developments in climate models, including better physical representation of sea ice, stratospheric effects and the effect of ocean currents, have an impact on storm tracks. The trends are masked by

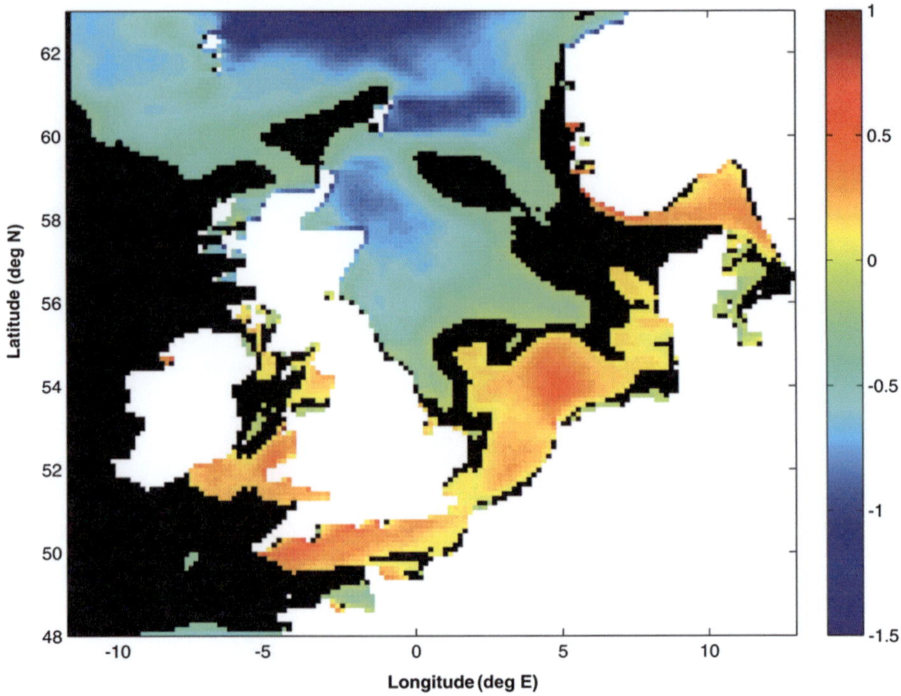

Fig. 2.8 Differences in extreme wave height; *dark-shaded* areas are where the differences are not statistically significant

decadal variability, partly natural and partly affected by global warming. In order to reduce the uncertainty in impact model projections, it is necessary to better understand the climate models, including further physical processes as necessary, and arrive at a consensus between the different climate models. It is to be expected that coupled ocean–atmosphere–wave regional models, more ensemble members and increased resolution will be possible with increasing computing power.

The end product of the offshore waves and surge projections developed here was to feed into look-up tables for the nearshore wave transformations in Chap. 3. The sea-level scenarios are used directly as explained in Chaps. 7, 8, and 9. With the HadCM3 forcing, the changes in surge and waves off the Norfolk coast were rather small, for different physical reasons: the wave height changes are larger to the north and south of an axis which approximately passes through East Anglia, whereas the projected changes in the surge are relatively small everywhere unless an assumption of increased storminess is used (this scenario was not implemented for the wave projections). The climate scenarios used in the Tyndall Coastal Simulator are summarised in Appendix A.

Future work is needed to resolve the discrepancies between climate models and better understand storm behaviour in a warming climate. The role of wind waves and storm surges in climate change science has been recognised as increasingly

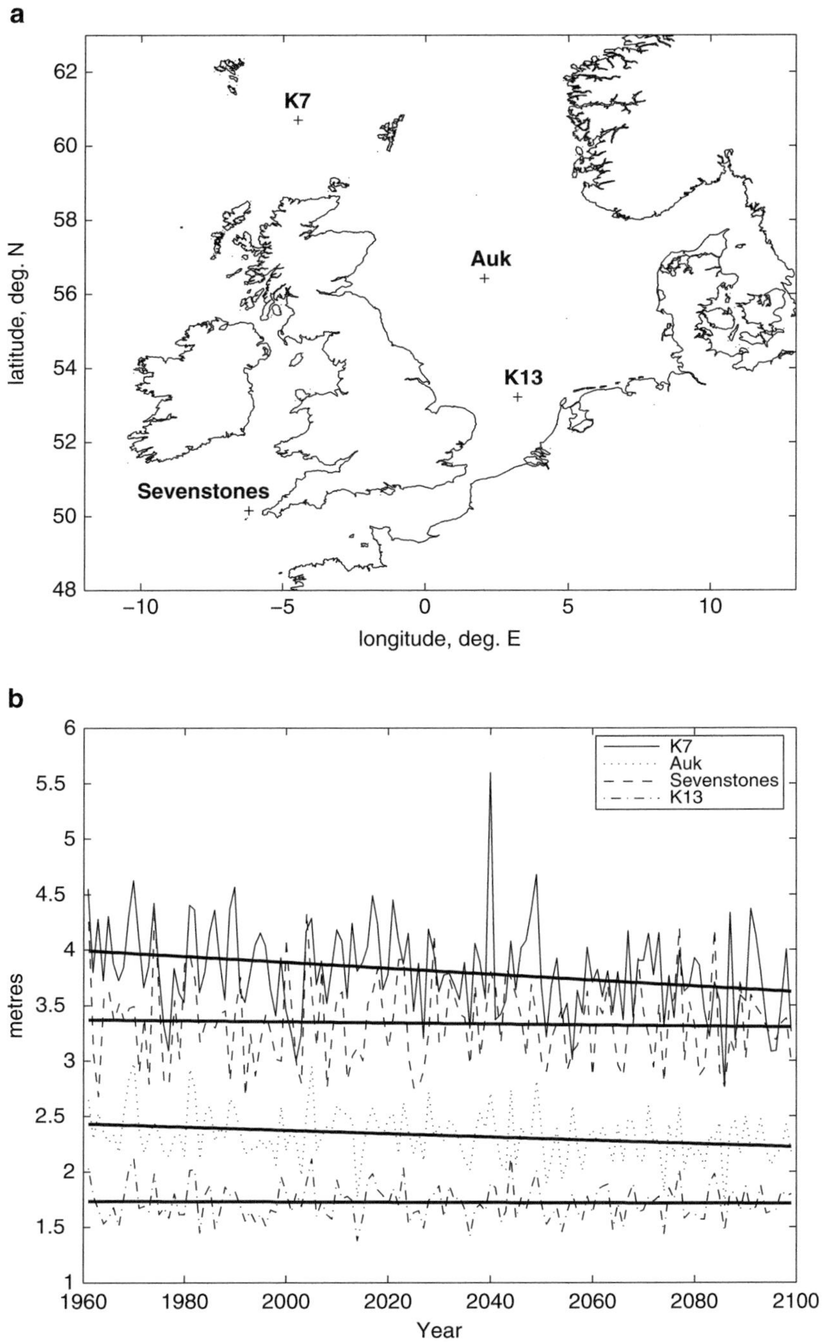

Fig. 2.9 Time series and trends of winter mean SWH at four locations around the UK

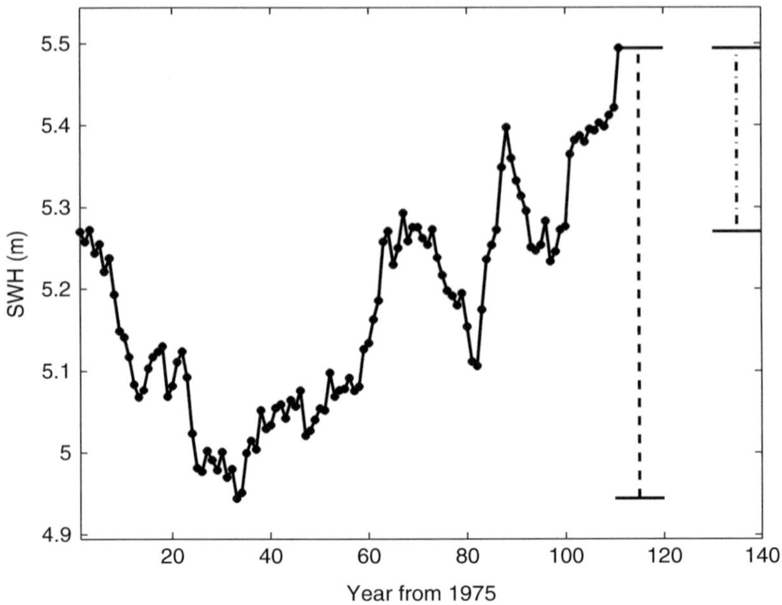

Fig. 2.10 A 30-year running mean of SWH in southern North Sea (k13: N 53°.17, E 3°.25). The *vertical bars* show the maximum (*dashed line*) and beginning to end (*dash-dotted line*) difference between 30-year running means

important in recent years (Lowe et al. 2010). The impact models discussed in Chaps. 7, 8, and 9 are by now rather robust, and the main sources of error are in the forcing conditions, although some benefits will come from increasing resolution of the impact models. Ensembles of atmospheric model forcing have been run for surges, which allow for a quantification of the uncertainty: similar techniques should be applied for the wave models. Future downscaling may be possible using coupled ocean–atmosphere–wave models.

Acknowledgements The authors would like to thank Professor Philip Woodworth for substantial input on the topic of sea-level rise.

References

Bindoff, N. L., Willebrand, J., Artale, V., Cazenave, A., Gregory, J., Gulev, S., Hanawa, K., Le Quéré, C., Levitus, S., Nojiri, Y., Shum, C. K., Talley, L. D., & Unnikrishnan, A. (2007). Observations: Oceanic climate change and sea level. In S. Solomon, D. Qin, M. Manning, Z. Chen, M. Marquis, K. B. Averyt, M. Tignor, & H. L. Miller (Eds.), *Climate change 2007: The physical science basis* (Contribution of Working Group I to the Fourth Assessment Report of the Intergovernmental Panel on Climate Change). Cambridge/New York: Cambridge University Press.

Boé, J., Terray, L., Habets, F., & Martin, E. (2006). A simple statistical-dynamical downscaling scheme based on weather types and conditional resampling. *Journal of Geophysical Research, 111*, D23106. doi:10.1029/2005JD006889.

Bradley, S., Milne, G., Shennan, I., & Edwards, R. (2011). An improved glacial isostatic adjustment model for the British Isles. *Journal of Quaternary Science, 26*, 541–552.

Brown, J. M., Souza, A. J., & Wolf, J. (2010). An investigation of recent decadal-scale storm events in the eastern Irish Sea. *Journal of Geophysical Research, 115*, C05018. doi:10.1029/2009JC005662.

Brown, J. M., Wolf, J., & Souza, A. J. (2012). Future extreme events in Liverpool Bay: Model projections from 1960–2100. *Climatic Change, 111*, 365–391. doi:10.1007/s10584-011-0145-2.

Caires, S., & Sterl, A. (2005). 100-year return value estimates for ocean wind speed and significant wave height from the ERA-40 data. *Journal of Climate, 18*(7), 1032–1048.

Church, J. A., Woodworth, P. L., Aarup, T., & Wilson, W. S. (2010). *Understanding sea-level rise and variability*. London: Wiley-Blackwell.

Church, J. A., Gregory, J. M., White, N. J., Platten, S. M., & Mitrovica, J. X. (2011). Understanding and projecting sea level change. *Oceanography, 24*(2), 130–143. doi:10.5670/oceanog.2011.33.

Collins, M., Tett, S. F. B., & Cooper, C. (2001). The internal climate variability of HadCM3, a version of the Hadley Centre coupled model without flux adjustments. *Climate Dynamics, 17*, 61–81. doi:10.1007/s003820000094.

de Vries, H., Breton, M., de Mulder, T., Krestenitis, Y., Ozer, J., Proctor, R., Ruddick, K., Salomon, J. C., & Voorrips, A. (1995). A comparison of 2-D storm surge models applied to three shallow European seas. *Environmental Software, 10*(1), 23–42. doi:10.1016/0266-9838(95)00003-4.

Flather, R. A. (2000). Existing operational oceanography. *Coastal Engineering, 41*(1–3), 13–40. doi:10.1016/S0378-3839(00)00025-9.

Gordon, C., Cooper, C., Senior, C. A., Banks, H., Gregory, J. M., Johns, T. C., Mitchell, J. F. B., & Wood, R. A. (2000). The simulation of SST, sea ice extents and ocean heat transports in a version of the Hadley Centre coupled model without flux adjustments. *Climate Dynamics, 16*(2–3), 147–168. doi:10.1007/s003820050010.

Hawkins, E., & Sutton, R. (2009). The potential to narrow uncertainty in regional climate predictions. *Bulletin of the American Meteorological Society, 90*, 1095–1107.

Heaps, N. S. (1983). Storm surges, 1967–1982. *Geophysical Journal of the Royal Astronomical Society, 14*, 331–376.

Holt, J. T. (2011). *Common set of forcing scenarios*. http://www.meece.eu/documents/deliverables/WP3/D3.1.pdf. Cited 22 Nov 2013.

Howard, T., Lowe, J. A., & Horsburgh, K. (2010). Interpreting century-scale changes in southern North Sea storm surge climate derived from coupled model simulations. *Journal of Climate, 23*, 6234–6247.

Jevrejeva, S., Moore, J. C., & Grinsted, A. (2010). How will sea level respond to changes in natural and anthropogenic forcings by 2100? *Geophysical Research Letters, 37*, L07703. doi:10.1029/2010GL042947.

Jones, P. D., Kilsby, C. G., Harpham, C., Glenis, V., & Burton, A. (2009). *UK Climate Projections science report: Projections of future daily climate for the UK from the Weather Generator*. Newcastle: University of Newcastle. http://ukclimateprojections.defra.gov.uk. Cited 22 Nov 2013.

Komen, G. J., Cavaleri, L., Donelan, M., Hasselmann, K., Hasselmann, S., & Janssen, P. E. A. M. (1994). *Dynamics and modeling of ocean waves*. Cambridge: Cambridge University Press.

Leake, J., Wolf, J., Lowe, J., Hall, J., & Nicholls, R. J. (2008, August 31-September 5). Response of marine climate to future climate change: Application to coastal regions. In *Proceedings of ICCE 2008*, Hamburg.

Lowe, J. A., Gregory, J. M., & Flather, R. A. (2001). Changes in the occurrence of storm surges around the United Kingdom under a future climate scenario using a dynamic storm surge model driven by the Hadley Centre climate models. *Climate Dynamics, 18*, 179–188.

Lowe, J. A., Howard, T., Pardaens, A., Tinker, J., Holt, J., Wakelin, S., Milne, G., Leake, J., Wolf, J., Horsburgh, K., Reeder, T., Jenkins, G., Ridley, J., Dye, S., & Bradley, S. (2009). *UK Climate Projections science report: Marine and coastal projections*. Exeter: Met Office Hadley Centre.

Lowe, J. A., Woodworth, P. L., Knutson, T., McDonald, R. E., McInnes, K. L., Woth, K., Von Storch, H., Wolf, J., Swail, V., Bernier, N., Gulev, S., Horsburgh, K. J., Unnikrishnan, A. S., Hunter, J. R., & Weisse, R. (2010). Past and future changes in extreme sea levels and waves. In J. A. Church, P. L. Woodworth, T. Aarup, & W. S. Wilson (Eds.), *Understanding sea-level rise and variability*. London: Blackwell.

Megann, A. P., New, A. L., Blaker, A. T., & Sinha, B. (2010). The sensitivity of a coupled climate model to its ocean component. *Journal of Climate, 23*, 5126–5150. doi:10.1175/2010JCLI3394.1.

Monbaliu, J., Padilla-Henandez, R., Hargreaves, J. C., Albiach, J. C., Luo, W., Sclavo, M., & Guenther, H. (2000). The spectral wave model, WAM, adapted for applications with high resolution. *Coastal Engineering, 41*, 41–62.

Murphy, J. M., Booth, B. B. B., Collins, M., Harris, G. R., Sexton, D. M. H., & Webb, M. J. (2007). A methodology for probabilistic predictions of regional climate change from perturbed physics ensembles. *Philosophical Transactions of the Royal Society A, 365*, 1993–2028.

Ostanciaux, É., Husson, L., Choblet, G., Robin, C., & Pedoja, K. (2011). Present-day trends of vertical ground motion along the coast lines. *Earth-Science Reviews*. doi:10.1016/j.earscirev.2011.10.004.

Pope, V. D., Gallani, M. L., Rowntree, P. R., & Stratton, R. A. (2000). The impact of new physical parameterizations in the Hadley Centre climate model - HadAM3. *Climate Dynamics, 16*(2–3), 123–146. doi:10.1007/s003820050009.

Prandle, D., & Wolf, J. (1978). The interaction of surge and tide in the North Sea and River Thames. *Geophysical Journal of the Royal Astronomical Society, 55*, 203–216.

Shennan, I., Milne, G., & Bradley, S. (2011). Late Holocene vertical land motion and relative sea-level changes: Lessons from the British Isles. *Journal of Quaternary Science*. doi:10.1002/jqs.1532.

Tamisiea, M. E., & Mitrovica, J. X. (2011). The moving boundaries of sea level change: Understanding the origins of geographic variability. *Oceanography, 24*(2), 24–39. doi:10.5670/oceanog.2011.25.

Wang, X. L., Zwiers, F. W., Swail, V. R., & Feng, Y. (2009). Trends and variability of storminess in the Northeast Atlantic region, 1874–2007. *Climate Dynamics, 33*, 1179–1195. doi:10.1007/s00382-008-0504-5.

Weisse, R., von Storch, H., Callies, U., Chrastansky, A., Feser, F., Grabemann, I., Günther, H., Pluess, A., Stoye, T., Tellkamp, J., Winterfeldt, J., & Woth, K. (2009). Regional meteorological-marine reanalysis and climate change projections: Results for northern Europe and potential for coastal and offshore applications. *Bulletin of the American Meteorological Society, 90*(6), 849–860. doi:10.1175/2008BAMS2713.1.

Wolf, J. (2008). Coupled wave and surge modeling and implications for coastal flooding. *Advances in Geosciences, 17*, 1–4.

Wolf, J. (2009). Coastal flooding – Impacts of coupled wave-surge-tide models. *Natural Hazards, 49*(2), 241–260.

Wolf, J., & Woolf, D. K. (2006). Waves and climate change in the north-east Atlantic. *Geophysical Research Letters, 33*, L06604. doi:10.1029/2005GL025113.

Woodworth, P. L., Teferle, N., Bingley, R., Shennan, I., & Williams, S. D. P. (2009). Trends in UK mean sea level revisited. *Geophysical Journal International, 176*, 19–30. doi:10.1111/j.1365-246X.2008.03942.x.

Woodworth, P. L., Gehrels, W. R., & Nerem, R. S. (2011). Nineteenth and twentieth century changes in sea level. *Oceanography, 24*(2), 80–93. doi:10.5670/oceanog.2011.29.

Woollings, T. (2010). Dynamical influences on European climate: An uncertain future. *Philosophical Transactions of the Royal Society A, 368*, 3733–3756. doi:10.1098/rsta.2010.0040.

Chapter 3
Broad-Scale Hydrodynamic Simulation, Wave Transformation and Sediment Pathways

Nicolas Chini and Peter Stansby

Abstract The stretch of coastline off East Anglia (UK) is characterised by the presence of large-scale seabed features, tidal sandbanks and shore-connected sand ridges, which determine tidal flow movement and wave propagation. This chapter describes the method used for transforming extreme offshore events into shallow waters which influences local flooding and erosion.

The climate projections presented in Chap. 2 are at a regional scale that allows the assessment of the influence of climate change on offshore wave and storm-surge conditions; in this case the effect is small. In order to propagate offshore wave conditions into shallow water and ultimately the coast, fine-resolution numerical wave and surge models are necessary to capture the spatial variability of the bathymetry and the physical processes associated with wave propagation in shallow water. Here, this is achieved by the TELEMAC hydro-informatics system. More precisely it is used to generate a look-up table to compute inshore waves from 1960 to 2099 taking account of the cumulative effect of climate change on atmospheric forcing and sea-level rise. Simulations take into account four scenarios of linear sea-level rise and a baseline climatic scenario for storm surges and offshore waves. Extreme event analysis is performed to estimate the effect of sea-level rise on nearshore wave height and the occurrence of extremes. For nearshore areas, it is shown that an increase of mean sea level leads to a reduction in the return period of extreme waves.

The wave simulations above are made assuming a fixed bed. Changes in wave climate and sea level potentially modify the sediment dynamics and the supply of sediment for the maintenance of offshore sandbanks. Here, the evolution of the sandbanks is analysed by simulating sediment transport and seabed morphodynamics using a process-based approach. The complexity of long-term residual flows and sediment pathways is shown by nearshore and sandbank interaction.

Wave transformation methods were used to provide the inputs to the erosion and flood analyses in Chaps. 7 and 8. The sandbank dynamics are important, but the understanding was too limited to include. This is an important topic for further research.

N. Chini • P. Stansby (✉)
Tyndall Centre for Climate Change Research, School of Mechanical, Aerospace and Civil Engineering, University of Manchester, Manchester M13 9PL, UK
e-mail: nicolas.chini@manchester.ac.uk; pr.k.stansby@manchester.ac.uk

R.J. Nicholls et al. (eds.), *Broad Scale Coastal Simulation*, Advances in Global Change Research 49, DOI 10.1007/978-94-007-5258-0_3

Keywords Open coasts • Sediment pathways • Wave propagation • Sea-level rise • Climate change

3.1 Introduction

For coastal management and planning, statistics of nearshore wave heights and water levels are the main input parameters for coastal structure design. These statistics, including extreme values, may be estimated using available data extracted from either field measurements or hindcast model results. Although methods for determining statistical estimates have improved, the temporal span of the available data still represents the main limiting factor. Moreover, statistical models based on historical data can be misleading in producing estimates for future changes due to lack of stationarity, i.e. changes in the long-term statistical characteristics. Estimation based on global-scale future scenarios is thus necessary as explained in Chap. 2, with a methodology to translate offshore conditions to the nearshore as developed in this Chapter.

Knowledge of nearshore sediment processes is also required by coastal planners in order to assess the natural protection offered by beaches and to design defences against coastal erosion and flooding. These sediment processes are directly linked to nearshore wave climate. This chapter aims therefore to assess the changes in nearshore wave heights and directionality.

The linkage between the offshore oceanographic climate, which was discussed in Chap. 2, and the nearshore conditions that affect coastline morphodynamics (Chap. 7) and habitats (Chap. 6) and induce coastal flooding (Chaps. 8 and 9) involves a wide range of physical processes that take place at various temporal and spatial scales. This occurs over a coastal domain where the water depths are less than 50 m. For this domain, temporal variability of water depth also determines shoreward wave propagation. This latter variability takes place at different timescales, either on a daily basis when water depth varies with storm surge and astronomical tides or on a larger timescale for which sea-level rise and bathymetric evolutions affect the water levels. With an increased mean sea level, offshore waves propagating towards the coast during a combination of high tide and surge levels potentially release more energy to the nearshore than the same waves with a lower sea level, due simply to the delay of depth-induced breaking and reduced bottom friction. This results in greater forces on natural and man-made sea defences.

State-of-the-art hydro-informatics systems such as DELFT3D, MIKE and TELEMAC represent the physical processes taking place in the coastal area and are able to transfer offshore wave and storm-surge predictions to the shoreline. Numerical modelling offers the option of not only covering a large coastal area but also of assessing different scenarios. However, one also needs to perform long-term simulations, and the use of process-based numerical models can be time consuming and hence impractical with all processes simulated concurrently. An alternative efficient method may be set up based on calculating a look-up table for wave characteristics at inshore locations, defined by relevant input parameters to enable rapid assessment of future changes inshore due to climate change scenarios and sea-level rise. This study of inshore wave-transformation is performed assuming a stationary seabed. This assumption is valid for hydrodynamic prediction provided that bed evolution is small compared to water depth, which is likely for small timescales (order 1 year). Historical charts can provide empirical information about the past mobility of the seabed. For the Norfolk case study, the seabed is variable with bed features ranging from sand ridges to elongated tidal sandbanks, reducing the water depth at their crests to about 5–10 m. The response of these seabed features to sea-level rise and future inshore wave climate is uncertain, and the future position of these sand volumes will in turn affect future wave climate. The adaptation of the sandbanks to the future conditions is, however, related to the exchange of sediment with the nearshore. Here, sediment pathways are determined by process-based numerical modelling. The model simulates the sediment transport induced by waves and currents taking into account sea-level rise. The residual, typically the annual average, sediment transport determines the sediment exchanges.

This chapter is structured as follows. Section 3.2 introduces the case study area located off East Anglia (UK) and the methodology used to assess wave climate and residual sediment transport. Section 3.3 presents the influence of sea-level rise and climate change scenarios on inshore wave climate. Changes both in extreme wave height and wave directionality are discussed. The residual sediment transport at the present time is presented and compared with the inclusion of sea-level rise in Sect. 3.4, while Sect. 3.5 concludes.

3.2 Methodology for Transferring Waves and Surge Towards the Coastline and Assessing Residual Sediment Transport

3.2.1 The Area of Interest

The continental shelf sea offshore of the East Anglia coast, between Sheringham and Southwold (Fig. 3.1), is considered in terms of its hydrodynamic conditions, its complex bathymetry and the nature of the coastline, which is subjected to erosion and coastal flooding. In this chapter we focus on the first two features.

Offshore of East Anglia, the tidal pattern is complex, being meso-tidal at Cromer and micro-tidal at Lowestoft. Tidal range for spring tides varies from 3 m in the north and reduces to 1.6 m in the southern part of the domain. Extensive wave monitoring is undertaken along the shore by the Environment Agency, but only one directional wave buoy was deployed off Cromer from 07/12/1985 to 30/06/1987 (Clayson and Ewing 1988). The location of this wave buoy is presented by a black triangle on Fig. 3.1. The statistics from that wave buoy show that waves are relatively

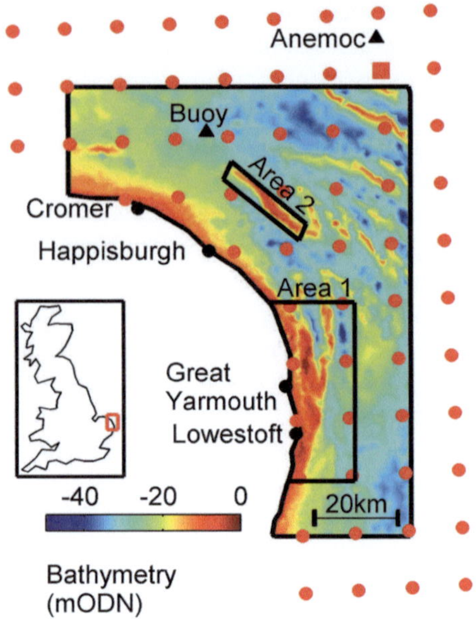

Fig. 3.1 The bathymetry of the computational domain, the nodes from the regional model (*red dots, one red square* represents the location of point of offshore analysis) and the position of the wave buoy (observations) and the Anemoc data (hindcast). The bathymetry is available under licence from Seazone (Ltd). Areas 1 and 2 are subdomains including the Great Yarmouth Sandbank system and Happisburgh sandbank, respectively

small on average, the mean observed significant wave height being 1 m, with waves predominantly coming from the north. From the collected data, the 50-year return wave height at this wave buoy is estimated to be 7.6 m (Clayson and Ewing 1988). A spectral analysis of the measured significant wave height reveals a semidiurnal modulation with an amplitude of 0.14 m (Chini et al. 2010a). Previous numerical tests showed that local wind speeds of less than 20 m/s do not influence the significant wave height (Kuang and Stansby 2004). In case of extreme winds, the significant wave height can be increased along the coast by about 0.6 m.

The sandbanks situated within this coastal domain have been widely surveyed and their mobility studied (Reeve et al. 2001; Park and Vincent 2007; Horrillo-Caraballo and Reeve 2008). The sandbanks system located off Great Yarmouth (see Area 1 on Fig. 3.1) has been monitored since the early nineteenth century, constituting a detailed database allowing Reeve et al. (2001) to show that the upper part of the sandbanks are accreting with a potential explanation related to sea-level rise. Further offshore, isolated and elongated sandbanks are noticeable. These sandbanks reduce the water depth from about 30 m to 5–10 m at their crest with implications for wave propagation. Area 2 focuses on the Happisburgh sandbank, which is the closest sandbank to the coast. The residual sediment pathways around these two types of sandbanks have been previously estimated. Those located off Great Yarmouth show a complex pattern with essentially the seaward side of the banks having a northward transport, with a transport in the opposite direction along the landward side (Park and Vincent 2007; Horrillo-Caraballo and Reeve 2008). For the offshore sandbanks that are not connected to the shore, clockwise residual circulation has been observed (Collins et al. 1995) and is explained by theoretical studies showing that this particular residual transport is induced both by the bottom friction and Coriolis acceleration (Huthnance 1982). The interconnection between the sandbanks is, however, not well understood, with the concept of 'stepping stones' being proposed for their maintenance (Stride 1988).

3.2.2 *The Modelling System*

To study the interactions between waves, tides, surge and sediment transport in the area under investigation, a numerical approach based on the state-of-the-art TELEMAC suite (Hervouet 2007) is adopted. The software contains three modules for simulating free-surface flows and bed dynamics using a finite-element method on an unstructured grid. The use of an unstructured grid allows fine spatial resolution to resolve seabed variability. The tidal flow and the surge motion are simulated using TELEMAC-2D that solves the depth-averaged non-linear shallow water equations. The third-generation wave spectral model, TOMAWAC (Benoit et al. 1996), is used to simulate wind–wave propagation. Sediment transport and bed evolution induced by waves, tides and surge are computed by the module SISYPHE (Villaret et al. 2013).

3.2.2.1 Long-Term Wave Climate Modelling

Simulating nearshore wave time series spanning several years using a third-generation wave spectral model such as TOMAWAC or SWAN (Booij et al. 1999) is time consuming. Several methods have been proposed to reduce the computational time, while maintaining the accuracy of numerical predictions (Stansby et al. 2006; Chini et al. 2010a; Camus et al. 2011). They are all based on the creation of a look-up table for the results of simulations, associated with an interpolation scheme. The principal difference between these methods lies in the way the look-up table is created. Here, a regular look-up table based on the four input parameters, i.e. water level, offshore significant wave height, offshore peak spectral wave period and offshore wave direction, is set up to both include the effect of varying water level on wave propagation and provide the integrated wave parameters at any point located within the computational domain. The wave modelling is performed assuming a fixed seabed, whose mobility is further discussed in the following section.

3.2.2.2 Sediment Transport Modelling

The coupling between the three modules of the TELEMAC suite is used to estimate bed evolutions over 1 year, and a schematic view of the coupling is presented in the flow chart in Fig. 3.2.

First, a 1-year TELEMAC-2D simulation is performed. For this first simulation, TELEMAC-2D is forced with the offshore water depth and velocity components $(h, u, v)_0$ provided by regional modelling, presented in the following subsection. This

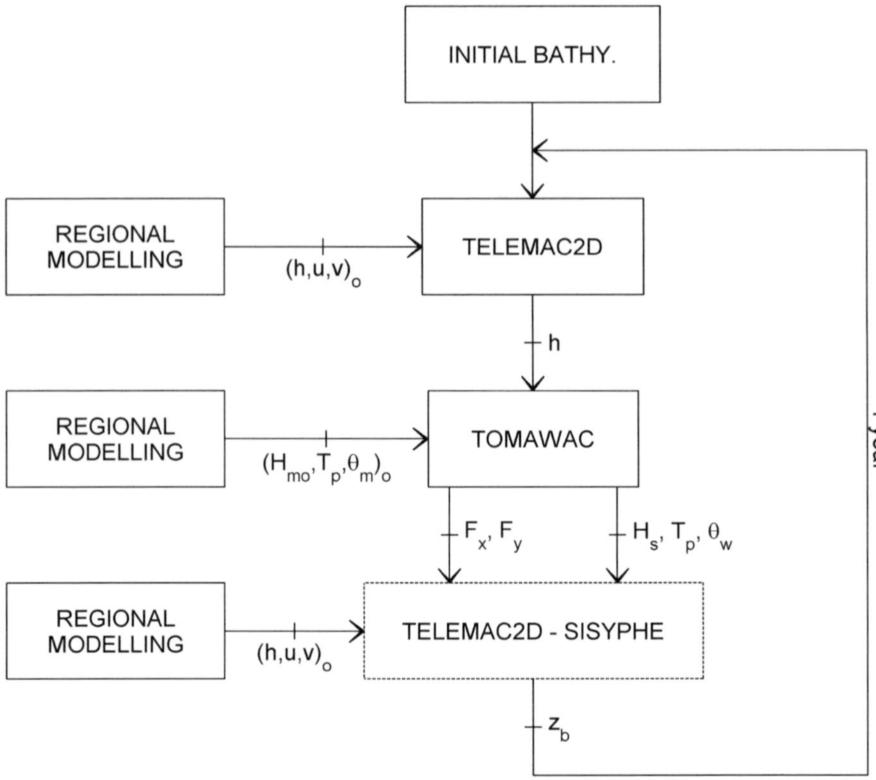

Fig. 3.2 Flow chart between the different modules for morphological computations

simulation provides TOMAWAC with the water depth, h. Refraction by currents can be included in TOMAWAC, but previous tests showed that wave refraction due to current is not significant in this area (Kuang and Stansby 2004). The integrated wave parameters (significant wave height, H_{m0}; the peak period, T_p; the mean wave direction, θ_w) and the radiation stresses (F_x, F_y) are then computed over 1 year by TOMAWAC and transferred to SISYPHE and TELEMAC-2D, respectively. At a final stage, SISYPHE and TELEMAC-2D are internally coupled to produce sediment transport and seabed evolution estimations.

The coupled system is used to assess the effect of sea-level rise and different climate change scenarios on residual sediment transport.

3.2.3 Offshore Boundary Conditions

Coastal modelling is forced along the offshore boundaries by climate change projections provided by UKCP09 as presented in Chap. 2. These projections are for one greenhouse gas emission scenario SRES A1B with three different regional modelling parameterisations. The projections are from 1960 to 2100 and provide water levels, current components and wave-integrated parameters. A bilinear interpolation is performed on the regional computational grid to provide values on the coastal grid. The locations of the regional nodes are shown in Fig. 3.1.

In order to assess the influence of sea-level rise on waves and sediment transport, linear trends in sea-level rise of different magnitude are superimposed on the water level. The time variation of sea-level rise is added to water levels along the offshore boundary for TELEMAC-2D. The increase in water depth could modify the currents. Here, only the magnitudes of the velocities are modified by assuming unchanged flux and direction along the boundary. Sea-level rise will also affect the propagation of inshore waves.

The assumption of sea-level rise superposition is considered reasonable for the projected sea-level rise rates (Lowe et al. 2001). Five sea-level rise rates are considered. The current sea-level rise at Lowestoft estimated by Haigh et al. (2009) is simulated. This sea-level rise is likely to be modified in the future, and the projected sea-level rise rates proposed by Meehl et al. (2007) are considered. These latter projections consider that the sea-level rise will accelerate in the future. However, here only a linear trend is assumed. These projections did not capture the effect of Arctic ice sheet melting and therefore a fourth sea-level rise leading to an increase in sea-level rise of 2 m in 2100 is included. Simulations with no sea-level rise are also performed. The following linear sea-level rise rates are thus considered: 0.0, 0.2, 0.6, 1 and 2 m by 2100.

3.3 Changes in Inshore Waves Due to Climate Change and Sea-Level Rise

The modelling system is used to assess the influence of climate change and sea-level rise on inshore wave parameters. For coastal management, two main changes in wave parameters are significant. The first is the change in extreme significant wave height, which may enhance breaching, cliff erosion and coastal flooding. The other is the change in nearshore wave directionality. Any change in nearshore wave direction can modify the littoral drift and consequently affect the stability of the shore. To understand the inshore wave climate, the offshore conditions are first presented and compared with hindcast data. The results of changes in inshore extreme significant wave heights and modifications in mean wave direction are then discussed.

3.3.1 Offshore Conditions

Table 3.1 presents the 50-year return wave height estimated from the regional modelling presented in Chap. 2. Here, we focus on a point located at 50-m depth in the north of the coastal domain (Fig. 3.1). The statistics are computed for two-time

Table 3.1 Simulation of offshore extreme wave heights (at the red square presented on Fig. 3.1)

Perturbed parameter ensemble	Emissions scenario	50-year return wave height (m)	95 % confidence intervals (m)
–	Baseline conditions	5.96	5.49–6.49
–	Anemoc hindcast	6.25	5.45–7.05
High	SRES A1B	5.21	4.82–5.64
Unperturbed	SRES A1B	6.36	5.72–7.07
Low	SRES A1B	5.54	5.12–6.02

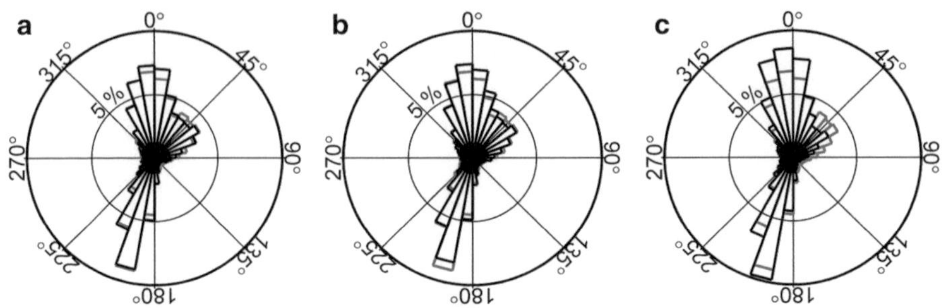

Fig. 3.3 Offshore mean wave directional occurrence. (**a**) low, (**b**) unperturbed and (**c**) high scenarios, baseline in *grey* and future in *black* at *red square* presented on Fig. 3.1

slices to infer any changes in offshore waves induced by climate change projections. The first time slice is called the baseline and it refers to the period lasting from 1960 to 1990. The second time slice presents the future projections estimated in the period lasting from 2070 to 2100. In order to give confidence to the baseline projections, results are compared with the Anemoc database (Benoit et al. 2008), which contains hindcast waves from 1979 to 2002 using re-analysed winds from ERA40 performed by the European Centre for Medium-Range Weather Forecasts. The location of the Anemoc datapoint is shown on Fig. 3.1. The statistics presented in Table 3.1 are estimated fitting a GEV (generalised extreme value) distribution (Coles 2001) to the annual maximum significant wave height extracted at the node from the regional model which is the closest to the Anemoc point (on Fig. 3.1, the node is represented by a square). While baseline estimations lie within the confidence interval of the Anemoc database, differences in the extreme offshore wave height for the three projections are noticeable. The high and low scenarios reduce the 50-year return significant wave height, whereas an increase in the extremes results from the unperturbed scenario. These results show the sensitivity of the projections to the parameterisation of the numerical modelling when transferring the SRES A1B scenario for marine projections around the UK and more particularly along the coast of East Anglia as highlighted in Chap. 2 (Fig. 2.7).

Offshore wave directionality can also be affected by changes in projected atmospheric conditions. Figure 3.3 presents the projected distribution of offshore mean wave direction for the baseline and the projections. The high and low scenarios induce some noticeable modifications with an increase in the occurrence of waves coming from the north and the south, jointly with a reduction of north-easterly waves.

3.3.2 Changes in Inshore Extreme Wave Heights

As for the offshore wave heights, the temporal series of inshore wave computed are analysed to estimate the extreme significant wave heights by fitting a GEV distribution to the annual maximum wave height, using maximum likelihood estimation (Coles 2001). Since the significant wave heights are sensitive to extreme local winds, the annual maximum wave heights at each location within the domain are computed by transferring selected offshore conditions including wind speed and direction. Firstly, the offshore conditions are distributed into bins of 45°. For each year the highest wave, within each bin, is selected. For each event, TOMAWAC is run for one tidal cycle assuming constant wind and offshore conditions.

The changes in the 50-year return significant wave height induced by the climate change projections are presented on Fig. 3.4. Figure 3.4a shows this significant wave height for the baseline scenario. The effects of the tidal sandbanks are noticeable as they induce a loss of wave energy due to bottom friction and wave breaking. As a result, a landward shadow area is generated where the extreme wave heights are reduced.

The modifications in the 50-year return significant wave height induced by climate change projections are presented in Fig. 3.4b–d. Consistently with the change in offshore conditions, the unperturbed scenario induces an overall increase. The low Scenario (Fig. 3.4d) induces on the other hand an expected decrease of the 50-year return level. The situation is less obvious for the high scenario, which leads to a slight increase of extreme wave height in the northern part of the domain and a reduction in the southern part. For that scenario, the results are dependent on wave directionality as the annual maximum wave heights propagating from the north-west are increasing.

Extreme wave height is also affected by sea-level rise. Sea-level rise induces an increase in the water depth, resulting in a reduction of the bottom friction and reduced-wave breaking. Figure 3.5 shows how the 100-year return significant wave height at two ports is modified by different sea-level rise. The first port is Great Yarmouth which is protected by the complex system of the tidal sand ridges. The second port is at Happisburgh in an open-coast location.

The 0.0035 m/year sea-level rise scenario has little effect on the extreme values at Great Yarmouth. For more pessimistic cases with sea-level rise higher than 0.007 m/year, the 100-year return wave height can increase by 12 % in 2100. For coastal structure stability, that means an increase of 24 % in wave energy. At Happisburgh, sea-level rise introduces a much smaller increase in the extreme wave height, no more than 6 % even in for the extreme sea-level rise rate leading to an increase of 2 m in water depth in 2100. Without bathymetric changes, sea-level rise

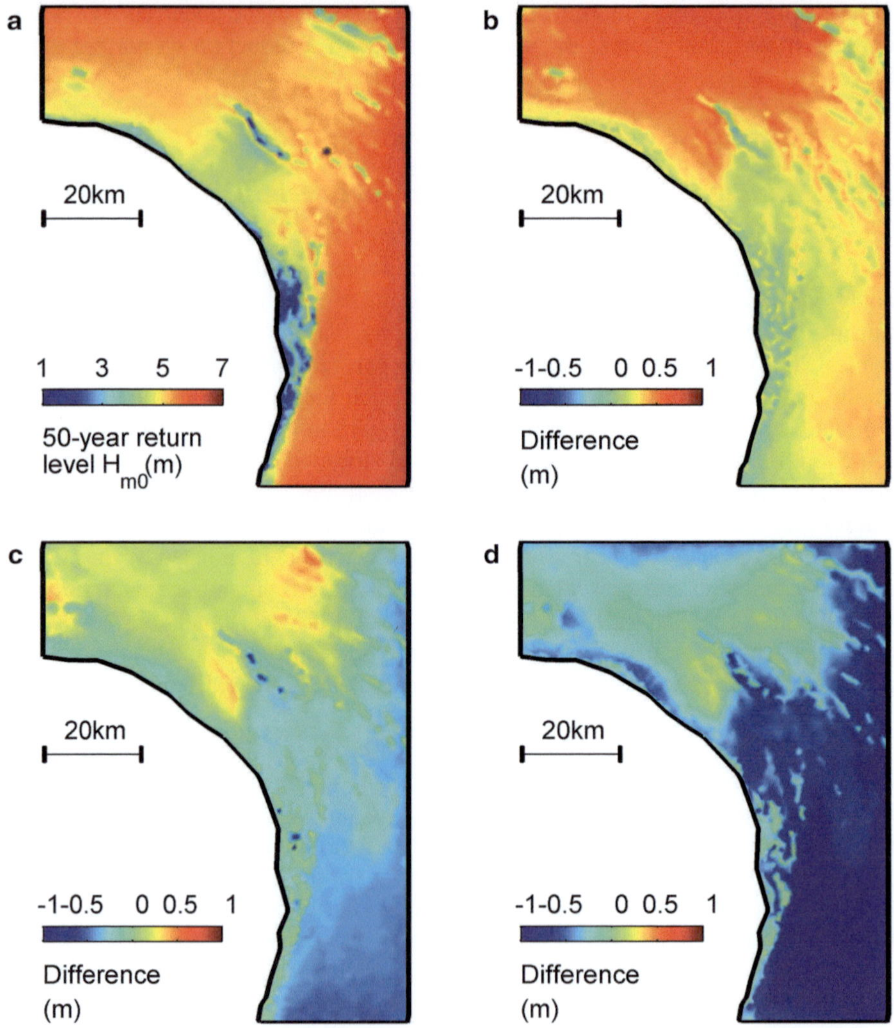

Fig. 3.4 50-year return level in significant wave height (**a**) from baseline and induced changes due projections unperturbed (**b**), high (**c**), and low (**d**). No sea-level rise included

causes areas currently protected by shore-connected sand ridges to be subject to a higher proportional increase in significant wave height than unprotected coastlines.

As wave height increases due to sea-level rise, the frequency of a given level is correspondingly increased. In other words, the 50-year wave height based on the baseline statistics becomes more frequent with an increased sea-level rise rate. Figure 3.6 presents the spatial distribution of the change in return period for the 100-year return period wave height from the baseline, for the case with a 0.0035 m/year sea-level rise. The influence of sea-level rise on the return periods is mainly in the shallower waters, nearshore and over the sandbanks. Consequently, this study

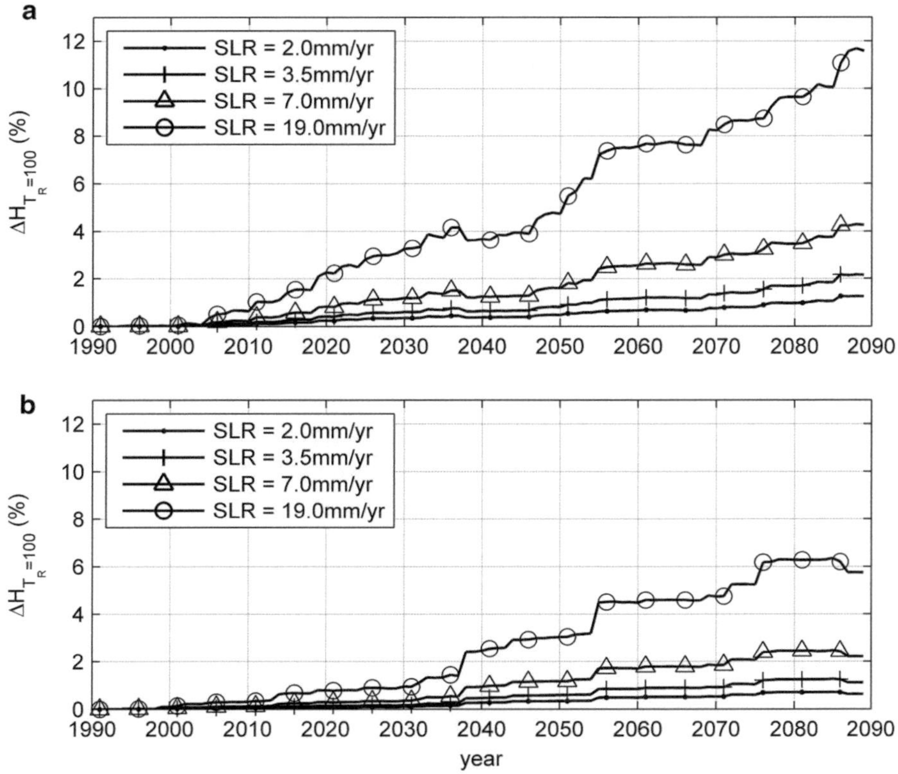

Fig. 3.5 Variation of the 100-year return wave height during the twenty-first century for different sea-level rise, using a running time slice of 30 years at (**a**) Great Yarmouth and (**b**) Happisburgh

shows that shorelines protected by sandbanks will experience an increase in nearshore wave heights.

3.3.3 Changes in Nearshore Wave Directions

Given the uncertainty related to climate change, the common assumption when assessing future nearshore sediment transport induced by waves is to consider that the current wave directions are all shifted by a constant angle. The use of the downscaling procedure based on the UKCP09 projections enables that assumption to be improved. As shown in the offshore wave directionality, there is no evidence of a single global shift in direction but a more complex redistribution (Fig. 3.3).

Here, the model, using the look-up table, is applied to transfer waves towards a particular point where the littoral drift has been studied and shows a clear south-eastward component (Vincent 1979). This longshore sediment transport is of importance for maintaining the sediment pathways between sediment input due to

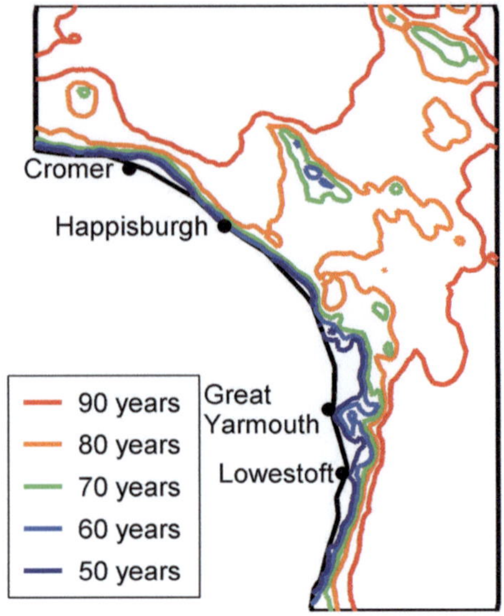

Fig. 3.6 Reductions in the return period of the 100-year return wave height from the baseline given 0.0035 m/year of sea-level rise

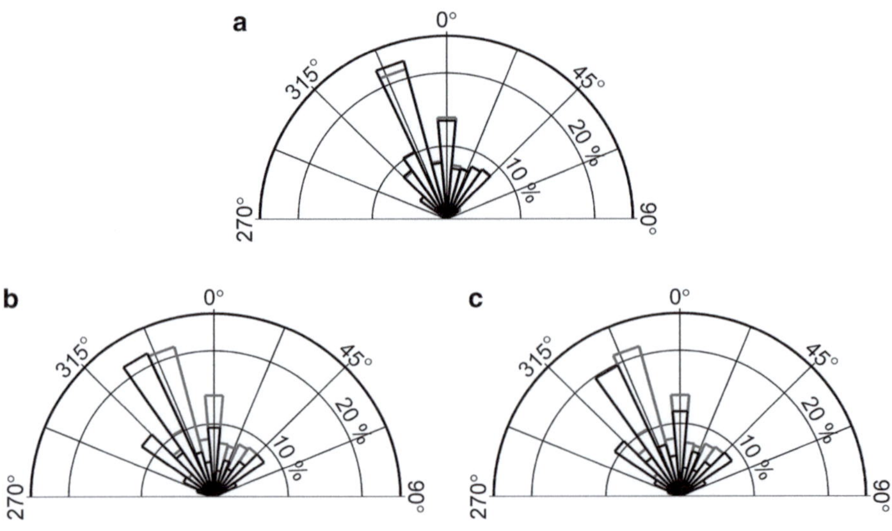

Fig. 3.7 Nearshore incident mean wave directional occurrence at Happisburgh. (**a**) unperturbed, (**b**) low and (**c**) high, baseline in *grey* and future in *black*

upstream cliff recession and downstream deposition on beaches with upper parts composed of dunes protecting low-lying land.

Figure 3.7 presents the variation in mean wave direction induced by the three climate projections. In this figure, the directions are represented relative to the normal to the shoreline. Consistent with the offshore conditions (Fig. 3.3), the unperturbed scenario does not induce much difference. However, the two other scenarios lead to more oblique waves resulting from an increase in offshore waves coming from the north and the south. Basically, these results show that the response of nearshore waves to offshore climate projections is more complex than a single global direction shift. These variations are related to the changes in offshore conditions. For the unperturbed scenario, one can notice an increase in the occurrence of waves coming from the north-north-west and a decrease in the number of waves from the north-east. For the two other scenarios, as well as an increase in the occurrence of waves reaching the coast more obliquely, there is a significant reduction in the number of waves coming from the north and the north-north-east. For the three scenarios, waves are predominantly reaching the coast from the north-north-east with a slight shift towards the west for the low and high scenarios. That would lead to an increase in the present longshore sediment transport.

3.4 Influence of Sea-Level Rise on Sediment Pathways

The annual residual sediment transport provides information about the long-term sediment transport by temporally filtering the effect of neap-spring tidal cycles and the intra-annual atmospheric variability. These residual transports are computed in order to assess the sediment pathways for different hydrodynamic conditions and sea-level rise. In order to understand the relation between the coastline and the sandbank system, sediment budgets are presented alongside the sediment pathways with focus put on two subdomains, the Great Yarmouth sandbanks system and the Happisburgh sandbank. First, the sediment pathways under present tidal conditions are considered. Then, the effects of waves are added according to the methodology presented in Sect. 3.2.2. Eventually, inclusion of sea-level rise is added in order to assess its influence on sediment pathways.

3.4.1 Baseline Situation

The modelling approach is used to assess the residual sediment transport defined as the averaged instantaneous sediment transport over 350 diurnal tides, broadly representing the mean over 1 year. The instantaneous sediment transport is driven by the flow conditions through a total load formula, for bed load and suspended load combined. In order to assess the sensitivity to the sediment transport formulations, two different semiempirical formulae for combined waves and currents, namely, the Soulsby–van Rijn (Soulsby 1997) and the Bijker (1968) formulations, are considered. These two formulations available in SISYPHE bound the range of possible magnitudes of sediment transport for waves and tides combined (Silva et al. 2009). Although the sediment transport takes place both as bed and suspended load, the use of total load was chosen given the large uncertainty in the suspended load computed using a depth-averaged model when waves are taken into account. The suspended load will be predominant for fine material. Here, we focus on fine to medium grain size. Tests were performed for two uniform spatial distributions of median grain size of 0.2 and 0.4 mm. The latter being the average grain size observed in the domain (Evans et al. 1998). Residual sediment transport direction and therefore the sediment pathways are found to be similar for the two different sediment transport formulations and for the two grain sizes. There are, however, differences in the magnitude of the sediment transport rates. The Soulsby–van Rijn formulation gives greater magnitudes than Bijker (1968), and the use of a smaller grain size increases the sediment transport rates.

Figure 3.8 presents the residual sediment transport computed by a simulation of 1 year when waves are omitted. The figure focuses on two areas centred on the Great Yarmouth sandbank system and on the Happisburgh sandbanks, respectively named as Area 1 and Area 2 (see Fig. 3.1), and it shows the complex circulation over these sandbanks. For the Great Yarmouth sandbanks, a divergence zone located in the centre of the area is noticeable in the deeper water, separated by the sandbanks. To the north, in deeper waters, the residual transport is northerly, and to the south, the transport is directed towards the south. On the other hand, over the sandbanks, the sediment transport direction is in opposition on each side of the sandbanks, in agreement with previous studies. Along the seaward part of the sandbank system, net northerly sediment transport can be observed. Within the sandbank system, the pattern is more complex with areas of divergence and convergence, with channels either ebb or flood dominated.

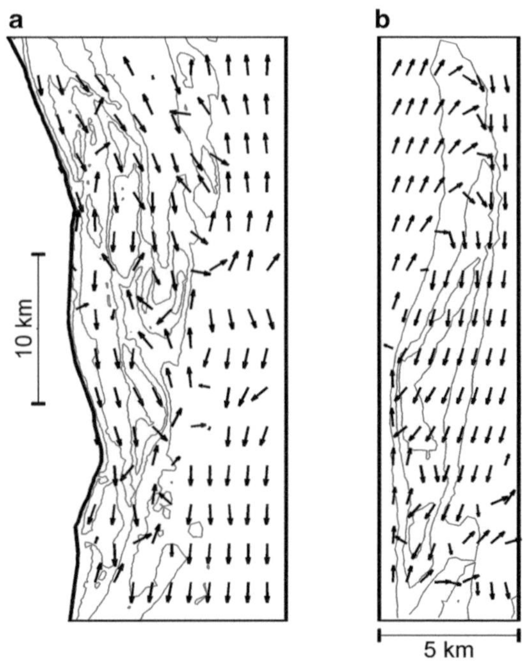

Fig. 3.8 Residual sediment transport direction induced by tide only in (**a**) Area 1 and (**b**) Area 2. *Grey line* shows −20, −10 and −5 m isobaths

Table 3.2 Sediment budget induced by tides only and tides and waves simulations

		Sediment budget (10^6 m^3/year)			
		Coastline	Area 1	Area 2	Total
Tides	Soulsby–van Rijn	−1.20	−2.76	5.56	−7.47
	Bijker	−0.05	−0.11	0.28	−0.40
Tides + waves	Soulsby–van Rijn	−9.20	3.71	2.34	−20.7
	Bijker	−0.14	−0.02	0.24	−0.62

For the Happisburgh sands (Fig. 3.8b), clockwise circulation is predicted by the modelling, with the residual deflected towards on the crest of the sandbank.

Although simulated bed evolutions are not directly compared with historical charts, the model is able to represent the residual sediment transport pattern which has been previously either estimated by measurements or by numerical modelling.

The simulations can also give information about the sediment budget within some specific areas of the computational domain by integrating the simulated bed evolutions. Three areas are considered. Along with Area 1 and Area 2, a third area representing the coastline is considered. This latter area follows the coastline with a width of about 250 m. For each area, the sediment budget is estimated by computing the volume changes. The result is then equivalent to the annual ingress of sediment within the considered area. Table 3.2 summarises the sediment budgets for the three areas and for the two sediment transport formulations. From this Table, it can be again seen that Soulsby–van Rijn provides much higher sediment transport rates than Bijker. However, both formulations show similar behaviour for the sediment budget at each location. Tidal simulations lead to a decrease in the sediment volume within the coastline area and in Area 1, whereas a net ingress is simulated for the offshore sandbank located in Area 2.

From the values in Table 3.2 and the direction of the residual sediment transport, it can be stated that the coastline provides a source of sediment for the shore-connected sand ridges. The connection between these two systems is, however, complex with recirculations along the nesses (coastal sedimentary features). Sediments travelling through Area 1 are subjected to pathways that will lead some to settle and others to exit the domain further offshore. The offshore sandbanks can be seen as attractors or sinks of sediment where sediments are deposited.

3.4.2 *Effect of Waves on Sediment Transport*

Waves play an important role in determining sediment motion. In the modelling presented here, waves affect the tidal circulation through their stresses induced over the scale of a wavelength, known as radiation stress. Therefore, the flow that generates and maintains sediment transport is modified both in magnitude and direction by waves. The effect of waves on the direction of sediment transport is determined indirectly by the flow. However, wave height and period are taken into account in sediment transport rates provided by the two formulae considered in order to represent the sediment pick-up induced by waves. The sensitivity of sediment transport rate and direction to waves is shown by comparing two simulations of sediment transport over 1 year, the first taking account of both waves and tides and the other with waves omitted. For each simulation, the annual residual sediment transport is computed. The effect of waves on the residuals rates is estimated by assessing the relative difference between the rates with waves and without waves. The effect on residual direction is estimated by computing the angle between the sediment transport vectors estimated for each simulation. Figure 3.9 presents the effect of waves on sediment transport rate and direction.

Residual sediment transport rates are affected most by waves where water is shallow, over the sandbanks and along the coastline. Waves can increase the residuals by a factor of up to 15. In deeper waters, waves have little effect. The direction of residual sediment transport is affected over the sandbanks and also in some areas in deeper water not protected by sandbanks.

As for the simulation with tides only, sediment budgets are estimated when waves are considered in the sediment transport modelling. The results are presented in Table 3.2, jointly with the ones from the tide-only simulations. Inclusion of waves tends to increase the loss of sediment along the coastline. This loss coincides with a net ingress of sediment in Area 1, shown by both sediment transport formulations. While waves are building up the volume within Area 1 or at least maintaining it, tides are shown to remove the sediment away from this particular location. Furthermore, waves make the ingress of sediment in Area 2 smaller, showing the distinctive behaviour of shore-connected sand ridges and offshore sandbanks when affected by waves.

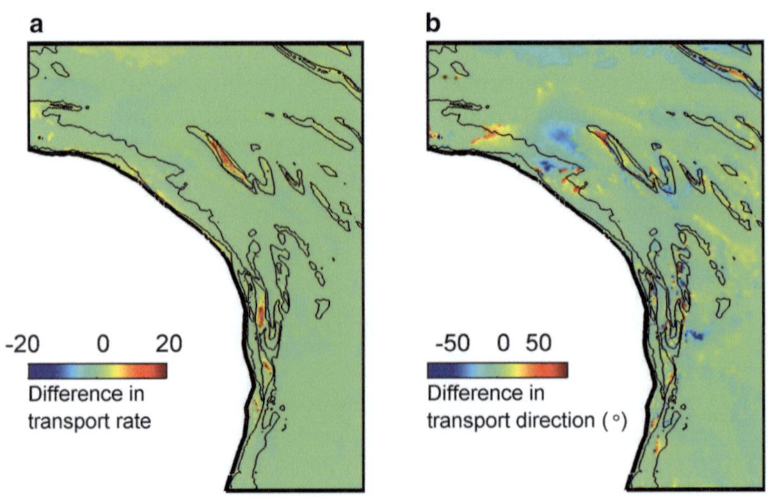

Fig. 3.9 The influence of waves on (**a**) residual sediment transport rates (as the relative difference between simulation with and without waves) and (**b**) sediment transport direction (in degrees)

For the results obtained for Area 1, historical charts surveyed from 1846 to 2002 are considered. This set of charts was analysed by Reeve et al. (2001). Here, this dataset is interpolated on to the computational grid in order to compare the historical bathymetries and to estimate the annual change of sediment volume contained in Area 1. Figure 3.10 presents the evolution of the ingress within Area 1 computed from the historical charts. The simulated sediment volumes are within the range of the historical volume changes. The volume of sediment entering the area is quite variable with a succession of periods with increasing and decreasing volumes. The numerical modelling is suggesting that net ingress of sediment in Area 1 occurred during years affected by substantial wave activity. On the other hand, the period of decrease of sediment in this area can be related to milder wave climate where sediment transport was mainly driven by tides. Figure 3.10b shows a comparison with the North Atlantic Oscillation (NAO) Index (Jones et al. 1997). This index provides information about the activity of storms and wind directionality in the Northern Atlantic. Positive indices indicate an increase in storminess and westerly winds in the North Atlantic, whereas negative indices are related to less intense storms. Even

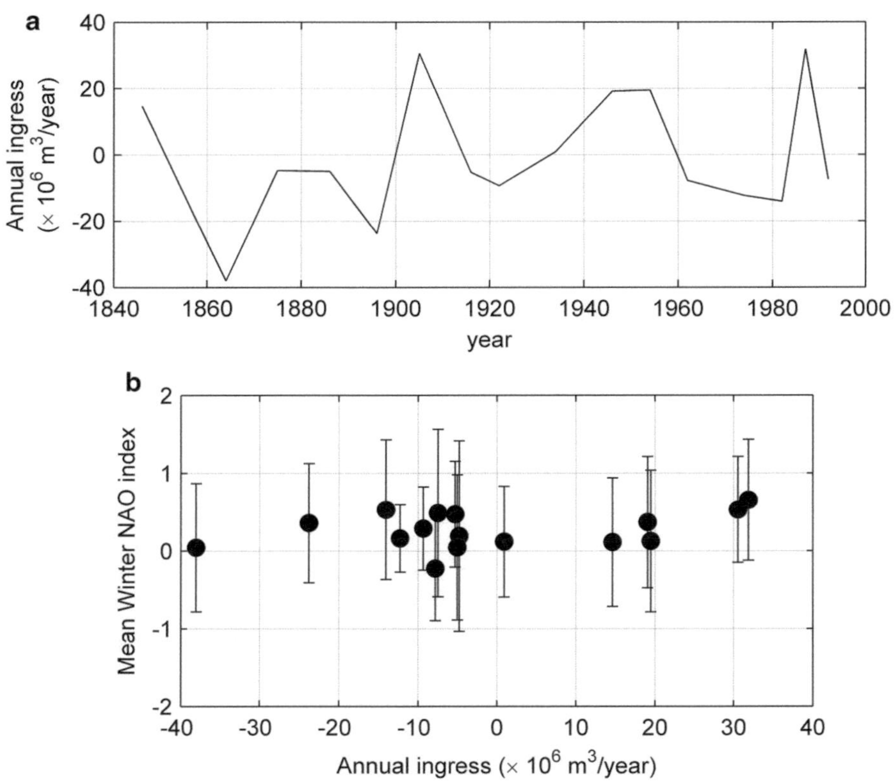

Fig. 3.10 Annual ingress of sediment from historical surveys in Area 1. (**a**) temporal series and (**b**) comparison with North Atlantic Oscillation index

if this index is not directly applicable for the North Sea, Fig. 3.10b shows that the extreme ingress of sediment in Area 1 occurred for extreme mean winter positive NAO indices. The error bars on Fig. 3.10b represent the standard deviation of the NAO index for the considered temporal periods, showing that during the extreme ingress, the NAO index remained predominantly positive over all the period between two bathymetric surveys. Decrease of the sediment volume in Area 1 can occur for positive NAO indices but the latter have smaller values or higher standard deviations.

3.4.3 *Effect of Sea-Level Rise on Sediment Transport*

Reeve et al. (2001) show a correlation between the volume of the sediment within the Great Yarmouth system and sea-level rise. Focusing on the upper part of the sandbanks, they found a linear trend in the temporal series of sediment volume that is similar to the actual sea-level rise rate. The same regression is performed for cumulative volume of sediment in Area 1 (Fig. 3.11). This shows according to historical charts that the volume of sediment is gradually increasing in the area over a period lasting 160 years with a rate of 0.18 million of m^3/year.

Sea-level rise is added to the numerical model in order to study its impact on the ingress of sediment within Area 1 and Area 2. It is analysed by comparing the volumes estimated with sea-level rise with those presented previously (see Table 3.2). The comparison is presented by computing the relative difference induced by the addition of different sea-level rise rates (Fig. 3.12). Results from the two sediment transport formulae are presented, applied with either tides only or the combination of waves and tides.

The influence of sea-level rise on tidally induced sediment transport is discussed first. The situation is compared for the two sediment transport formulae and for

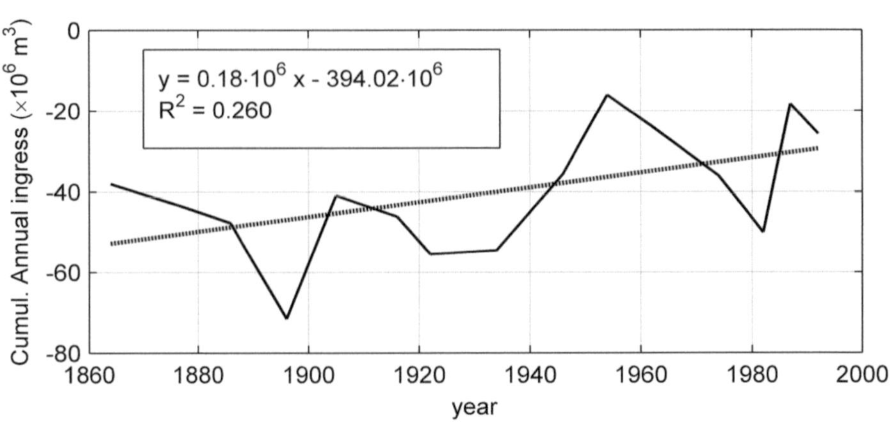

Fig. 3.11 Cumulative annual ingress of sediment in Area 1 (Fig. 3.1)

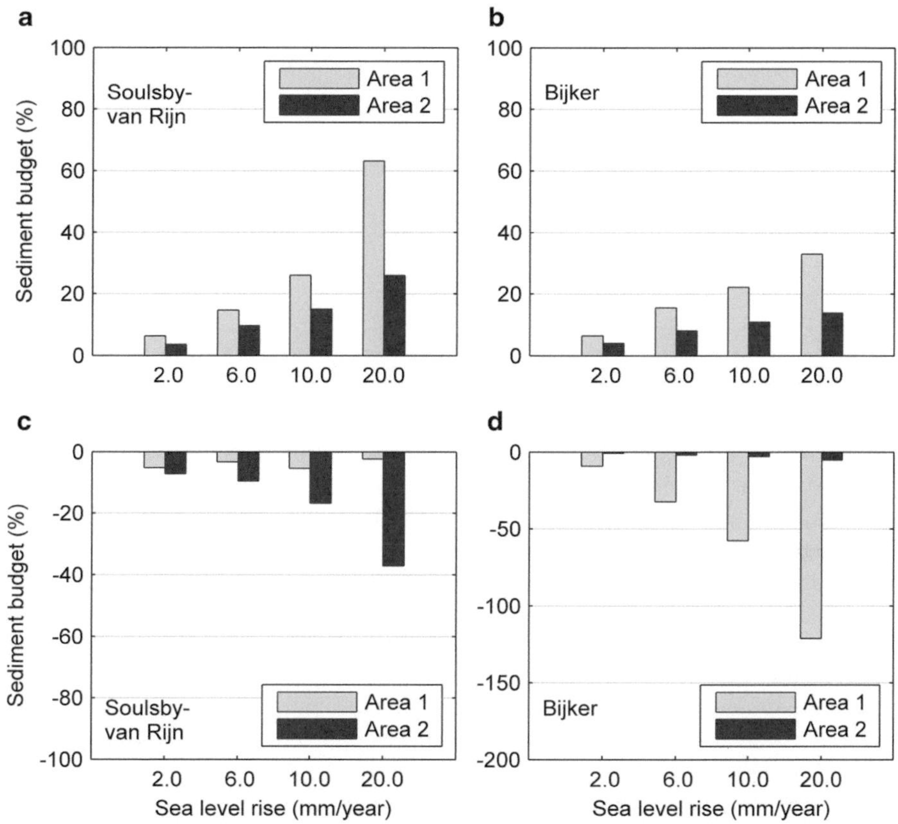

Fig. 3.12 Evolution of the sediment budget when including sea-level rise. (**a**) and (**b**) when tides are only considered and (**c**) and (**d**) when the sediment transport is driven by the combination of waves and tides. On panel (**d**), scale is increased

Areas 1 and 2. It can be seen that increased sea level induces ingress of sediment within Area 1 and Area 2. This result for Area 1 is in agreement with the results presented by Reeve et al. (2001). Moreover, for the Soulsby–van Rijn formula, the inclusion of the sea-level rise that took place during the twentieth century leads to an increase in the volume by 0.17 million m^3/year, which compares with the regression presented in Fig. 3.11. Similar agreement in the magnitude of the ingress is not found for the Bijker formula with much smaller sediment transport rates. However, both formulae show an increased ingress with a sea-level rise up to 2 m. Sandbank volumes are thus sensitive to the water level and tend to adapt to an increased sea level by increasing their volume.

The situation when sea-level rise is added to the water level in simulations combining waves and tides is more complex than when tides are only considered. The overall picture is that sea-level rise reduces the ingress of sediment within both Area

1 and Area 2. However, the situation in these two areas is quite different for the two sediment transport formulae. The Soulsby–van Rijn formula tends to show that sea-level rise has little impact on the ingress in Area 1 and significantly reduces the amount of sediment depositing in Area 2. On the other hand, sea-level rise has a greater influence when the Bijker formula is used in Area 1, where waves were found to stabilise the volume, and hardly modifies the volume in Area 2. For the results obtained with the Bijker formula, it should be born in mind that transport rates are small in comparison with the ones simulated with the Soulsby–van Rijn formula.

3.5 Conclusions and Next Steps

Coastal managers require future nearshore hydrodynamic conditions in order to plan appropriate policies for coastal erosion and coastal inundation. With sea-level rise and future atmospheric forcing projections, the present conditions will be modified and present-day management approaches may no longer be effective in preventing future coastal risks. The nearshore hydrodynamic conditions are, however, related to large-scale oceanographic and atmospheric phenomena that are generated offshore and propagate over varying water depths before reaching the coastline. During their shoreward propagation, waves and surges are subjected to complex physical processes that modify their characteristics. In order to take into account these processes, a downscaling procedure can be applied to transfer offshore projections towards the coastline.

Climate change and sea-level rise impact on inshore waves and sediment transport are analysed here using process-based numerical modelling. The modelling system described in this chapter is forced with offshore climate change scenarios to assess inshore water levels, wave-integrated parameters and residual sediment transport. The modelling is applied to a coastal area with complex bathymetry in the southern North Sea.

Changes in extreme inshore significant wave height are expected to occur in this area due to the climate change scenarios. The spatial pattern is complex with differences depending on climate scenarios. Sea-level rise leads to an increase in the occurrence of extreme wave heights of a given magnitude and for extreme sea-level scenarios to a significant increase in the 100-year return wave height along the coastline protected by sandbanks. Coastal flood risk is thus increased with sea-level rise due to increasing water level and extreme wave height.

Changes in wave directionality due to climate change projections are also discussed. Projections cause complex variations in offshore wave directions that are not a simple global shift, but a partial reduction or increase in the number of waves coming from a given direction. These offshore changes induce modifications to nearshore wave directionality. This affects littoral drift and consequently the stability of the coastline and coastal erosion. The transferred wave-integrated parameters are used to force nearshore models of morphodynamics and inundation (Chaps. 7

and 8) in order to assess shoreline stability and inform different adaptation policies.

Finally, insight on sediment pathways is presented in terms of residual sediment transport and sediment budgets. The simulations are performed for two different sediment transport formulae that tend to give qualitatively similar results, although the rates predicted by the two formulae can differ by a factor of 10. The sandbanks tend to adapt to sea-level rise by capturing more sediment. Waves also play an important role in the sediment dynamics. When they are included in the sediment transport modelling, the sediment transport rates are increased where the water is shallow and the volume of sediment in the sandbank areas is either maintained or increased. The addition of sea-level rise tends to reduce the latter. The sediment input from the eroding coastline (Chap. 7) appears to be the source maintaining the structure of the sandbanks. In the future, more explicit coupling of the coastal morphodynamics with the analysis here would allow these effects to be better quantified.

Further research should be undertaken to give more quantitative evidence of the link between nearshore sediment processes and the maintenance of the sandbanks offshore. Ultimately, the analysis of seabed morphodynamics and wave-transformation needs to be linked to provide more realistic simulations of future conditions. The proposed downscaling enables the assessment of nearshore hydrodynamic processes that affect coastline stability and induce coastline flooding. The methodology is implemented in order to perform large-scale long-time simulations of waves and surge in the upper part of the continental shelf and to take into account a range of climate change and sea-level rise scenarios. The output from the downscaling could be used to drive shoreline modelling, as shown in Chap. 7 and Chini et al. (2010b), and therefore translate this nearshore forcing into shoreline recession, inundation and protection costs.

References

Benoit, M., Marcos, F., & Becq, F. (1996). *Development of a third generation shallow-water wave model with unstructured spatial meshing*. 25th International Conference on Coastal Engineering, Orlando.

Benoit, M., Lafon, F., & Goasguen, G. (2008). Constitution et exploitation d'une base de données d'états de mer le long des côtes françaises par simulation numérique sur 23 ans. Base ANEMOC en Atlantique – Manche – Mer du Nord. *European Journal of Environmental and Civil Engineering, 12*, 35–50.

Bijker, E. W. (1968). *Littoral drift as function of waves and currents*. 11th International Conference on Coastal Engineering, London.

Booij, N., Ris, R. C., & Holthuijsen, L. H. (1999). A third-generation wave model for coastal regions 1. Model description and validation. *Journal of Geophysical Research, 104*, 7649–7666.

Camus, P., Mendez, F. J., & Medina, R. (2011). A hybrid efficient method to downscale wave climate to coastal areas. *Coastal Engineering, 58*, 851–862.

Chini, N., Stansby, P., Leake, J., Wolf, J., Roberts-Jones, J., & Lowe, J. (2010a). The impact of sea level rise and climate change on inshore wave climate: A case study for East Anglia (UK). *Coastal Engineering, 57*, 973–984.

Chini, N., Stansby, P., Walkden, M., Hall, J., Wolf, J., Lowe, J., & Nicholls, R. (2010b). *Modelling long term implication of climate change projection on shore morphology of North Norfolk, UK, combining TOMAWAC and SCAPE.* 32nd International Conference on Coastal Engineering, Shanghai.

Clayson, C. H., & Ewing, J. A. (1988). *Directional wave data recorded in the southern North Sea.* Wormley: Institute of Oceanographic Sciences - Deacon Laboratory.

Coles, S. (2001). *An introduction to statistical modelling of extreme values.* London: Springer.

Collins, M. B., Shimwell, S. J., Gao, S., Powell, H., Hewitson, C., & Taylor, J. A. (1995). Water and sediment movement in the vicinity of linear sandbanks: The Norfolk Banks, southern North Sea. *Marine Geology, 123*, 125–142.

Evans, C., Crosby, A., Wingfield, R., James, J., Slater, M., & Newsham, R. (1998). *Inshore seabed characterisation of selected sectors of the English coast* (British Geological Survey technical report WB 98/45). Nottingham: British Geological Survey.

Haigh, I., Nicholls, R., & Wells, N. (2009). Mean sea level trends around the English Channel over the 20th century and their wider context. *Continental Shelf Research, 29*, 2083–2098.

Hervouet, J. M. (2007). *Hydrodynamics of free surface flows modelling with the finite element method.* Chichester: Wiley.

Horrillo-Caraballo, J. M., & Reeve, D. E. (2008). Morphodynamic behaviour of a nearshore sandbank system: The Great Yarmouth Sandbanks, U.K. *Marine Geology, 254*, 91–106.

Huthnance, J. M. (1982). On the formation of sand banks of finite extent. *Estuarine, Coastal and Shelf Science, 15*, 277–299.

Jones, P. D., Jónsson, T., & Wheeler, D. (1997). Extension to the North Atlantic Oscillation using early instrumental pressure observations from Gibraltar and South-West Iceland. *International Journal of Climatology, 17*, 1433–1450.

Kuang, C., & Stansby, P. (2004). Efficient modelling for directional random wave propagation inshore. *Maritime Engineering, 157*, 123–131.

Lowe, J. A., Gregory, J. M., & Flather, R. A. (2001). Changes in the occurrence of storm surges around the United Kingdom under a future climate scenario using a dynamic storm surge model driven by the Hadley Centre climate models. *Climate Dynamics, 18*, 179–188.

Meehl, G. A., Stocker, T. F., Collins, W. D., Friedlingstein, P., Gaye, A. T., Gregory, J. M., Kitoh, A., Knutti, R., Murphy, J. M., Noda, A., Raper, S. C. B., Watterson, I. G., Weaver, A. J., & Zhao, Z. C. (2007). Global climate projections. In *Climate change 2007: The physical science basis* (Contribution of Working Group I to the Fourth Assessment Report of the Intergovernmental Panel on Climate Change). Cambridge/New York: Cambridge University Press.

Park, H. B., & Vincent, C. E. (2007). Evolution of Scroby sands in the East Anglian coast, UK. *Journal of Coastal Research, 50*, 868–873.

Reeve, D., Li, B., & Thurston, N. (2001). Eigenfunction analysis of decadal fluctuations in sandbank morphology at Great Yarmouth. *Journal of Coastal Research, 17*, 371–382.

Silva, P. A., Bertin, X., Fortunato, A. B., & Oliveira, A. (2009). Intercomparison of sediment transport formulas in current and combined wave-current conditions. *Journal of Coastal Research, 56*, 559–563.

Soulsby, R. (1997). *Dynamics of marine sands.* London: Thomas Telford.

Stansby, P., Zhou, J., Kuang, C. P., Walkden, M., Hall, J., & Dickson, M. (2006). *Long-term prediction of nearshore wave climate with an application to cliff erosion.* 30th International Conference on Coastal Engineering, San Diego.

Stride, A. H. (1988). Indications of long term episodic suspension transport of sand across the Norfolk Banks, North Sea. *Marine Geology, 79*, 55–64.

Villaret, C., Hervouet, J. M., Kopmann, R., Merkel, U., & Davies, A. G. (2013). Morphodynamic modelling using the Telemac finite-element system. *Computers & Geosciences, 53*, 105–113.

Vincent, C. E. (1979). Longshore sand transport rates – A simple model for the East Anglian coastline. *Coastal Engineering, 3*, 113–136.

Chapter 4
Land Use Dynamics and Coastal Management

Corentin M. Fontaine, Mustafa Mokrech, and Mark D.A. Rounsevell

Abstract Changes in climate risk are driven by a range of socio-economic factors, as well as the climatic drivers considered in Chaps. 2 and 3. Exploring how the built environment might change over time is an essential element for assessing changing risks. This chapter presents and discusses two algorithm-based approaches used in the Tyndall Coastal Simulator to analyse local changes in residential and other urban land uses.

The first approach uses multi-criteria analysis (MCA) to calculate spatial weights based on a number of attracting features such as transport, existing development, flood risk and proximity to the coast to identify the development patterns under different socio-economic futures. The second method uses agent-based modelling (ABM) to examine interactions between residential households and local planners as a demand–supply process that produces possible development patterns under different socio-economic futures. The core elements of this model are the location preferences of (changing) individual residential agents and the constraints imposed by planners through land-use policy.

The MCA method is simpler than the ABM method and useful for developing and realising socio-economic scenarios because of its flexibility and the possibility of quick implementation and adjustment. As a result, the MCA method is easy to implement, and it provides flexible tools that can be used for first-step socio-economic scenario development. While the MCA method requires fewer input variables, the ABM method is better able to account for feedbacks between governance

C.M. Fontaine (✉)
Namur Centre for Complex Systems and Research Group in Sustainable Development, Department of Geography, University of Namur, 61 Chaussée de Bruxelles, 5000 Namur, Belgium
e-mail: corentin.fontaine@gmail.com

M. Mokrech
Environmental Institute of Houston, School of Science and Computer Engineering, University of Houston Clear Lake, Houston, TX 77058, USA
e-mail: mokrech@uncl.edu

M.D.A. Rounsevell
Institute of Geography and the Lived Environment, School of Geosciences, University of Edinburgh, Edinburgh EH8 9XP, UK
e-mail: mrounsev@staffmail.ed.ac.uk

R.J. Nicholls et al. (eds.), *Broad Scale Coastal Simulation*, Advances in Global Change Research 49, DOI 10.1007/978-94-007-5258-0_4

processes, individual choices and external changes such as migration. The ABM is more sophisticated and more detailed and so provides a comprehensive approach that considers both location preferences and planning policies. The ABM method captures local behaviour and the iterative characteristics of urban dynamics. For both models, the key outputs are plausible distributions (scenarios) of urban dwellings in East Anglia over the twenty-first century under four different socio-economic storylines (see also Appendix A). Both methods are relevant to the process of risk assessment of climate change, and both have much scope for improvement, e.g. including brownfield sites and change in vertical as well as horizontal urban densities.

For the Tyndall Coastal Simulator, these model outputs provide the distribution of property that is vulnerable to coastal erosion and flooding, as described in the risk analysis in Chap. 9. More generally these types of algorithm-based models of socio-economic parameters have great potential for inclusion in a wide range of future change analyses.

Keywords Urban development • Land-use change • Urban system dynamics • Algorithm-based modelling • Agent-based modelling • Multi-criteria analysis

4.1 Introduction

Coastal areas are attractive places for people to live. Scenic value, perceptions of health benefits and economic assets all contribute to people wanting to live at the coast, and the consequences of this have for residential development pressures. However, residential development comes with landscape management challenges, including flooding and erosion control and mitigation. Cliff stability can be threatened by erosion leading to the potential loss of property to the sea (see Chap. 9). Likewise, low-lying land is at a greater risk of flooding due to the proximity of the sea and major rivers and their estuaries. These problems are likely to be exacerbated under the future changes in climate and sea-level rise described in Chaps. 2 and 3. Thus, evaluating future risks needs to also take account of socio-economic drivers, especially as these affect the location of new buildings.

Socio-economic drivers include the growth and distribution of population and the built environment, as well as land management and land-use policy (see Fig. 4.1). Land Use Science is an emerging discipline that addresses this class of question. Researchers design models of land-use change with a landscape management perspective (Lambin and Geist 2001; Verburg 2006). The objective is to identify the mechanisms of change in order to better understand the implications of these complex space–time relationships for the wider environment into the future. Assessing risk in a coastal area means understanding the spatial dynamics of land use with a focus on both built-up areas and sensitivity to flood/erosion areas. This includes in particular the accelerating development of new residential housing outside of urban areas, driven by residents preferring to move from areas with high to low urban densities (Fontaine and Rounsevell 2009). These new developments may occur on

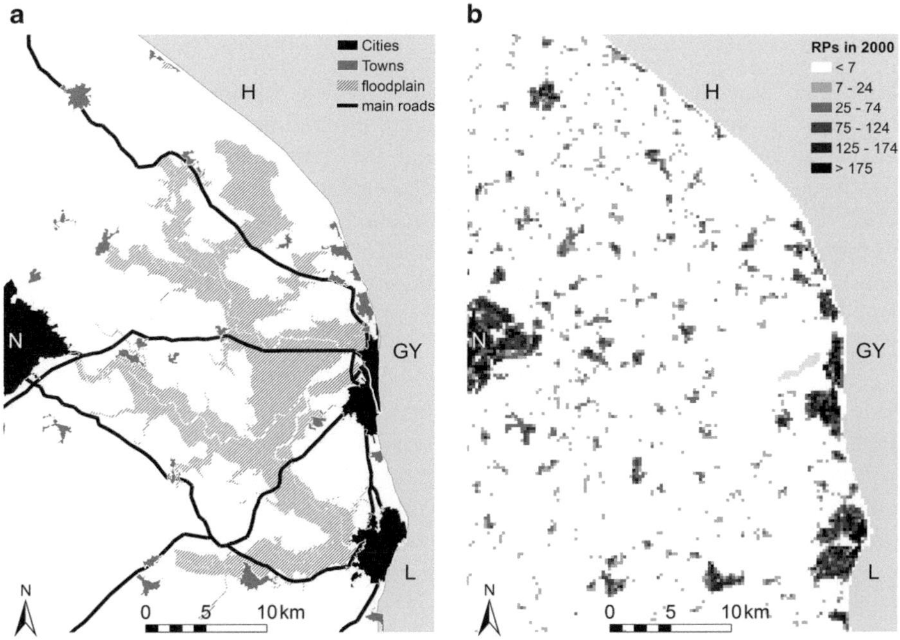

Fig. 4.1 Main features of the socio-environmental context in East Anglia. (**a**) Location of major settlements (cities and towns) and the main roads network around the principal floodplain in subcell 3b. (**b**) Detailed distribution of residential properties (*RP*) densities (number of properties per cell of 250 m by 250 m) in subcell 3b. *N* Norwich, *L* Lowestoft, *GY* Great Yarmouth, *H* Happisburgh

low-lying land areas within flood plains, or close to top cliffs. This process of peri-urbanisation relates to the individual and societal preference for more personal space and "greener" local amenities (Cavailhes et al. 2004; Brown et al. 2008). With climatic change, flood plains could extend, and cliffs could erode with increasing numbers of people put at risk. These risks are likely to increase with land-use change through time.

The objective of this chapter is to present two approaches used in the Tyndall Coastal Simulator to estimate urban land-use change: (1) multi-criteria analysis (Sect. 4.3) and (2) agent-based modelling (Sect. 4.4). Both methods consistently and transparently analyse local changes in land use for residential and other urban activities, based on an interpretation of socio-economic storylines embedded in scenarios and climate change projections. In Sect. 4.2, we discuss the implication of, and the challenges for, land-use modelling in the context of coastal management. This includes the interpretation of scenario storylines. Storylines describe what could happen in the future, in terms of social and economic development paths in qualitative terms (see Chap. 1 for details). Thus, in simulating the spatial dynamics of urban land use, it is necessary to transform these qualitative narratives into quantitative model inputs, as detailed in Sects. 4.3 and 4.4. Finally, in Sect. 4.5 we discuss some key results and lessons learned.

4.2 Modelling Land Use Change in Coastal Areas

Land Use Science aims to understand the complex, dynamic and non-linear relationships between human activities and our environment. The research focus is on the interactions between society and the natural environment (dubbed "socio-environmental system" or "socio-ecological system") and the outcome observed in the landscape. Hence, socio-economic drivers may be at least as important as climate change in shaping coastal areas in the future. Moreover, coastal zone change may result from interactions between different environmental and socio-economic change drivers at different geographical scales and/or at several decisional levels.

Empirical models formalising the links within socio-environmental systems provide a comprehensive approach to the understanding of land-use change (e.g. see Parker et al. 2003). Nonetheless, the majority of land-use models are designed within an environmental management perspective. Consequently, urban growth is only taken into account when buildings are considered to have an impact on agricultural, seminatural or natural land uses. In most cases, urbanisation is usually interpreted as a passive diffusion process that is controlled exogenously by "land-use planning". In reality, however, urbanisation is the result of human actions and decisions that cannot be regarded as a monotonous physical process or as a constant. The socio-economic context in which urban planning decisions are made will impact enormously on landscape dynamics in the long term and will modify critically the spatial distribution of flooding risks in a coastal context. In addition to the land use change research community, studies in urban economics and urban geography focus on the mechanisms leading to the emergence of cities and their internal organisation (Batty 2008). Urbanisation is often seen as the most important, complex and persistent process of change affecting the natural environment, society and individual people, from local to global scales (Batty 2005; Reginster and Rounsevell 2006; Batty 2008). Various urban growth theories and models have been developed over time with different levels of complexity (Batty 2005). With today's computing power, modellers have moved from analysing static distributions to developing dynamic models that do not seek equilibrium solutions, but that acknowledge the intrinsic unstable state of cities.

4.2.1 Challenges

Few studies have investigated how the built environment might evolve in the future in the context of environmental impacts and risk assessment. The interdependent challenges for research in urban land-use change, in the context of sustainable development, are (1) taking a forward-looking stance (i.e. prospective modelling; e.g. the storyline of a scenario), (2) incorporating both urban and environmental features in the analysis (i.e. considering more than just the direct environment of urban area; e.g. including climatic change in the urbanisation process), (3) analysing urban change as well as possible environmental consequences and (4) considering several possible development pathways since the "real" future is unknown by definition (e.g. a set of maps presenting different scenarios of urbanisation). In order to meet these challenges, socio-economic scenarios provide a starting point as they contain descriptions of urban development in broad terms. These qualitative storylines can be translated into (more) quantitative outcomes (also named "projections") using models and computer-based simulations. This process involves the interpretation of model input parameters using expert judgement, deviation and/or continuation of current trends or the results of other scenario studies (Rounsevell and Metzger 2010).

Some studies have applied socio-economic storylines to urbanisation directly based on expert judgement (e.g. Holman et al. 2005a, b), while others have produced scenarios of urban land-use change using either algorithm-based approaches (e.g. Mokrech et al. 2012) or decision-based approaches (e.g. Fontaine and Rounsevell 2009). Scenarios built with expert judgement usually combine qualitative knowledge, expressed in the storylines, with a more quantitative approach, such as magnitude and location of change to determine where land-use changes are more likely to occur (Rounsevell and Metzger 2010). It can be difficult to include all aspects of the problem domain, which may lead to gaps and possible inconsistencies (Parker et al. 2003). As an example, an expert judgement method tends to over simplify the planning processes which may bias the outcome.

4.2.2 Spatial Dynamic Modelling: Some Principles

An alternative to the expert judgement "top-down" analysis is "bottom-up" simulation techniques for analysing urban spatial processes. In either algorithm-based or decision-based approaches – they follow the same spirit – spatial dynamics are driven by explicit rules, applied locally and iteratively. Most rules consider that objects which are closer to one another have a greater direct influence. Using similar data inputs, these approaches only differ in the nature of the "rules" and to which object they apply.

For example, the most common tool to model urban growth is cellular automata (CA). In these models, urban growth is interpreted as a strict diffusion process simulated with neighbourhood functions. Space is usually represented by a regular grid of cells, which always take one of a finite number of states ("urban" and "not urban" in the simplest versions, but can be any land-use type or even levels, e.g. urban densities). Time is represented by a discrete iteration of steps (most often, one step refers to 1 year). At each time step, each cell may change state according to a set of rules relating to the configuration of the targeted cell and the surrounding cell states (the neighbourhood) observed at the previous time step. Typically, the rules are the same for all cells, do not change over time and are applied to the whole grid simultaneously, but independently (i.e. the rule refers to the spatial configuration in the previous time step). In sophisticated CA, the set of rules can be a combination of detailed calculations (e.g. mathematical algorithms evaluating the local urban density) and logical operators (e.g. a restriction to build on waterscape or a constraint to maintain urban hierarchy) relating to, for example, exogenous land-use planning.

Another tool that is being used increasingly to model urban growth is agent-based modelling (ABM). This approach concentrates on describing and reproducing the decision process followed by land-use agents (another "layer" of autonomous objects considered mobile on top of the grid cells) that ultimately lead to urban sprawl. In these models, urban growth is interpreted as the observable consequence of agents' effective actions. In other words, land-use change is the tip of the iceberg only. In most ABM modelling, the representation of space and time is the same as in CA for computational reasons, but other representations are also used, depending on the research objectives (e.g. polygons when the key spatial structure is administrative boundaries, a monthly time step when the key temporal structure varies greatly within a year, etc.). The main difference is in the inclusion of autonomous entities, called agents (e.g. individuals, households, etc.), which not only impose their decisions on cells (e.g. "build a new house here"), but also interact with one another (e.g. "get married") without any direct impact on the landscape. Nonetheless, these behaviours may have delayed and indirect consequences for land-use dynamics (e.g. the future child will eventually leave the parents and build its own house). The nature and the form of the behavioural rules are so varied that they cannot be described in detail here. In practice, there are an infinite number of ways to formalise agent decisional processes, but the principle is to keep the number of rules to a minimum in order to adequately represent the process under investigation (as for CA).

4.2.3 Socio-Economic Storylines

The simulations of urban land-use dynamics presented in Sects. 4.3 and 4.4 require specific socio-economic information in the form of scenarios. These inputs were derived from a two-stage adaptation of the national scenario storylines described in Appendix A to the spatial extent of East Anglia. First, the "ForeSight" socio-economic storylines were qualitatively downscaled to the level of coastal management in East Anglia. With an objective of exploring long-term changes in risk, we focused on four well-established and internally consistent UK scenarios (Table 4.1). Secondly, these regional storylines were translated into the quantitative indicators used in the simulations (see Sects. 4.3.2 and 4.4.2).

Table 4.1 The four storylines for UK and East Anglia in a nutshell

The UK "ForeSight" storyline	Storyline for East Anglia (EA)
World Market (*WM*)	
A world defined by an emphasis on private consumption and highly developed and integrated world trading systems.	The integrated trading systems lead to more development of the main urban centres in EA, which are still shadowed by London. They attract more workforces, at the expense of more remote locations. The increasing economy and wealth boosts private consumption and materialistic personal well-being, testified by larger new houses and more second homes. There will be a continued pressure to build more houses, and these are attracted to coastal and floodplain locations. Consequently, significant developments would occur in all areas including the floodplain and cliff areas.
Global Sustainability (*GS*)	
A world in which social and ecological values are considered in economic decisions and in which strong collective action through global institutions are taken.	The "think global, act local" motto encourages people to reduce their ecological footprint and to invest in their social network rather than in materials and goods. The economy grows at a lower path than today. The main urban centres tend to delegate to smaller towns part of the activities so services are closer to everyone and size differences are lessened. Population and household numbers increase; new developments will be mostly confined to existing urban areas. Less new development on coasts and floodplains arises partly through more planning and also through smaller growth in the number of households. Floodplain occupancy is discouraged under this scenario.
Local Stewardship (*LS*)	
A world of highly devolved political structures where actions are decided on and implemented at the community and neighbourhood level. Conservation of natural resources is a key concern. Labour-intensive activities are preferred to capital-intensive ones in order to conserve fuel, provide work and boost local economies.	Political actions are put in place to boost the local economy which increases job provision. The concern about natural resources and the environment pushes political actions to be taken to better inform on the negative consequences of new development in general and second-home ownership in particular. Most people are receptive to these concerns and information. The local economic activities maintain a sustainable viability of every community which are as autarchic as possible and mainly maintain contacts with the surrounding communities. A conservationist ethic and lower demand for new development contribute to the continuing dominance of traditional housing. Floodplain occupancy is kept stable. The urbanisation of the countryside comes to a halt as planning controls are tightened and strong land-use control avoids flood prone areas.

(continued)

Table 4.1 (continued)

The UK "ForeSight" storyline	Storyline for East Anglia (EA)
National Enterprise (*NE*)	
A world in which people aspire to personal independence and material wealth within a nationally rooted cultural identity. Liberalised markets, together with a commitment to build capabilities and resources to secure a high degree of self-reliance and security, are believed to best deliver these goals. Political and cultural institutions are strengthened to buttress national autonomy in a more fragmented world.	Liberalised markets do not greatly shift the gravity centre of economy towards the region but maintain its activities to today's levels. Population, households, urban areas and second homes increase largely in East Anglia. However, the increase will not be as high as that of the World Market scenario since there are lower levels of investment in housing. As well as the issue of urban sprawl, the increase in the number of households may be exacerbated by a coastal attraction and people relocating in coastal regions from large urban areas like London. Planning controls are relaxed, allowing more buildings and more people to live in floodplains.

See Appendix A for full scenario details

4.3 Translating Socio-Economic Storylines with an Algorithm-Based Approach

Mokrech et al. (2012) proposed an accessible/flexible, algorithm-based approach to construct scenarios of the future built environment using a GIS-based, multi-criteria analysis to interpret socio-economic scenarios. The objective was to facilitate broadscale flood and erosion risk analysis, with the following specifications:

- Quick to operationalise
- Flexible enough to explore many planning scenarios
- Accessible and easily interpreted by non-technical policy makers
- Based on readily available and standard datasets
- Provides spatial outputs at a broadscale
- Auditable and transparent

Here, we focus on the method used to estimate the spatial change in the built environment. The three steps taken were: (1) constructing projections of regional growth, (2) defining spatial attractors of growth, including their weights and (3) spatial allocation of new development.

4.3.1 Quantifying Total Change in the Built Environment

The first step builds projections of regional growth by quantifying the total number of new residential properties (RP) and non-residential properties (NRP) over the twenty-first century in East Anglia and under each socio-economic scenario (see Table 4.2). Current trends and the storylines of socio-economic scenarios provided the basis to estimate most change rates in RP and NRP from the baseline year. The change in the number of NRP is obtained by applying a correction factor to the change of RP based

Table 4.2 The percentage growth of residential properties (*RP*) and non-residential properties (*NRP*) and the total numbers of new properties created under each socio-economic scenario in subcell 3b by 2080s

Subcell 3b	Change in RP (%)	NRP correction (%)	Change in NRP (%)	Total no. of new RP	Total no. of new NRP
Local Stewardship	+28	−10	+18	62,607	4,069
Global Sustainability	+44	0	+44	98,158	9,939
National Enterprise	+61	0	+61	136,053	13,785
World Markets	+77	+10	+87	171,790	19,662

Mokrech et al. (2012), Nicholls et al. (2005)

Table 4.3 Summary of the attraction factors under the socio-economic scenarios

	Existing settlements	Transport networks	Coast	Floodplains
Local Stewardship	−−	−	-	−−
Global Sustainability	+	0	+	−
National Enterprise	0	0	+	+
World Markets	0	+	++	++

on expert judgement. This reduces relative growth of NRPs under Local Stewardship and increases growth under World Markets, which is consistent with the storylines. The correction factor is a decrease and increase of 10 % from the RP rate, which translates into an increase in NRP of +18 % and +87 %, respectively. NRPs under Global Sustainability and National Enterprise are assumed to follow the RP trends.

4.3.2 Identifying Triggers of Change in the Built Environment

In a second step, four attraction factors were selected as relevant to the spatial strategy, general process of urbanisation and planning policies of interest: (1) existing settlements, (2) transport networks, (3) the coastline and (4) floodplain. The number of factors was kept to a minimum to facilitate interpretation of the results. The socio-economic storylines were interpreted based on expert judgement to derive the relative importance of each attraction factor and the effect it would have on the distribution of people and employment in the future (see Table 4.3). As a result, urbanisation patterns under the Local Stewardship scenario tend to avoid both floodplains and existing settlements and move away from the coast and transport networks. This contrasts to the World Markets scenario, where urban development expands across much of the countryside, floodplain and the coastal zone, with a corresponding desire to be close to transport networks for easy commuting. The Global Sustainability governance structure lends itself to stronger planning regulation, which is parameterised in the model in terms of a reduction in permitted housing developments in the floodplain. Conversely, the utility of land-use planning in the World Market scenario is reduced by a more laissez-faire attitude to governance.

4.3.3 Simulating Spatial Change in the Built Environment

In the spatial allocation of new development, the single objective multi-criteria decision-making technique (Eastman et al. 1993) was implemented in order to determine the locations of both RP and NRP. The technique combined the attraction factors with additional constraints (used to exclude some areas, e.g. inland water bodies) in order to produce a single quota map. The map was then used to extract the number of new properties in each postcode polygon, standardised by their area and weighted according to their proximity to major settlements, roads and coastline (see Mokrech et al. 2012 for equations and more details).

As the attractors here were applied at the postcode scale, it is worth noting that spatial complexity occurring at smaller scales (e.g. attraction/repulsion of railway stations) is not captured in the modelling. Also, the possibility of properties being abandoned is not considered because of the underlying population growth and demand for development under all scenarios, so only new properties are created. The numbers of new NRP for each particular business type were scaled according to changes in the economy associated with each scenario (e.g. a greater proportion of the new NRP are financial services under the World Markets scenario). NRP were subsequently allocated to polygons with similar attributes leading to natural expansion trends in business types.

4.3.4 Results and Key Findings

The spatial distributions of new properties show consistency with the urbanisation policies under the four socio-economic scenarios and the attraction factors described in these storylines. For example, under the World Markets scenario, there is significant urban development across the entire region including floodplains, coastal areas and the countryside, while under the Local Stewardship scenario, there is limited development in the floodplains, in the coastal areas and within the existing major settlements. The same pattern can be observed for non-residential properties under these two scenarios. All the scenarios show new properties in the floodplains: the highest percentage increase is under the World Markets scenario, while the lowest is under the Local Stewardship scenario. The high floodplain development in the World Market scenario is due to the combination of all attractions – a desire to live within the floodplain as properties may cost less than elsewhere, as well as other attractants such as the coast and transport networks. Conversely, in the National Enterprise scenario, only the coast is an attraction, which increases the proportion of new properties within the tidal floodplain.

Planning policies have changed much over the last century highlighting the need to explore a range of socio-economic scenarios in the course of considering long-term climate risks. Thus, while the model does not predict future urban growth or address local design issues, it provides an important tool for exploring long-term changes to climate risks over a broad scale. This information can be used to inform

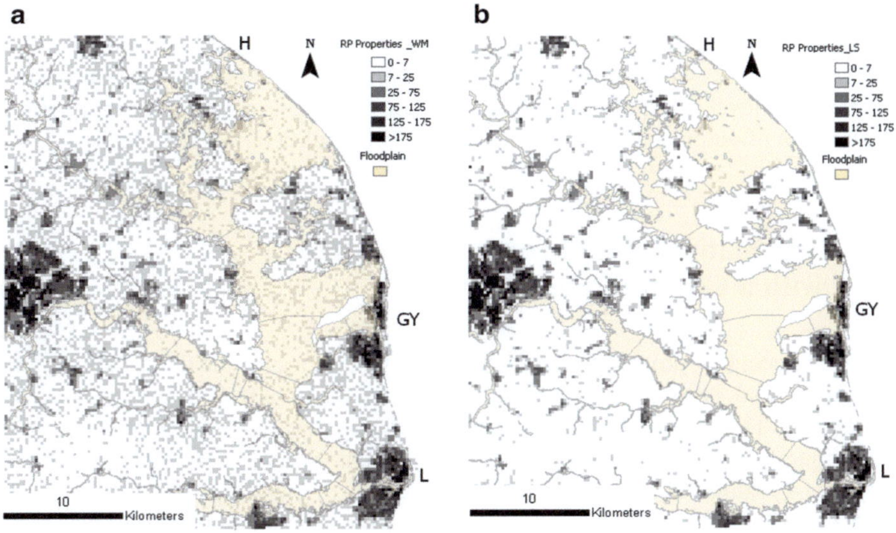

Fig. 4.2 The spatial distribution of total residential properties under the World Market (**a**) and the Local Stewardship (**b**) scenarios in the Norfolk Broads over the twenty-first century. *L* Lowestoft, *GY* Great Yarmouth, *H* Happisburgh

and set the context for the local planning process and underpin understanding of the wider and longer-term impacts of development plans on climate risks.

While the expert judgement method tends to increase urban patterns uniformly leading to a constant pattern of exposure, the multi-criteria analysis links the spatial pattern of exposure to the storyline interpretation of various futures allowing a wider range of evolution pathways and hence changing risks. The method described here is an algorithm-based approach capable of providing feasible and flexible tools for developing and exploring plausible future socio-economic scenarios for impact and risk assessment. It has the advantage that the scenario assumptions are explicit, are reproducible and can be easily changed. This allows for an in-depth investigation of stakeholder views concerning possible futures and their implications in terms of climate risk. Although the model has proven itself fit for the purposes of this study, there are a number of improvements that could be made. These include capturing both horizontal and vertical urban changes (e.g. converting single properties into multistorey buildings). Likewise adding other attraction factors such as brownfield sites would enable additional planning tensions to be explored. Furthermore, the use of a linear distance attraction and linear multi-criteria summation of total attraction could be improved through the development of logit or other value functions or the application of a spatial interaction model. However, this either requires additional information within the scenarios or a further elicitation process to parameterise such functions.

The results (see examples in Fig. 4.2) from this method were influenced by the assumptions, especially the number of attraction factors and the scaling scores of these factors. Different weight scales are expected to generate similar generic urban patterns, but these may vary at the local scale. It is noted that the inclusion of more factors and details will make the model more difficult to implement. Thus, a balance between the level of detail incorporated in these broadscale analyses and the appropriate level of outputs is key in maintaining the flexibility of this approach.

4.4 Translating Socio-Economic Storylines with a Choice-Based Approach

Another recent approach to estimate the distributions of new residential and non-residential properties was proposed by Fontaine (2010) using agent-based modelling. The approach was designed to investigate the impacts of plausible behavioural shifts in resident and planner interactions and decisions about urban land-use dynamics in a polycentric region, i.e. where there is no central dominant city. The objective was to present alternatives in the spatial distribution of new residential development resulting from a multi-scale evolution of society in terms of demographic dynamics (macro-geographical level), planning strategies (meso-geographical level) and resident location decisions (micro-geographical level). In order to encompass all three levels of processes, the whole region of East Anglia was used as a case study (see Fig. 4.3).

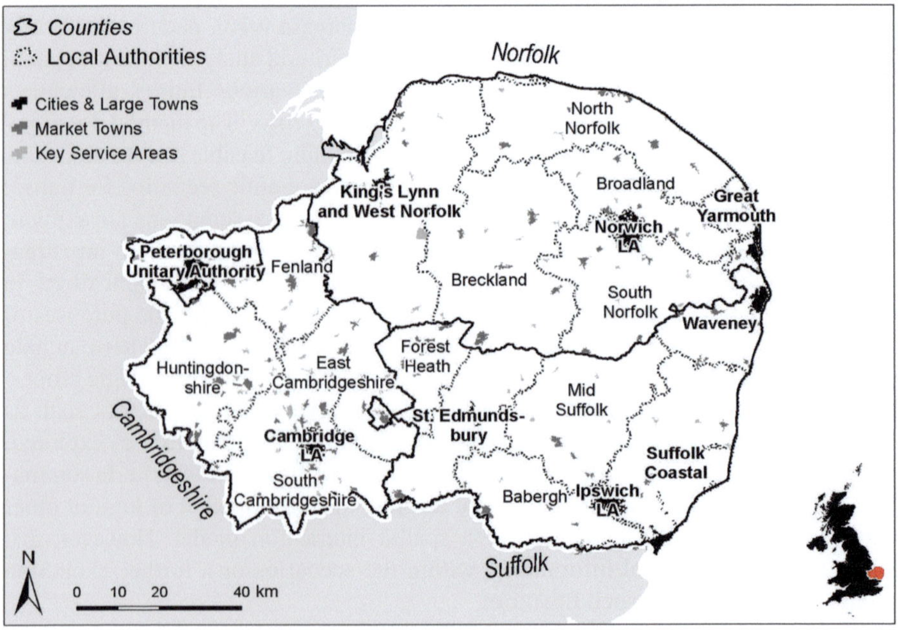

Fig. 4.3 Administrative boundaries in East Anglia, UK

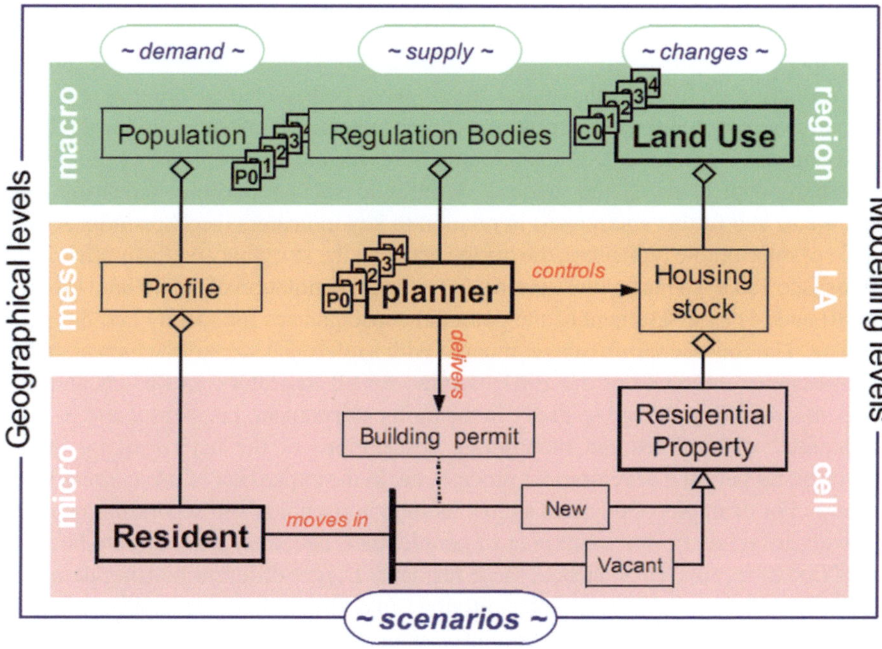

Fig. 4.4 Design of the decision-based modelling framework. The links between objects comply with UML notations. Processes happen at three geographical levels, represented by three modelling levels in the model (cell = 250 m square; *LA* = local authority boundary; region = EA borders). The demand (*left-hand side column*) is driven by population dynamics (macro) but also by each household (micro) wanting to move house. The demand (*middle column*) is met either by existing available properties ("vacant") or by new houses built with permission of the local planner (meso) if the demand complies with the planner's housing stock target. Consequently, the new properties (micro) modify the local housing stock (meso), producing urban dynamics that changes the regional land use (macro)

Residential land-use dynamics are seen as the spatial outcome of interactions between the decisions of individuals for locations and new housing development at the micro-geographical level (see Fig. 4.4). The challenge is to create a coherent link between the scenario narrative and the operational framework used to simulate *a priori* unknown outcomes, in which both local heterogeneities and diverse agents can coexist. The simulation model also needs to consider time within the storyline (residential development is a temporal process) as well as its space extent and resolution (residential development occurs at specific locations). Thus, the model has to output disaggregated information useful for an analysis of how the future may unfold at a fine resolution. Moreover, analysis of a complex system requires the incorporation of feedbacks in both the storyline and the model as agents interact and adapt to changes in the state of the system (Batty 2005; Verburg 2006; Crooks et al. 2008). In other words, the current residential distribution is influenced by previous planning and resident decisions, and these will influence future changes.

4.4.1 Scenario Indicators and Output Indicators

Four scenarios of future behavioural trends were constructed in order to estimate alternative residential land-use futures and their potential consequences for the landscape and the environment of East Anglia. The scenarios comprised (1) a written storyline, adapted from the generic ForeSight socio-economic scenarios (see Sect. 4.2.3) and further interpreted in relation to key indicators (see Appendix A), (2) a table of measurable indicators, the bridge between the storyline and the model simulations (see Table 4.4. as an illustration), (3) a spatial simulation of future land use (the model) and (4) an assessment of the potential consequences for society and the environment. The four scenarios were compared with a reference scenario, which assumed steady-state evolution from the 2001 baseline for all indicators. In this chapter, we focus on simulating change spatially in the built environment, i.e. steps 2 and 3.

Qualitative storylines can incorporate descriptions of the heterogeneity of the agents taking part in a development process, i.e. in the expansion of urban areas studied here. For example, residential agents include households and second-home owners, with different attributes (number of people, their age, etc.) and different housing needs (location, size, etc.). Hence, these residents trigger different housing demands in specific places at specific moments in time. In addition, housing supply is largely controlled by different policy makers and planners (particularly in the UK), managing different areas (urban, rural, coastal, etc.) at several levels within the governance

Table 4.4 Planning meso-indicators for the ABM for each scenario

	Ref	WM	GS	LS	NE
Hierarchy of priority areas[a]					
Urban planner	✓	✗	✓	✓	✗
Rural planner	✗	✗	✓	✗	✗
Building restrictions[b] in sensitive zones					
Protected areas		↘↘	↗↗	→	↘↘
	−1	None	2	−1	None
Flooding areas		↘↘	↗↗	↗	→
	0[c]	None	2	1	0
Shoreline areas		↘↘	↗↗	↗	→
	1[c]	None	3	2	1

[a]Changes in planning strategy were evaluated with a distinction between urban and rural local authorities (*LA*). In 2000, the nine urban LA (out of 20) were those including no cities or large towns sheltering more than 25,000 people. The hierarchy indicates a ranking amongst the cells, giving a development priority: brownfields in urban areas first, then the periphery of urban areas and then greenfield sites. When the LA contains several levels of urban areas, the more populated areas come first (e.g. brownfields in cities have a higher priority than those in towns).

[b]When existing restrictions are applied to the cells that belong to a buffer drawn around the corresponding spatial feature. The buffer size is measured as a number of cells: 0 means the exact extent of the feature.

[c]As observed in 2000, there is a tolerance for small increases within cells that are already urbanised even if they are within the corresponding restricted feature.

Table 4.5 The simulated percentage growth of residential properties (*RP*) and the total numbers of new properties created under each socio-economic scenario in East Anglia by 2100

East Anglia	Total no. of new RP	Change in RP (%)
Local Stewardship	586,936	+61
Global Sustainability	401,123	+42
National Enterprise	614,776	+64
World Markets	686,337	+72

hierarchy (local, subregional, regional, national, supranational). Consequently, they respond to residential demands for new housing in specific places only. Beside these two key categories of agents, housing dynamics are also influenced by contractors (who build houses, blocks of apartments, etc.), community groups and lobbyists. Lobbyists attempt to influence planners in making decisions that are beneficial to their own interests (e.g. regional builders and investors, environmentalists, local neighbourhoods, etc. – after Barker 2006). In the work presented here, residents and planners are considered explicitly because these agents provide the central interaction that drives residential land-use dynamics, each agent representing one side of the supply and demand relationships for new housing. Other groups are considered implicitly through their indirect influence on the residents and planners. This approach has the virtue of simplicity (Verburg 2006; Zellner et al. 2008).

The key elements of the generic socio-economic storylines were first qualitatively interpreted (see Appendix A, Table 4.5). They were then translated into quantitative input parameters for the model runs: one or several cardinal or ordinal indicators were first identified per storyline as part of "Demography and Migration" (macro-indicators), "Planning Strategy" (meso-indicators, see Table 4.4 for an example) and "Location preferences" (micro-indicators). These indicators were either regional variables (e.g. migration rates) or agent state variables (e.g. probability of death). Secondly, a qualitative change was assigned to each indicator based on an interpretation of a storyline narrative (the arrows in Table 4.4). Finally, the value or the level of an indicator observed in 2001 was adapted in order to reflect the qualitative directional change for each scenario (the numbers in Table 4.4).

As with the model output indicators, key values were compiled at three geographical levels for any number of time steps. While information at the micro-level can be illustrated with maps, graphs of trends are more appropriate for rendering the information at more aggregated levels. Hence, a choice is needed between more spatially explicit (maps) and more temporal-explicit (graphs) results (see output illustrations in Fig. 4.5 and in Sect. 4.4.3).

4.4.2 Residential Land Use Change Simulations

Changes in future land use were simulated with a model integrating two modules. The first module is a framework for modelling and simulating, in space and time, residential demand (left-hand side of Fig. 4.4) for new housing driven by Household

Interactions through LIFE cycle stages (the HI-LIFE model, see Fontaine and Rounsevell 2009 for more technical details). The core principle of the model is that households have specific location preferences relating to the age and composition of its members. As the household structure evolves through time (e.g. from marriage, newborn, retirement, etc.), its current location, which was adequate for a previous configuration, may, at that moment, no longer suit the household requirements (e.g. the property is too small/big, not close enough to work, to the coast, etc.). Unsatisfied households trigger a relocation process that may result in a demand for new housing at a specific location when no properties are already available at that location.

This demand is handled by the second module, a framework designed to stylise spatial and local interactions influencing meso-scale planning (the SLIM-P model, see Chap. 5 in Fontaine 2010 for more details; and middle column of Fig. 4.4). Local planners may or may not grant a building permission, depending on the number of local demands, the level of priority for the area or the planner's housing targets.

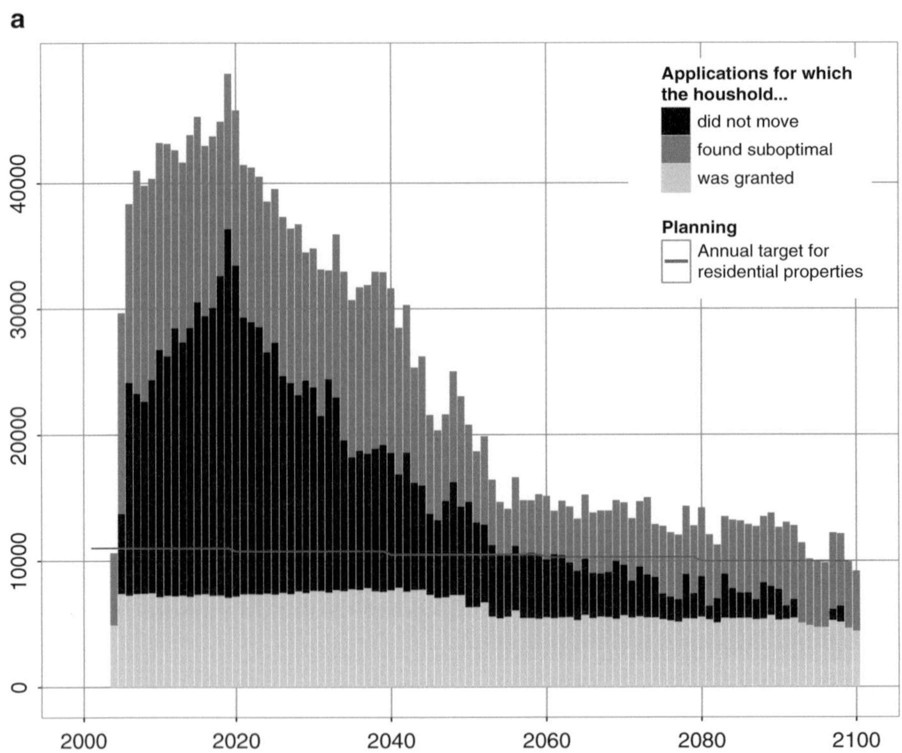

Fig. 4.5 Example of graphical output at (**a**) regional and (**b**) meso-geographical levels for planning: evolution of applications and annual targets for the National Enterprise scenario. Target: the number of new properties the local planner should permit every year. The (**a**) tile shows that the regional target is decreasing although some "urban" local planners increase (strongly) their own target as seen for e.g. Cambridge in (**b**) tile

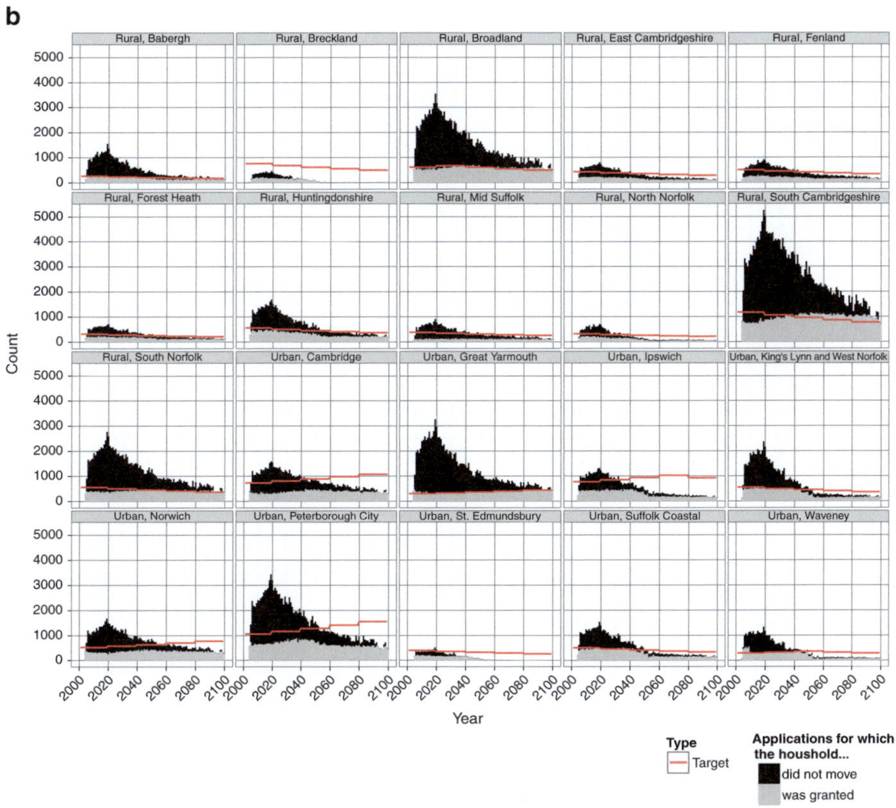

Fig. 4.5 (continued)

The outcome of these models is a dynamic simulation of the evolving patterns of residential land use and changes in the future regional landscape. Other (non-residential) buildings follow the trends caused by residential development.

4.4.3 Results and Key Findings

There are a number of unexpected similarities between the scenarios, which depend on the scale of analysis. At the macro-level, the storylines depicted distinct societal choices. Most model input parameters were given different values in each scenario, but even so, these led to similar outcomes. For example, in demography (see Fig. 4.6), only the World Market scenario led to a significant deviation in the number of households compared to the Reference trend. However, at the micro-level, the four scenarios showed distinct differences in the spatial change of residential development.

Fig. 4.6 Simulated and aggregated residential population (macro-level output), relative to the Reference (BU) scenario. The relative evolution of the count is defined as the ratio between the absolute number for one measure at one date for one scenario and the absolute number for the same measure for the Reference scenario

Conversely, the response trends at the macro- and meso-geographical levels are very similar for the World Markets and Local Stewardship scenarios (although different in magnitude). For example, LS planners grant permissions at levels that are more similar to WM than GS (see Table 4.5). This was a surprise since the LS and WM storylines have very different basic assumptions. WM describes a society-embracing globalisation through economic development. The LS storyline describes a self-sustaining world of local communities, concerned for the security of people and the environment. However, the spatial distributions of new housing development at the micro-level are very different between the storylines. In LS, urban development is mostly concentrated in the largest urban centres (see Fig. 4.7d), while WM leads to urban sprawl (see Fig. 4.7a). This suggests greater changes in organisation and infrastructure of cities and hence in the overall region, under a LS scenario (even though the absolute increase was lower than for WM).

The close similarity between the Global Sustainability scenario and the Reference scenario through time and space was unforeseen (with the exception of the greater restrictions on building in sensitive areas for GS, see Fig. 4.7b). The magnitude of the values for total population and most household stages (see Fig. 4.6), targeted housing stocks, vacant properties, unsatisfied demand and effective urban growth were different, but trends were essentially the same, at all scales. For example, the

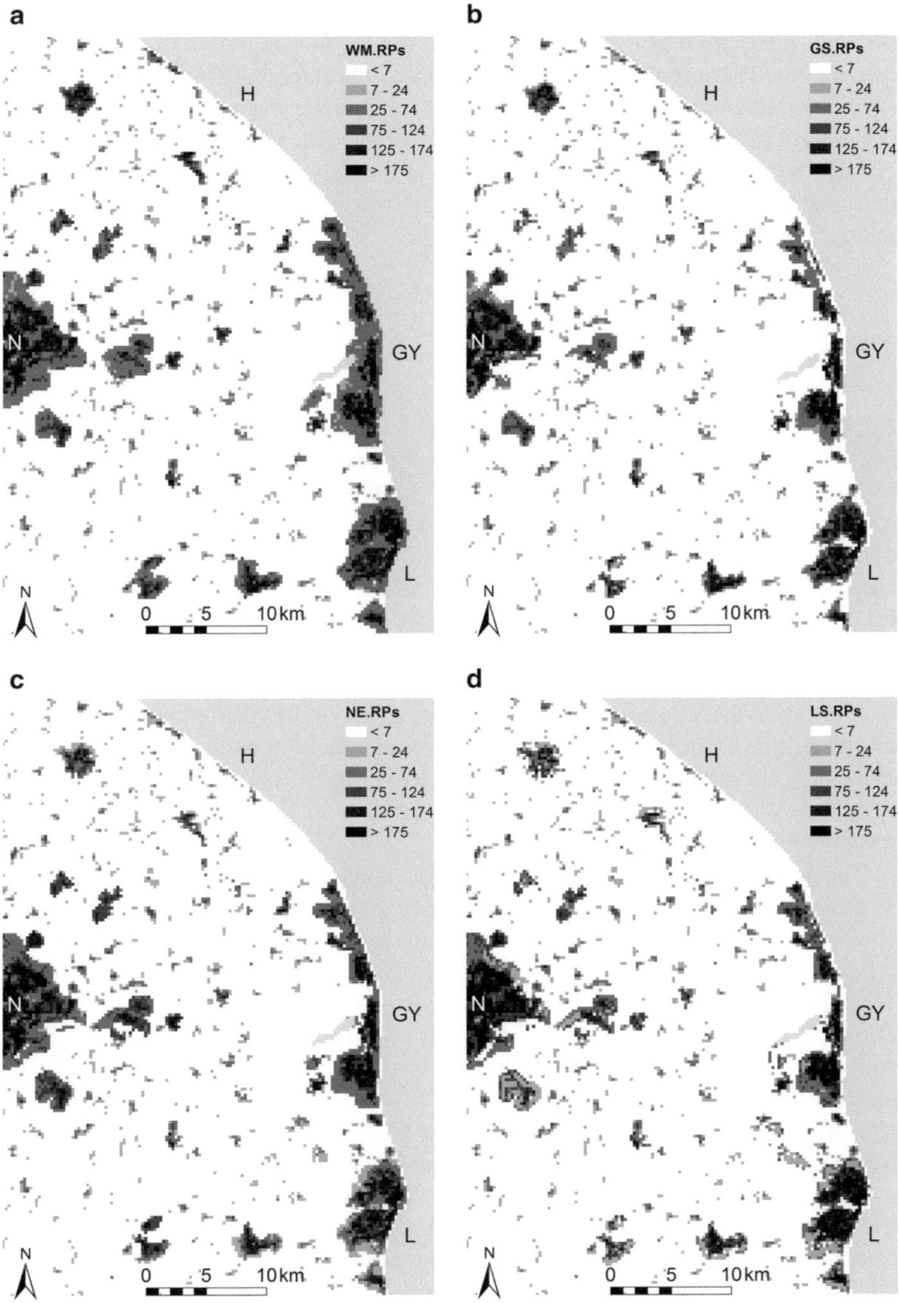

Fig. 4.7 Residential land-use distribution simulated with the ABM approach, until 2100, under the four socio-economic scenarios. *N* Norwich, *L* Lowestoft, *GY* Great Yarmouth, *H* Happisburgh. (**a**) World Market scenario. (**b**) Global Sustainability scenario. (**c**) National Enterprise. (**d**) Local Stewardship

resulting trend in the number of households is similar between GS and the Reference scenario, although the birth rate is lower and the outflow migration of young people cancels out. The principal difference, limited to the three most westerly rural LAs (Fenland, Huntingdonshire and South Cambridgeshire), was in the evolution of the GS annual planning target that increased once (by 2.5 % in the second planning period), but not in the Reference scenario. Ultimately, these observations indicate that East Anglia might already be on a path towards sustainable urban land-use management.

4.5 Conclusions and Next Steps

Both the multi-criteria analysis (MCA) and the agent-based modelling (ABM) frameworks have pros and cons in terms of model complexity, data collection and analysis and interpretation. These were illustrated when the two modelling approaches were used to inform the Tyndall Coastal Simulator about urban land-use dynamics. For both methods, socio-economic storylines for East Anglia were the basis for quantifying potential urban development, with an explicit consideration of existing settlements, transport networks, the coastline and floodplains. While Mokrech et al. (2012) developed an urbanisation algorithm based directly on these components, Fontaine (2010) added an extra "layer" depicting agent behaviours that rely upon these components. The advantage of considering human behaviour in the modelling process is to allow for non-monotonic evolution in the land-use dynamics, i.e. specific feedback at specific locations for specific moments in time. In other words, the total number of new properties is not decided a priori, but results from iterations and system dynamics.

The multi-criteria analysis can generate rapidly urban development patterns reflecting the storylines of the socio-economic scenarios, which are appropriate for conducting impact assessments of climate change at the regional scale. Conversely, the ABM approach focuses on the interactions between households and planners. Thus, in the ABM, the agents are treated as autonomous entities, evolving in relation to the scenario storyline and reacting to the need for household mobility. Although simulations generate similar aggregated results for both methods and for most scenarios (compare Tables 4.2 and 4.5), some differences can be identified; e.g. the relative increase in residential properties under the LS scenario doubles between both methods (MCA: +28 %; ABM: +61 %), although this has to be interpreted carefully since the spatial extent of the case study areas was not the same. Moreover, the ABM generated detailed results at the local level to support the identification of societal choices in planning strategies, location preferences and demography that could lead to sustainable landscape management. The results and outputs of this work were applied to the integrated assessment through the risk analysis presented in Chap. 9.

Both methods are transparent about the assumptions made and flexibility in tuning the influence of different drivers. This creates the potential to explore stakeholder views interactively about future urbanisation scenarios in the context of climate change risk assessment. Thus, if the modelled results were to be examined by practitioners, the causes of these results could be quickly ascertained from the known model input values. Subsequently, alternative tuning of these inputs could be further tested. This closer interaction with stakeholders should be the next step for research in this field, aiming to build relationships between model simulations and the planning process. This would allow planners to influence model development towards specific policy questions.

References

Barker, K. (2006). *Barker review of land use planning: Final report recommendations*. London: The Stationary Office.

Batty, M. (2005). *Cities and complexity: Understanding cities with cellular automata, agent-based models and fractals*. Cambridge: MIT Press.

Batty, M. (2008). The size, scale, and shape of cities. *Science, 319*(5864), 769–771.

Brown, D. G., Robinson, D. T., An, L., Nassauer, J. I., Zellner, M., Rand, W., Riolo, R., Page, S. E., Low, B., & Wang, Z. F. (2008). Exurbia from the bottom-up: Confronting empirical challenges to characterizing a complex system. *Geoforum, 39*(2), 805–818.

Cavailhes, J., Peeters, D., Sekeris, E., & Thisse, J. F. (2004). The periurban city: Why to live between the suburbs and the countryside. *Regional Science and Urban Economics, 34*(6), 681–703.

Crooks, A. T., Castle, C. J. E., & Batty, M. (2008). Key challenges in agent-based modelling for geo-spatial simulation. *Computers, Environment and Urban Systems, 32*(6), 417–430.

Eastman, J., Kyem, P., Toledano, J., & Jin, W. (1993). GIS and decision making. In *Explorations in geographic information systems technology* (Vol. 4). Geneva: United Nations Institute for Training and Research (UNITAR).

Fontaine, C. M. (2010). *Residential agents and land use change modelling*. Thesis, School of GeoSciences, Faculty of Science and Engineering, The University of Edinburgh, Scotland (UK).

Fontaine, C. M., & Rounsevell, M. D. A. (2009). An agent-based approach to model future residential pressure on a regional landscape. *Landscape Ecology, 24*(9), 1237–1254.

Holman, I. P., Rounsevell, M. D. A., Shackley, S., Harrison, P. A., Nicholls, R. J., Berry, P. M., & Audsley, E. (2005a). A regional, multi-sectoral and integrated assessment of the impacts of climate and socio-economic change in the UK Part I. Methodology. *Climate Change, 71*, 9–41.

Holman, I. P., Nicholls, R. J., Berry, P. M., Harrison, P. A., Audsley, E., Shackley, S., & Rounsevell, M. D. A. (2005b). A regional, multi-sectoral and integrated assessment of the impacts of climate and socio-economic change in the UK Part II. Results. *Climate Change, 71*, 43–73.

Lambin, E. F., & Geist, H. J. (2001). Global land-use and land-cover change: What have we learned so far? *Global Change Newsletter, 46*, 27–30.

Mokrech, M., Nicholls, R. J., & Dawson, R. J. (2012). Scenarios of future built environment for impact assessment of climate change using a multi-criteria approach in GIS. *Environment and Planning B: Planning and Design, 30*, 120–136.

Nicholls, R. J., Mokrech, M., Richards, J., Bates, P., Dawson, R., Hall, J., Walkden, M., Dickson, M., Jordan, A., & Milligan, J. (2005). *Assessing coastal flood risk at specific sites and regional*

scales: Regional assessment of coastal flood risk (Tyndall Centre technical report no. 45). Norwich: Tyndall Centre for Climate Change Research.
Parker, D. C., Manson, S. M., Janssen, M. A., Hoffmann, M. J., & Deadman, P. (2003). Multi-agent systems for the simulation of land-use and land-cover change: A review. *Annals of the Association of American Geographers, 93*(2), 314–337.
Reginster, I., & Rounsevell, M. (2006). Scenarios of future urban land use in Europe. *Environment and Planning B: Planning and Design, 33*(4), 619–636.
Rounsevell, M. D., & Metzger, M. (2010). Developing qualitative scenarios and storylines. *Wiley Interdisciplinary Reviews: Climate Change, 1*(4), 606–619.
Verburg, P. H. (2006). Simulating feedbacks in land use and land cover change models. *Landscape Ecology, 21*(8), 1171–1183.
Zellner, M. L., Theis, T. L., Karunanithi, A. T., Garmestani, A. S., & Cabezas, H. (2008). A new framework for urban sustainability assessments: Linking complexity, information and policy. *Computers, Environment and Urban Systems, 32*(6), 474–488.

Chapter 5
Evaluating Broadscale Morphological Change in the Coastal Zone Using a Logic-Based Behavioural Systems Approach

Susan Hanson, Jon French, Tom Spencer, Iain Brown, Robert J. Nicholls, William J. Sutherland, and Peter Balson

Abstract Climate change will have pervasive effects on the world's coasts, but at broad scales these changes have typically proven difficult to analyse in a quantifiable manner. Consequently, individual management decisions are often taken without consideration of the wider, regional coastal system with its physical linkages between geomorphological elements.

S. Hanson (✉)
Faculty of Engineering and the Environment, University of Southampton,
Southampton SO17 1BJ, UK
e-mail: s.e.hanson@soton.ac.uk

J. French
Coastal and Estuarine Research Unit, Department of Geography, University College London,
Pearson Building, Gower Street, London WC1E 6BT, UK
e-mail: j.french@ucl.ac.uk

T. Spencer
Cambridge Coastal Research Unit, Department of Geography, University of Cambridge,
Downing Place, Cambridge CB2 3EN, UK
e-mail: ts111@cam.ac.uk

I. Brown
The James Hutton Institute, Craigiebuckler, Aberdeen AB15 8QH, UK
e-mail: iain.brown@hutton.ac.uk

R.J. Nicholls
Tyndall Centre for Climate Change Research, Faculty of Engineering and the Environment,
University of Southampton, Southampton SO17 1BJ, UK
e-mail: r.j.nicholls@soton.ac.uk

W.J. Sutherland
Conservation Science Group, Department of Zoology, University of Cambridge,
Cambridge CB2 3EJ, UK
e-mail: w.sutherland@zoo.cam.ac.uk

P. Balson
Formerly of British Geological Survey, Kingsley Dunham Centre,
Nicker Hill, Keyworth, Nottingham NG12 5GG, UK
e-mail: peterbalson@outlook.com

R.J. Nicholls et al. (eds.), *Broad Scale Coastal Simulation*, Advances in Global Change Research 49, DOI 10.1007/978-94-007-5258-0_5

In this chapter, this concern is addressed using an outcome-driven deductive methodology. This provides a qualitative analysis based on geomorphological principles and expert knowledge. A coastal classification, based on generic coastal elements, is used to define the active coastal system with and without coastal management. Potential outcomes (futures) for these elements are then described based on the current understanding of geomorphological processes which are then linked with external drivers using a Driver-Pressure-State-Impact-Response (DPSIR) framework. The uncertainty in future geomorphological change is then described using likelihoods for these possible responses based on individual or consensus expert judgement captured within a simple, transparent matrix.

The methodology provides a whole system level of analysis. It allows different views on system dynamics to be articulated and captured in terms of broadscale patterns of potential geomorphological futures. It is also beneficial as a precursor to more quantitative models as it provides a framework for interaction between actors with interests in the coastal zone, promoting a more strategic shared understanding of the coastal system. This simplification of the coastal system to key responses and their relationship to external drivers also has advantages for communication with non-experts, promoting confidence in the decision-making process. The methodology is demonstrated in an analysis of part of the English coastal sediment cell 3 which covers the coastline of Norfolk and Suffolk, on the east coast of the UK.

In terms of the Tyndall Coastal Simulator, this study was used to characterise the broad behaviours of the coastal geomorphic system in the region around our main focus in North Norfolk. The method allows the main geomorphic processes and interactions, which need to be considered quantitatively, to be identified. This stage of assessment, which has often been omitted, is an important step for future studies. The methods used here also link strongly to the wetland and habitat analysis described in Chap. 6.

Keywords System-based behavioural approach • Outcome-driven modelling • Likelihood matrix • Regional coastal classification • Coastal evolution • Coastal management policy

5.1 Introduction

The understanding of coastal morphological change is at the root of all coastal investigations, planning and management decisions. However, translating the changes in physical drivers described in Chaps. 2 and 3 into changes in the coastal geomorphological system is a complex issue; extrinsic alterations force change in coastal landforms, whilst intrinsic system state dynamics are also influential. Anthropogenic activities, such as coastal protection and flood defence interventions, are likewise important. In addition the magnitude, mode and timeframes of morphological response are likely to vary between different geomorphic components. In England and Wales, shoreline management plans (SMPs) require that management decisions be based upon an assessment of coastal change over 100 years for areas defined by sediment cells and subcells (e.g. Cooper et al. 2001; DEFRA 2006). Over such time and spatial scales, it is problematic to rely on conventional process-based numerical modelling, which is difficult to scale up due to changing boundary conditions, is often specialised and computationally demanding and may become invalid once one of the underlying assumptions is updated (e.g. Chau 2006). These models can also be difficult to comprehend for the non-specialist, and this can lead to questioning of model outcomes.

Within the Tyndall Coastal Simulator, a systems-based behavioural approach was therefore developed to provide an analytical framework at the regional scale and a context for the more detailed modelling described in Chaps. 7–9. Systems analysis has become increasingly recognised as a means of overcoming difficulties associated with detailed modelling as they are less data demanding, focusing on key interconnections between elements and their combined responses to driving forces (e.g. Burgess et al. 2004; EstSim Consortium 2007; HR Wallingford et al. 2006; Whitehouse et al. 2009). They can be particularly useful as drivers of change are modified beyond historical trends, for example, through an accelerated rate of sea-level rise (Hanson et al. 2007, 2010). This chapter therefore presents a qualitative (descriptive) to semi-quantitative deductive methodology that adapts and expands a systems-based behavioural approach to the analysis of broadscale coastal response to external drivers. Using geomorphological principles supported by expert opinion, this more descriptive method provides a systematic description of the geomorphic state and possible futures, which experts can debate, and is more understandable by the non-coastal expert. This allows the strategic planner to recognise the wider context of management decisions and facilitate easier communication between specialists and stakeholders, as well as providing a framework for the subsequent development of more appropriate quantitative models.

5.2 Capturing Current Knowledge and Understanding

Analysing and communicating complex coastal geomorphological response to external drivers in a clear, transparent and understandable way is essential for assessing the impacts of, and appropriateness of adaptations to, projected climate change. Often, the compunction is to attempt to generate more precise predictions of geomorphological change, without considering the context in which predictions will be used or the difficulty of making such predictions in the face of uncertain futures. These uncertainties include (1) the non-linearity of morphology-process feedback linkages in geomorphic systems, (2) the possibility of emergent (or unanticipated) behaviour or responses due to the complexity of the system structure and (3) the interaction between natural and socio-economic systems. This is particularly important at regional scales and above. Here, many policy and strategic decisions are made by those not necessarily knowledgeable about the intricacies of geomorphological change. At this scale, such details are less important than the understanding of the general trends and consequences that must be taken into account in proactive policy-making. There is thus a clear niche for a more qualitative but structured approach to analysing coastal geomorphological change and building on the long history of coastal geomorphological investigation in the UK and elsewhere (e.g. Bird 2000; King 1972; Pethick 1984; Steers 1946). The benefits of such an approach include (1) its ability to assimilate diverse sources of knowledge on the functioning of the coastal system, (2) the clarification and simplification of geomorphological principles and therefore (3) the clear communication of those principles to those both inside and outside the scientific community.

The deductive methodology advocated here aims to identify important patterns, themes and interrelationships in coastal evolution. This follows an increasing trend in modelling to look at the likelihood of broad outcomes rather than attempting prediction of specific values (e.g. Freckleton et al. 2011; Sutherland 2006; Taylor and Hasting 2004). The likelihood of outcomes can be assessed within a framework of ideas from Bayesian analysis, probabilistic reasoning and fuzzy logic (e.g. Bouma et al. 2011), allowing the exploration of landscape form and its behaviour over broad scales of time and space. Whilst this approach may produce findings that differ between experts, the confidence in the results can be assessed by a range of techniques. These include replication across a range of individual experts, consensus building workshops, comparison with findings from previous research, feedback from participants in the research and feedback from users of the research findings.

Complex systems do not consistently evolve to the same end state in response to a change in the driver. Rather, a range of potential responses may exist. An outcome-driven approach focuses on these potential futures, describing the system, making it clear which aspects of the system are able to respond to change and the nature of the responses which might be expected (see Deng et al. 2011). The qualitative nature of the methodology is also compatible with the principles of fuzzy logic, which incorporates facts with a degree of confidence and allows for the possibility of multiple outcomes (Nguyen 1997). It also allows for a level of abstraction and generalisation, commensurate with a regional scale of coastal analysis. This is because the methodology uses a collection of fuzzy sets (classes with inexact boundaries) and rules to simplify and reason about data within a structured framework. Qualitative terminology can also be used judiciously to describe non-explicit relationships between variables, for example geomorphological drivers and behaviours, which can then be organised within a rule-based ('if…then'), open, reasoned and adaptable structure. In addition, fuzzy analysis does not preclude the use of quantitative data that can be added, where appropriate, to contribute to the collection, classification and analysis of data. In essence, the methodology described here aims to:

- *Provide* broadscale understanding of geomorphological responses to changes in external drivers. This understanding should capture general principles, current knowledge; condense extensive and varied information into a brief, summary structure; and provide a convenient and efficient way of analysing qualitative data.
- *Identify* key characteristics of the coastal system and its drivers that can be used to illustrate these potential responses.
- *Acknowledge* that there may be more than one possible response to any change in drivers reflecting recognised behavioural patterns and accepting that conclusions are not absolute, as shown in the use of structured language and descriptors such as probably, possibly and (un)likely.

5.2.1 *Structuring the Knowledge*

A number of methods exist for structuring qualitative knowledge. The Driver-Pressure-State-Impact-Response (DPSIR) framework is well recognised in impact studies, underlying much recent environmental research (e.g. Atkins et al. 2011; Borja 2006; EEA 2007; Karageorgis et al. 2006; Omann et al. 2009). This has been shown to offer a logical and efficient way to present information in a well-structured and user-friendly manner. It is particularly appropriate when scientific process knowledge has to be translated into knowledge for policy support and for improving communication with stakeholders (Helming et al. 2011).

In this research, the DPSIR framework used here was simply defined, being based on four assumptions: the premises that (1) only a small number of physiological and functional properties are required to explain the basic working of a geomorphic coastal system; (2) analysis can be based on discrete and interacting physiographic units as used in shoreline management planning; (3) a tendency towards dynamic equilibrium exists between contemporary surface processes and the external forces that influence behaviour; and (iv) a systems approach, incorporating inputs, throughputs and outputs, can be used to characterise coastal geomorphological change. The steps used in the methodology are illustrated in Fig. 5.1.

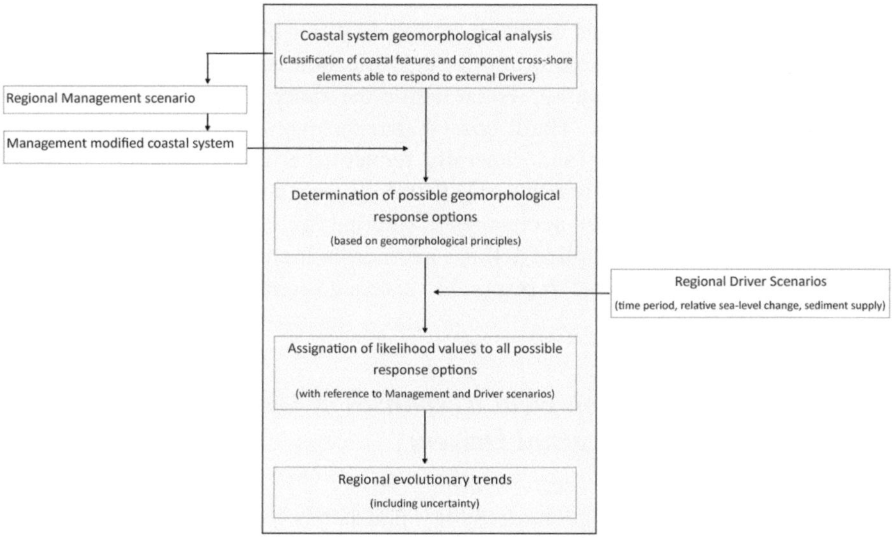

Fig. 5.1 Methodology used to determine broadscale coastal change in the Tyndall Coastal Simulator

5.2.2 Generic Coastal Geomorphological Elements

As with many deductive approaches, a qualitative classification was developed to form the basis for subsequent analysis. A simple modular classification provides a systematic way of both mapping geographical variation in morphology and a framework for understanding the implications of specific management options. Existing coastal typologies provide the basis for a classification based on the identification of generic geomorphological elements and combinations of these elements (e.g. Finkl 2004; Greenlaw et al. 2011; Mikhailova 2008). The purpose of this classification is to identify the components of the coastal system able to adjust their morphology in response to environmental drivers. It is explicitly formulated to describe coasts with and without engineered defences, recognising that defences are an integral part of the coastal system.

In shore-normal cross section, individual coastal geomorphological elements were identified from changes in lithology, sediment type, localised slope, relative relief, exposure to waves and tides and cross-shore position. The same factors can also be used to define shore-parallel longshore limits for the geomorphological elements. Described generically (e.g. reef, foreshore, cliff, barrier), these elements then form the building blocks of a simplified regional coastal classification producing, in combination, a number of widely identifiable profile types (see Fig. 5.2). These profile types can be broadly grouped into barrier and non-barrier coastal types, with barrier types including a geomorphological element (barrier) that usually prevents the direct impact of wave action and modified tidal influence on other geomorphological elements to its landward side.

Applied to a coast that has not been modified by coastal defences, the landward boundary of the coastal system using this classification is defined by the inherited natural geology and topography, which define the space within which wave and tidal processes can operate. 'Hard' coastal defences can be viewed as playing a similar role. They redefine (and generally reduce) the scope and complexity of coastal geomorphological elements and therefore commonly reduce the range of options available for response to changes in external drivers. In either situation, the same classification can be applied to those geomorphological elements that are considered to be 'active' (i.e. able to respond to external forcing).

5.2.3 Key Geomorphological Responses of Generic Elements to External Drivers

A key principle of coastal geomorphology that needs to be incorporated in this methodology is that mutually interdependent landform changes are largely controlled by sediment exchanges. The latter are mediated both by sediment supply and sediment transport processes. Landform elements are therefore mutually dependent on the sediment budget system where sediment stores exist in dynamic equilibrium

a

	Element						
	Hinterland	Backshore	Backbarrier		Barrier	Foreshore	Offshore
Cross-shore profile	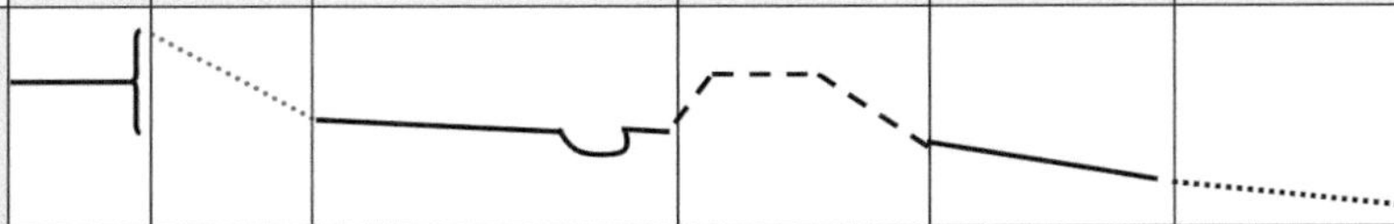						
Description	Area landward of tidal and wave influence	Area at limit of tide and wave influence	Area within the tidal flood plain and landward of a barrier. May include a tidal channel		Area above regular tides which protects landward areas	Area subject to regular tide and wave influence May be vegetated	Area below tidal limits
Geomorphology (landform)	Various (n/a)	Cliff face Rising ground	a) Protected from direct marine processes Lowland Coastal grazing marsh	b) Direct marine processes active due to channel network Saltmarsh Sand flats Mud flats Lagoons	Beach ridge Spit Barrier island Berm Ness Sand dunes	Upper foreshore Lower foreshore Saltmarsh Mud flats Sand flats	Shore platform

b

Coastal type	Profile class		Cross-shore element combination
Non-barrier	Non-barrier(low)		Foreshore
	Non-barrier (cliffed)		Backshore; foreshore
Barrier	Barrier-backbarrier	a)With channel	Back-barrier; barrier; foreshore
		b)Without channel	Back-barrier; barrier; foreshore
	Fringing barrier		Backshore; barrier; foreshore

Fig. 5.2 (**a**) Generic cross-shore elements and (**b**) possible element combinations in the UK

with the hydrodynamic processes acting upon them (Cooper et al. 2001; Cowell et al. 2003). Changes in this dynamic equilibrium will result in generally well-understood alterations to the size, shape and cross-shore position of a given element (e.g. Pethick 1984; Woodroffe 2002). Onshore-offshore migration is the anticipated response to changes in mean water level as coastal landforms maintain their position within the coastal system relative to dynamics; changes in sediment supply result in adjustments in the cross-shore profile geometry of any geomorphological element. Changes in geomorphological elements can therefore be represented in terms of movement (migration) and changes in shape (geometry). These changes can be described relative to the initial reference state (which may vary according to the research question being posed, but would logically be the presently active coastal configuration) as comparative descriptions. In broad terms, changes can be represented by three classes relative to the reference state: greater than (>), equal to (=) or less than (<).

Consequently, migration can be defined as significant landward movement, no significant movement or significant seaward movement relative to the initial position of the geomorphological element. For changes in geometry, if the coastal elements are seen as idealised triangular prisms, geometric rules can be used to relate descriptive changes to cross-shore width. The three geometric classes are significantly widened, are significantly narrowed or remain geometrically constant. Migration and geometric classes can then be combined to provide a simple matrix of potential future outcomes. Similarly for changes in the external drivers, the future rate of sea-level rise can be described as greater than the current rate (accelerating), equal to the current rate (no change) or less than the current rate (decelerating). Changes in sediment supply can be classed as greater than current supply (positive), equal to current supply (neutral) or less than current supply (negative). Combining all these potential changes leads to the generation of a potential outcome matrix, which can be used as a framework for incorporating expert opinion.

It is also possible to incorporate the influence of shoreline management within the methodology. In many densely populated and economically developed areas, management is primarily concerned with the maintenance of a predetermined defensive line by (1) hard defences or (2) the manipulation of natural processes (e.g. beach nourishment) in order to sustain a preferred coastal configuration. An important advantage of this methodology therefore is that it promotes a deeper understanding of the interdependency of management policy and coastal dynamics, especially amongst non-specialist stakeholders and interested lay audiences.

5.2.4 Incorporating Expert Knowledge

At this stage of the analysis, the use of expert knowledge of those with an understanding of the coastal system becomes a key aspect of the methodology for dealing with areas of uncertainty, in this case for determining how its landform elements will change under different conditions. It is important however, due to the range of geomorphological response times, to clarify the time period over which the analysis takes place. To keep the methodology simple, one of the geomorphological elements within each of the profile types is used as an indicator of overall system change. The selection of the most applicable element may vary, but the element chosen should be that considered to be the most responsive to changes in sediment supply and water levels within the profile type. In general, this will usually be the barrier element in barrier profile types and the foreshore element in non-barrier types. For these two geomorphological elements, migration descriptions can be defined by the position of the crest line of the barrier or centre line of the foreshore relative to its initial position.

Completion of the outcome matrix is based on current geomorphological understanding using a two-step process:

1. The elimination of outcomes considered geomorphologically impossible under the given changes in external drivers
2. The qualitative relative likelihood of occurrence between remaining possible outcomes (i.e. high, medium, low)

The matrices of potential outcomes thus capture expert judgement systematically using the answer to a series of 'if...then' questions. For example, if the rate of sea-level rise accelerates (>) and sediment supply is negative (<), then, using the definitions of change described earlier, a barrier could not move seaward or maintain/increase its volume, eliminating five of the potential futures. This would leave only four possible system responses. Of these, the likelihood would be high that the barrier would (1) narrow in width and (2) move landward relative to its initial position. The other possible responses can either be considered low or medium in likelihood, with the relative assessment of each outcome depending on the knowledge of system dynamics. These matrices are a transparent and flexible means of containing these answers (or rules) and can be altered/deleted/added to according to the coastal system being investigated or as knowledge develops. Completion of the matrices can be undertaken by individuals, but increased confidence in the likelihood values can be achieved using contributions from a number of experts.

5.3 Illustrative Application to the East Anglian Coast, UK

For the Tyndall Coastal Simulator, the East Anglian coast (English counties of Norfolk and Suffolk) of the UK was used to identify broadscale coastal evolutionary responses to changes in mean sea level, sediment supply and coastal defence (see Fig. 5.3 and Hanson et al. 2007). This coastal region corresponds to Coastal Sediment Cell 3 in the scheme of Motyka and Brampton (1993) and is characterised by a variety of both open coast and estuarine geomorphic settings, although only the open coast elements are analysed here. The region also has a long history of coastal erosion and flooding, which has culminated in an extensive legacy of artificial defences and numerous ongoing management challenges.

The stretches of open coast are relatively self-contained in terms of sediment transport, with eroding cliff sections thought to provide significant contemporary sediment supply to neighbouring sand and gravel beaches and offshore bank systems (e.g. Brooks and Spencer 2010). The wave-driven movement of this beach-grade material provides a strong longshore link between adjacent profile types (through the geomorphological elements used as indicators) with changes in the direction of this longshore movement (based on HR Wallingford 2002) defining profile groupings. It is then possible to determine the changes in sediment availability for each profile grouping. It is recognised that there are more sediment interrelationships in

Fig. 5.3 East Anglian region of the UK used for broadscale analysis

coastal geomorphology than those considered here, but on this coast those relationships appear less direct than the cliff/beach interaction. For other coastal types, the cliff/beach relationship can be replaced by other dominant linkages or the approach expanded to consider these additional processes.

5.3.1 Coastal Classification

The coastal classification was applied to the East Anglian coast under a full spectrum of coastal management policies ranging from Total Defence (TD; continuous defences at mean high water (MHW)) to No Defences (ND) (see Fig. 5.4). The classification clearly indicates the potential future changes to the character of the coastal landscape and the coastal elements able to respond to changes in external forcing. For example, in comparison to the No Defence (natural) profile classification, current defences (Maintain the Existing Line or baseline management; column 3 in Fig. 5.4) have largely reduced the active coastal system to a non-barrier low type (the only active geomorphological element being the foreshore). The main consequence of this management strategy has been to significantly constrain the range of

Management policies	
ND	No defences
TD	Continuous defence ant Mean High Water
MEL	Current defence line maintained
SSR	Small scale Realignment—all settlements defended
LSR	Large Scale Realignment—major settlements only

Profile class	
	Non-barrier (low)
	Non-barrier (cliffed)
	Fringing barrier
	Barrier-backbarrier
	Barrier-backbarrier with channel
*	Ness

Fig. 5.4 Illustrative section of the East Anglian (sediment *subcells b* and *c*) coast showing the change in the active coastal configuration under the five management options

possible coastal responses to future sea-level changes – sediment supply will decrease as the cliffs are known to be a major source of sediment, and the fixed defences will limit shoreline movement. Consequently, the response of the remaining foreshore element will be to narrow, which may in turn have implications for the effectiveness of the defences and the protection that they provide. Conversely, the introduction of a management policy that defends major settlements only (Large-Scale Realignment; column 5 in Fig. 5.4) would reintroduce a significant number of non-barrier (cliffed) profile types to the currently active coastal system. The future erosion of these sections would supply additional sediment to the coastal system, and this would increase the set of the potential futures that could be considered. Realignment management policies also show that backbarrier areas need to be included in the active coastal system. Compared to the Maintain the Existing Line management option, the more targeted Large-Scale Realignment option restores active backbarrier areas to an area comparable to an undefended shoreline; the more piecemeal Small-Scale Realignment management option (column 4 in Fig. 5.4) also increases active backbarrier areas. These sites are characteristic of the North Norfolk (sediment subcell 3a, see Fig. 5.3) and Suffolk coasts. This has important implications for policy-makers as these areas are commonly saline lagoons, salt marsh and mudflat habitats, and obligations regarding the loss/gain of such habitat areas are included in international biodiversity legislation (see European Commission 1992; Spencer and Brooks 2012).

5.3.2 Potential Futures: Likelihood Matrix Completion

To capture expert opinion about the possible futures for the geomorphological elements found on the East Anglian coast, 24 coastal experts from a range of organisations and with varying interests in coastal landforms, geomorphological processes and management were invited to participate in a workshop held in London. A consensus approach, with a facilitator from the project members assigned to each of two subgroups, was used at the workshop. This was because open debate and discussion were considered appropriate to the development of the methodology at this point. However, issues regarding selective participation would need to be considered in depth in any future use of the methodology.

Matrices were completed for unrestricted barrier and foreshore geomorphological elements under a range of sea-level rise and sediment supply scenarios. Changes in sediment supply were described as (1) positive (increase in input), which results in an internal volume increase; (2) neutral (no change in input), indicating no internal change in sediment volume; and (3) negative (decrease in input), where an internal volume decrease occurs. The rate of sea-level rise was described as (1) accelerating – an increase compared to a defined historic rate; (2) no change –rate is maintained; or (3) decelerating – a decrease compared to the historic rate. Table 5.1 compares the matrices completed by the two groups during the workshop for one scenario combination.

Table 5.1 Example of a likelihood change matrix completed for an unconstrained barrier by two subgroups of experts at the workshop

		Migration			
	Geometric change	Landward	No movement	Seaward	Breach
Group 1	Widen	None	High	Medium	Low
	No change	None	None	None	
	Narrow	None	None	None	
Group 2	Widen	None	High	Low	None
	No change	None	None	None	
	Narrow	None	None	None	

The scenario assumes no change in the rate of sea-level rise but increased sediment supply

For barrier elements, an additional geomorphological possibility of major interest to the coastal manager is a break in longshore continuity due to breaching of the element during storm conditions (e.g. Hartley and Pontee 2008). On this coast, numerous such breaches occurred during the well-documented 1953 storm surge (e.g. McRobie et al. 2005; Muir Wood and Bateman 2005). Some breaches may be temporary and will naturally seal; others may remain as new tidal inlets, effectively reconfiguring the active coastal system and the geomorphological elements they contain (Bruun and Gerritsen 1960). In this model, permanent breaches were considered more likely under particular future states. For example, a narrowing barrier, which in accordance with geometric rules is also experiencing lowering of its crest, is more likely to breach than a widening barrier, irrespective of height (e.g. Orford et al. 1995). Breaching was therefore included as a conditional possibility within the likelihood matrices.

The group discussions held at the workshop confirmed that the potential outcomes presented encompass the range of possible geomorphological responses. The groups also discussed whether a set time span for the projected geomorphological changes was desirable. The majority thought that the inclusion of a time span would be necessary as any potential outcome is not likely to be perpetual and at a future point change will have been such that a new baseline would be needed. It was therefore decided, for the purposes of this exploratory process, that a time span of 50 years would be considered.

Once the two groups had completed the sample matrices, these were compared both to each other and to a set previously completed by the research team. It was found that there was an agreement between the groups in terms of which future outcomes were considered possible and the outcome considered to have the highest likelihood of occurrence. Differences in opinion between workshop participants arose in the likelihood values assigned to outcomes considered possible but less likely (medium and low); it was recognised, however, that these still needed to be included. This general consensus instilled confidence that the matrices represent a way of capturing geomorphological principles that are simple and broadly applicable.

5.3.3 Assessing the Future

Using the active coastal system identified for each management policy and the knowledge that, on the East Anglian open coast, coarse beach-grade sediments (sands and gravels) are dominant (e.g. Pontee 2005), potential changes in the selected representative elements (foreshore and barrier) can be established using existing studies of longshore transport, sediment cells/subcells and their budgets (e.g. HR Wallingford 2002). Existing longshore sediment studies (e.g. HR Wallingford 2002) were used to identify linked geomorphological elements (and therefore profile types) with the updrift and downdrift limits and drift direction for each linked group. Where longshore drift is not known, or believed not to be a dominant process, onshore-offshore sediment transport with a neutral longshore flux was assumed.

Sediment supply scenarios were determined within each group of profiles linked by longshore sediment drift. The effect of management on cliff sediment supply (the dominant source of sediment for this coast) was then assessed. For example, if an active non-barrier cliffed profile type in an undefended coast is reclassified as a non-barrier low-profile type following the selection of a particular management scenario, then downdrift profiles will experience a change to a negative sediment budget due to the 'removal' of the cliff sediment supply from the active coastal system. This sediment deficit is assumed to propagate to the end of the drift section or until an intermediate active non-barrier cliffed profile type is reached. Conversely, if the management option is to remove a defence in front of a cliff, a non-barrier (low) profile will switch to a non-barrier (cliffed) profile, and the downdrift sediment budget will become positive, due to the reintroduction of sediment supply to the coastal system. This reasoning highlights two aspects of the active coastal system that require further, more detailed investigation: first, the location of any drift divide (in terms of its effect on the source and direction of sediment supply) and second, the relative longshore lengths of adjacent profiles (e.g. to what distance can a reduced length of active non-barrier (cliffed) profile maintain downdrift geomorphological elements?). A more quantitative (and possibly physically based) analysis of these issues might result in modification of the 'if…then' rules of sediment supply which could then be easily applied.

Using these 'if…then' rules within the Tyndall Coastal Simulator, coastal elements and profiles were linked with relevant likelihood matrices from the expert workshop to produce geomorphological trend predictions relative to the current coastal configuration (reference state) under the different sea-level change and sediment supply scenarios (e.g. see Fig. 5.5).

In addition to the individual coastal sites, the general tendency of regional coastal evolution under the different management options and environmental conditions can be assessed. For example, under a scenario where the rate of sea-level rise was assumed to remain constant, the management option determines the balance between barrier and non-barrier profile types. In particular, the foreshore becomes increasingly the more dominant element type as management policy becomes more

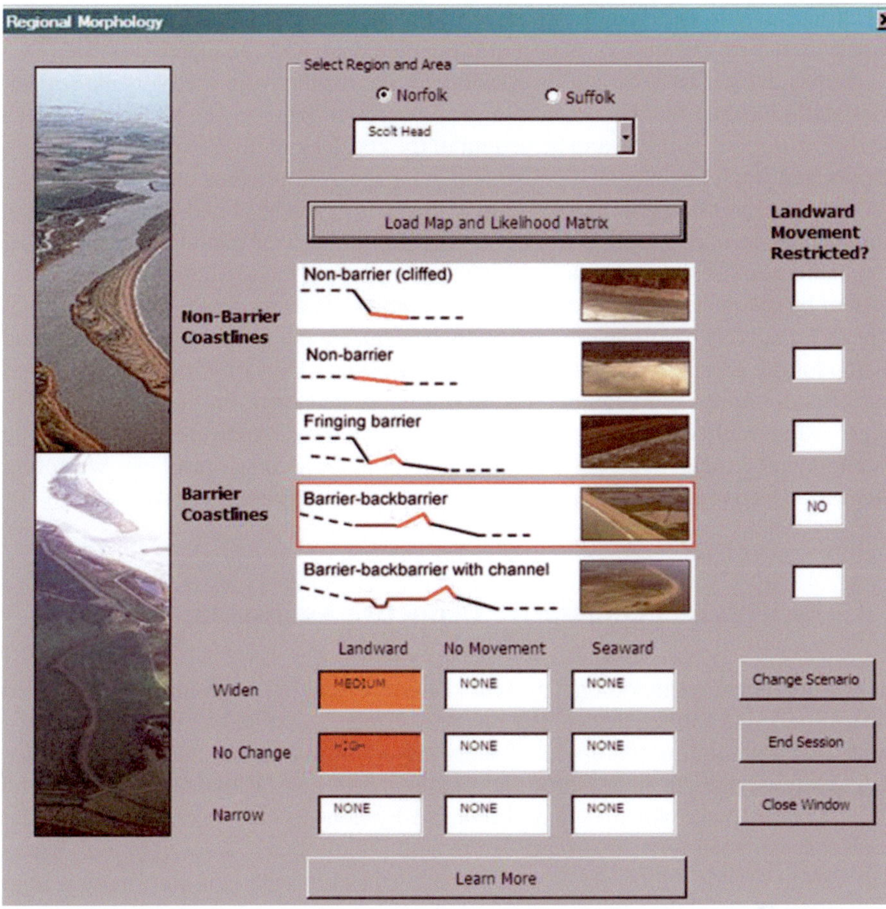

Fig. 5.5 Example of profile identification and outcome matrix for Scolt Head Island, North Norfolk (sediment *subcell 3a*), under an accelerated rate of sea-level rise and MEL (Maintain the Existing Line) management option to the 2050s

protection orientated. For the Maintain the Existing Line (MEL) and Total Defence (TD) management options, the active coastal system is effectively restricted to a single non-barrier (low) profile. The exclusion of the cliff elements leads to negative sediment supply using the 'if…then' rule structure. Consequently, with a rise in sea level, the outcome with the highest likelihood for the active coastal zone can be expected to be landward movement of the centre line along with a narrowing of the foreshore. This, in turn, has implications for the viability of defence structures as the dissipation of wave and tidal energy by the foreshore is lessened. In comparison, for the realignment and undefended options when compared to the MEL scenario, positive sediment scenarios develop as 'reactivated' non-barrier (cliffed) profiles act as sediment sources within the active system. Active barriers and foreshores therefore have the potential to increase in sediment volume, and alternative future states, including seaward extension of these elements, are considered possible.

Across the sea-level scenarios considered, the results from the Norfolk/Suffolk case study indicate that the positioning and length of defences are major influences on the future configuration and behaviour of the coastal system. By altering sediment supply and imposing physical constraints, management policies can determine the ability of the coastal system to respond to changes in sea level and be more influential than sea-level rise itself as a determinant of future coastal evolution. This finding also indicates that the opportunity exists for improving the resilience of the coastal system by using appropriate management policies. Addressed at the strategic level where broadscale relationships between coastal elements can be identified, the methodology can provide a framework for local decisions which otherwise might not consider wider implications. For example, related work on the North Norfolk coast using multi-criteria assessment found that the 'compromise' Small-Scale Realignment management option linked also to the rate of sea-level rise was often the least effective in maintaining an aggregated range of benefits (Brown 2006).

5.4 Benefits of a Qualitative Deductive Assessment

In the analytical methodology presented here, evidential reasoning is combined with knowledge formalisation. The coastal system is distilled to its basic components and potential morphological change based on commonly adopted principles that acknowledge the intangibles and unknowns of coastal dynamics. Within the DPSIR framework, which is understandable by non-experts, this depiction of the coastal system can be readily modified to assess the impacts associated with external forcing and/or policy decisions. It can also incorporate new information (e.g. Oliver 2002). Crucially, the outcome-based emphasis of the methodology also highlights the existence of multiple possible futures, even though one will often be considered the most likely.

Simplifying a complex system in this manner is not unproblematic and is likely to generate (and indeed requires) some debate over the degree of abstraction that is appropriate. However, this can also be seen as an advantage since meaningful analysis is not dependent on the amount of data that would be needed to calibrate and parameterise a process-based model. The methodology is thus consistent with the challenges of managing complex systems with limited data during a time of major change in key drivers. It is easy to incorporate advances in understanding and assess the effects they may have on the coastal system. The qualitative, expert-based, geomorphological reasoning employed allows the methodology to be consistent across geomorphological elements for which detailed quantitative data are unavailable. Building a consensus on the understanding of coastal systems, especially at broader scales, can, of course, be a precursor to more detailed process modelling. The more complete the a priori knowledge incorporated into the formulation and implementation of quantitative process-based models, the better the predictive results are likely to be.

The methodology also lends itself to exploring different views and controversies and how sensitive future coastal evolution might be to alternative models of controlling factors. In this sense, its value should be judged not by the precision of the estimates it generates, but rather by how well it represents potential futures in the broader regional sense. This allows a decision-maker to quickly anticipate and plan for change by formalising existing forms of knowledge (qualitative data) rather than waiting for more detailed quantification. The outcome-based approach can also accommodate stakeholder preferences for different outcomes and test the robustness of these preferences under a range of varying conditions; this can be further investigated using an inverse modelling process evaluating the pathways to different outcomes.

In addition, the transparent characterisation of uncertainty (as likelihoods) facilitates adaptive management based upon the incremental incorporation of new knowledge as the system continues to evolve, either in dynamic equilibrium or in transition to a new state. Conventional coastal process models usually have their assumptions embedded within computer coding. This makes them less easy to adapt either to changes in system understanding or to natural reconfigurations of the system over time.

5.5 Conclusions and Next Steps

The methodology described here has successfully provided an underlying understanding of the broadscale context for the more detailed coastal modelling and has informed the development of the quantitative habitat method described in Chap. 6. In a basic sense, it has defined the coastal geomorphic system including the role of management. From this system understanding the response to various drivers can be explored in a class-based system of geometric response. This provides a good basis for different experts to discuss and debate their understanding of these trends. Further, it is a good foundation for quantitative modelling including priority effort. The outcome-driven, deductive approach could also be used for other complex policy-making issues.

Ultimately, coastal management requires a broad view of the entire interconnected regional coastal system and its response to external forcings including future climate change. To help maintain a sustainable coast, it is better to use the qualitative knowledge available than to simply not consider particular issues because of either a lack of quantitative datasets or an inability of process-based models to handle the problem in question. Whilst treating qualitative assessment as a component of a larger management process raises questions that then demand further empirical evaluation, such assessments are nevertheless capable of indicating expected trends and identifying and prioritising the information needed to enable a coastal management decision to be made. In addition, a conceptually simple scheme founded on a simplified vocabulary has communication value across disciplinary boundaries as it is more likely to be understood and capable of being questioned. Communication of complex issues with non-experts is also essential if confidence in the decision-making process is to be encouraged: this is likely to make it more trusted.

References

Atkins, J. P., Burdon, D., Elliott, M., & Gregory, A. J. (2011). Management of the marine environment: Integrating ecosystem services and societal benefits with the DPSIR framework in a systems approach. *Marine Pollution Bulletin, 62*(2), 215–226.

Bird, E. C. F. (2000). *Coastal geomorphology: An introduction*. Chichester: John Wiley and Sons.

Borja, A. (2006). The European Water Framework Directive and the DPSIR, a methodological approach to assess the risk of failing to achieve good ecological status. *Estuarine, Coastal and Shelf Science, 66*(1–2), 84–96.

Bouma, J. A., Kuik, O., & Dekker, A. G. (2011). Assessing the value of Earth Observation for managing coral reefs: An example from the Great Barrier Reef. *Science of the Total Environment, 409*(21), 4497–4503.

Brooks, S. M., & Spencer, T. (2010). Temporal and spatial variations in recession rates and sediment release from soft rock cliffs, Suffolk coast, UK. *Geomorphology, 124*(1–2), 26–41.

Brown, I. (2006). Modelling future landscape change on coastal floodplains using a rule-based GIS. *Environmental Modelling and Software, 21*(10), 1479–1490.

Bruun, P., & Gerritsen, F. (1960). *Stability of tidal inlets*. Amsterdam: North Holland Publishing Company.

Burgess, K., Jay, H., Hosking, A., & Pinto, F. T. (2004). Futurecoast: Predicting the future coastal evolution of England and Wales. *Journal of Coastal Conservation, 10*(1), 65–71.

Chau, K. (2006). A review on the integration of artificial intelligence into coastal modeling. *Journal of Environmental Management, 80*(1), 47–57.

Cooper, N. J., Hooke, J. M., & Bray, M. J. (2001). Predicting coastal evolution using a sediment budget approach: A case study from southern England. *Ocean and Coastal Management, 44*(11–12), 711–728.

Cowell, P., Stive, M. J. F., Niedoroda, A. W., De Vriend, H. J., Swift, D. J. P., Kaminsky, G. M., et al. (2003). The coastal-tract (part 1): A conceptual approach to aggregated modelling of low order coastal change. *Journal of Coastal Research, 19*(4), 812–827.

DEFRA. (2006). *Shoreline management plan guidance*. London: Department of Environment, Farming and Rural Affairs (DEFRA).

Deng, Y., Sadiq, R., Jiang, W., & Tesfamariam, S. (2011). Risk analysis in a linguistic environment: A fuzzy evidential reasoning-based approach. *Expert Systems with Applications, 38*(12), 15438–15446.

EEA. (2007). *The DPSIR framework used by the EEA*. Retrieved January 5, 2011, March 2012, from http://root-devel.ew.eea.europa.eu/ia2dec/knowledge_base/Frameworks/doc101182

EstSim Consortium. (2007). *Development and demonstration of systems-based estuary simulators (EstSim)*. London: Department of Environment, Food and Rural Affairs (DEFRA) and the Environment Agency (EA).

European Commission. (1992). *Habitats directive*. Brussels: European Commission.

Finkl, C. W. (2004). Coastal classification: Systematic approaches to consider in the development of a comprehensive scheme. *Journal of Coastal Research, 20*(1), 166–213.

Freckleton, R. P., Sutherland, W. J., Watkinson, A. R., & Queenborough, S. A. (2011). Density-structured models for plant population dynamics. *American Naturalist, 177*(1), 1–17.

Greenlaw, M. E., Roff, J. C., Redden, A. M., & Allard, K. A. (2011). Coastal zone planning: A geophysical classification of inlets to define ecological representation. *Aquatic Conservation: Marine and Freshwater Ecosystems, 21*(5), 448–461.

Hanson, S., Nicholls, R. J., Balson, P., Brown, I., French, J. F., Spencer, T., et al. (2007). *Capturing coastal morphological change within regional integrated assessment: An outcome-driven fuzzy logic approach* (Working Paper, 113). Norwich: Tyndall Centre for Climate Change Research.

Hanson, S., Nicholls, R. J., Balson, P., Brown, I., French, J. R., Spencer, T., et al. (2010). Capturing coastal geomorphological change within regional integrated assessment: An outcome-driven fuzzy logic approach. *Journal of Coastal Research, 26*(5), 831–842.

Hartley, L., & Pontee, N. (2008). Assessing breaching risk in coastal gravel barriers. *Proceedings of the ICE - Maritime Engineering, 161*(4), 143–152.

Helming, K., Diehl, K., Bach, H., Dilly, O., Koenig, B., Kuhlman, T., et al. (2011). Ex ante impact assessment of policies affecting land use, Part A: Analytical framework. *Ecology and Society, 16*(1), 27.

HR Wallingford. (2002). *Sediment transport report. Report EX4526* (Report no. EX 4526). Wallingford: HR Wallingford.

HR Wallingford, ABPMer, & Pethick, J. (2006). *Review and formalisation of geomorphological concepts and approaches for estuaries*. London: Department of Environment, Food and Rural Affairs (DEFRA) and the Environment Agency (EA).

Karageorgis, A. P., Kapsimalis, V., Kontogianni, A., Skourtos, M., Turner, K. R., & Salomons, W. (2006). Impact of 100-year human interventions on the deltaic coastal zone of the Inner Thermaikos Gulf (Greece): A DPSIR framework analysis. *Environmental Management, 38*(2), 304–315.

King, C. A. M. (1972). *Beaches and coasts*. Gordonsville: Palgrave Macmillan.

McRobie, A., Spencer, T., & Gerritsen, H. (2005). The big flood: North Sea storm surge. *Philosophical Transactions of the Royal Society A: Mathematical, Physical and Engineering Sciences, 363*(1831), 1263–1270.

Mikhailova, M. (2008). Hydrological and morphological features of river mouths of different types (the Columbia estuary and the Fraser delta as examples). *Environmental Research, Engineering and Management, 46*(4), 4–1111.

Motyka, J. M., & Brampton, A. H. (1993). *Coastal management: Mapping of littoral cells* (Report no. SR328). Wallingford: HR Wallingford.

Muir Wood, R., & Bateman, W. (2005). Uncertainties and constraints on breaching and their implications for flood loss estimation. *Philosophical Transactions. Series A, Mathematical, Physical, and Engineering Sciences, 363*(1831), 1423–1430.

Nguyen, H. T. (1997). Fuzzy sets and probability. *Fuzzy Sets and Systems, 90*(2), 129–132.

Oliver, I. (2002). An expert panel-based approach to the assessment of vegetation condition within the context of biodiversity conservation: Stage 1: The identification of condition indicators. *Ecological Indicators, 2*(3), 223–237.

Omann, I., Stocker, A., & Jaeger, J. (2009). Climate change as a threat to biodiversity: An application of the DPSIR approach. *Ecological Economics, 69*(1), 24–31.

Orford, J. D., Carter, R. W. G., McKenna, J., & Jennings, S. C. (1995). The relationship between the rate of mesoscale sea-level rise and the rate of retreat of swash-aligned gravel-dominated barriers. *Marine Geology, 124*(1–4), 177–186.

Pethick, J. (1984). *An introduction to coastal geomorphology*. Baltimore: Edward Arnold.

Pontee, N. I. (2005). Management implications of coastal change in Suffolk. *Proceedings of the ICE - Maritime Engineering, 158*(MA2), 69–83.

Spencer, T., & Brooks, S. M. (2012). Methodologies for measuring and modelling change in coastal saline lagoons under historic and accelerated sea-level rise, Suffolk coast, eastern England. *Hydrobiologia, 693*(1), 99–115.

Steers, J. A. (1946). *The coastline of England and Wales*. Cambridge: Cambridge University Press.

Sutherland, W. J. (2006). Predicting the ecological consequences of environmental change: A review of the methods. *Journal of Applied Ecology, 43*(4), 599–616.

Taylor, C. M., & Hasting, A. (2004). Finding optimal control strategies for invasive species: A density-structured model for Spartina alterniflora. *Journal of Applied Ecology, 41*(6), 1049–1057.

Whitehouse, R., Balson, P., Beech, N., Brampton, A., Blott, S., Burningham, H., et al. (2009). *Characterisation and prediction of large-scale, long-term change of coastal geomorphological behaviours*. Bristol: Environment Agency.

Woodroffe, C. D. (2002). *Coasts. Form process and evolution*. Cambridge: Cambridge University Press.

Chapter 6
Coastal Wetland Habitats: Future Challenges and Potential Solutions

Mustafa Mokrech, Sarah Gardiner, Robert J. Nicholls, Andrew R. Watkinson, and William J. Sutherland

Abstract Coastal habitats are being widely degraded and squeezed around the world due to human pressure and competition for space in the coastal zone. Rising sea levels and other aspects of climate change are and will continue to exacerbate these problems. At the same time, the value of these habitats is increasingly recognised through policies such as the EU Habitats Directive and the UK National Ecosystem Assessment, in terms of the ecosystem services they provide. As a result, management goals such as 'no-net-loss' targets have been implemented, requiring habitat creation or restoration projects to be included in coastal management plans.

Modelling provides a powerful approach to better understanding and addressing these threats. Simulations of future changes as part of an integrated assessment methodology can identify the key drivers of change and assess the effectiveness of potential response options. This goes beyond GIS analyses of habitat futures and begins to address relevant system dynamics and capture pertinent uncertainties including climate drivers considered in Chap. 2. Coastal simulations that aim to quantify future coastal habitats are however subject to many sources of uncertainty

M. Mokrech (✉)
Environmental Institute of Houston, School of Science and Computer Engineering, University of Houston Clear Lake, Houston, TX 77058, USA
e-mail: mokrech@uncl.edu

S. Gardiner
Faculty of Engineering and the Environment, University of Southampton, Southampton SO17 1BJ, UK
e-mail: scgardiner@hotmail.com

R.J. Nicholls
Tyndall Centre for Climate Change Research, Faculty of Engineering and the Environment, University of Southampton, Southampton SO17 1BJ, UK
e-mail: r.j.nicholls@soton.ac.uk

A.R. Watkinson
School of Environmental Sciences, University of East Anglia, Norwich NR4 7TJ, UK
e-mail: a.watkinson@uea.ac.uk

W.J. Sutherland
Conservation Science Group, Department of Zoology, University of Cambridge, Cambridge CB3 3ET, UK
e-mail: w.sutherland@zoo.cam.ac.uk

R.J. Nicholls et al. (eds.), *Broad Scale Coastal Simulation*, Advances in Global Change Research 49, DOI 10.1007/978-94-007-5258-0_6

including incomplete scientific knowledge and understanding of environmental processes and uncertain future drivers such as the magnitude of sea-level rise.

This chapter describes two lines of investigation to address these challenges: (1) analysing the impacts of sea-level rise and climate change on coastal wetlands from the perspective of coastal management and land use and (2) analysing and managing the implications of uncertainties on wetland changes and losses. The first, conventional, method provides a deterministic set of outputs for a given input, where future impacts can be quantified for management decisions. The second, more novel, method uses outcome-driven modelling to produce probabilistic results reflecting the uncertainties. This method narrows down the impacts and possible management responses to a limited number of outcomes, including their associated likelihood values. Both approaches have their utilities and allow users to explore the potential impacts of environmental change on coastal wetlands; they are illustrated here using examples drawn from the south coast of England that are typical of northwest European coastal wetlands.

In the Tyndall Coastal Simulator, this work does not currently feed directly into the other models, as important habitats are not represented in the Norfolk case study. However, these insights are vital for managing our diverse coasts. Inclusion of habitat models with the erosion and flood models in Chaps. 7 and 8 should be a priority in future research.

Keywords Coastal wetlands • Sea-level rise impacts • Habitat change • Uncertainty management • Outcome-driven modelling

6.1 Introduction

Coastal wetlands are naturally located within the tidal frame giving them both terrestrial and aquatic characteristics. They have a richer flora and fauna than many other natural environments and are also valuable in terms of the ecosystem goods and services that they provide (Jones et al. 2011). These include supporting wildlife habitats, food-web support for fish and shellfish, flood buffers, shoreline erosion control, recreation and waste assimilation (Hopkinson et al. 2008). However, inundation and erosion present major threats to coastal wetland habitats and are already resulting in widespread global losses (Nicholls et al. 2008, 2012). These will be exacerbated by future sea-level rise and human activities, particularly when combined with increasing coastal development, infrastructure and coastal protection works, leading to coastal squeeze (Lee 2001). Another factor that is known to influence coastal wetland loss is sediment supply. A decrease in sediment availability, which often results from human interventions to control rivers and main streams as well as reduce coastal erosion, will contribute to wetland loss as the lack of sediment supply inhibits habitats, such as salt marshes, responding vertically to the rising sea.

Assessing both human and climate (including sea-level rise) impacts is important for identifying viable management policies for coastal wetlands, as these events will shape the structure, species composition and function of coastal ecosystems from local to regional scales. There is, however, considerable uncertainty in the extent to which sea levels will rise in the future and an incomplete understanding of associated environmental processes (Sutherland 2006). Conventional approaches for analysing habitat losses and changes due to sea-level rise do not provide any likelihood estimates for future change. Instead a limited number of sea-level rise scenarios are usually selected and investigated, and a range of corresponding outputs are produced. Furthermore, the projections of climate change and sea-level rise scenarios are continuously changing due to an improving understanding of the climate and earth systems. As these changes happen periodically (e.g. linked to the Intergovernmental Panel on Climate Change assessment process), any modelling results based on earlier scenarios become redundant, and a new assessment effort is required. The consequences of these new predictions and their policy implications then also have to be reconsidered.

The UK National Ecosystem Assessment (Jones et al. 2011) provides an assessment of the states and trends of ecosystem services provided by coastal habitats and also highlights how important regulatory frameworks have been in protecting and enhancing coastal ecosystem services. The Habitats Directive and the Birds Directive together form the basis of the European Union's nature conservancy policy (Natura 2000 Network).[1] This has led to the development of the Natura 2000 network, which consists of protected sites for wildlife and habitats of European importance. Coastal habitats, such as salt marshes, mudflats, maritime cliffs, sand dunes, vegetated shingle, coastal grazing marsh and saline lagoons, are designated within this framework, which requires that member states protect, maintain or enhance these areas. In order to deliver flood and coastal defence schemes which comply with the requirements of the Habitats Directive, in the UK Coastal Habitat Management Plans (CHaMPS) were formed. They quantify habitat changes and recommend on how to prevent future losses and identify compensation where designated habitats are impacted.

In the UK, Shoreline Management Plans were introduced in 1994 and have been developed as non-statutory documents aimed at producing sustainable policies for coastal defence and development (see Chap. 1). They are large-scale assessments of the flood and erosion risks associated with coastal processes (such as tides, wave height and direction and the movement of beach material) and aim to help, from a strategic perspective, in reducing the risks to people and the developed, historic and natural environments. In so doing they guide the future management of the shoreline through assessing the interrelationships between coastal processes, defences, land use and the natural environment. It is essential that they draw on a firm evidence base not only of the risks associated with coastal processes but also of the impacts on coastal habitats.

Effective management of future wetland changes requires predictive methods for estimating impacts, together with an assessment of the uncertainties involved and

[1] Coastal habitats often have multiple designations linked to the Habitat Directive. These are the European designations of Special Protection Areas (SPAs) and Special Areas of Conservation (SACs), national designations of Special Sites of Scientific Interest (SSSI), while many wetlands across the world are designated under the Ramsar Convention.

the policy response options to the expected changes. This approach should be able to address a number of important questions in relation to adaptation to future changes and spatial planning such as:

1. Are there opportunities for implementing policies that minimise wetland losses?
2. Are the suggested policy and management responses applicable?
3. What are the spatial extents of these policy and management responses?
4. What is the time frame for implementing these policies and achieving effective outcomes?
5. How should uncertainty be managed in the context of setting future policies?

This chapter will address the above questions by discussing possible impacts of sea-level rise in the context of coastal management and the uncertainty associated with sea-level rise projections. It will analyse how coastal wetlands respond to sea-level rise including vertical accretion and horizontal migration under a range of sea-level rise scenarios. Possible impacts of land use and coastal management on habitat responses including coastal squeeze will be described. Then, the issue of uncertainty associated with sea-level rise and its effects on setting future management policies will be discussed. Predicting the impact of climate change can be complicated due to the range of possible interacting effects (Mustin et al. 2008). The outcome-driven modelling approach is presented as an option for incorporating uncertainties and providing a range of most likely outcomes for setting future management policies. This is beneficial for the management community as it supports more flexibility and preparedness for future adjustment in the context of developing future responses and policies for adaptation. The investigations consider coastal habitats relevant to northwest European coasts, but the principles apply to a broad range of environmental and other problems.

6.2 The Tyndall Coastal Simulator Approach

Future impacts on biodiversity in coastal zones due to climate change (including sea-level rise) and coastal management decisions have been investigated extensively during the development of an integrated coastal simulator (Ridley et al. 2006; Gill and Sutherland 2006; BRANCH Partnership 2007). The investigations have included deterministic (scenario-based) and probabilistic methods.

The first uses conventional models involving hard computing to provide a deterministic view of future changes. The work of the BRANCH Partnership (2007) is presented here as an example of this method, where the impact of sea-level rise on intertidal habitats and spatial planning is investigated. Throughout the BRANCH project, stakeholder engagement activities were carried out, ranging from informal discussions with local coastal users to the European and international level. Presenting findings and gaining feedbacks through workshops involving local spatial planners and coastal managers allowed the project results to be more relevant to end users and to highlight issues with current coastal management and legislation.

One of the main concerns raised by stakeholders was the robustness of scenarios used and the uncertainties of whether these will remain applicable in the future (BRANCH Partnership 2007). These issues and questions over the interpretation and understanding of scenarios also emerged as important in the stakeholder activities described in Chap. 12.

The second method uses a probabilistic approach to account for the uncertainty associated with future projections of sea-level rise and the incomplete knowledge and understanding of environmental processes. The outcome-driven modelling is presented here as a potential approach for addressing uncertainty and providing probabilistic answers, which represent more realistic views of future environmental changes. The underlying philosophy is to consider broad states – as initially adopted in studies of plant populations where densities were grouped into broad categories (Taylor and Hasting 2004). The logic is that it is a delusion to imagine that we can predict say the exact loss of salt marsh or the exact level a weed population will reach so it is better to classify into broad categories and consider under what condition each final state is likely. This approach of considering broad states has a number of advantages including that practitioners just have to consider their responses to these broad states, and these do not change if the assumptions of the model change – only the likelihood of each changes. They do not provide excessive provision (the classic problem of a predictive line on a map that typically generates undue respect), and it is possible to combine a range of information to determine how likely each outcome is. These two methods are explained further in the following sections.

6.2.1 Deterministic Coastal Wetland Simulations

This section describes the methodology implemented in the BRANCH project for modelling habitats under future sea-level rise. The methodology is based on habitat zonation within the tidal frame due to inundation, tolerance of salinity and topography. It also considers horizontal erosion and vertical accretion of salt marshes.

The study site used here to illustrate the methodology is located along 9 km of the north Solent shore from Hurst Spit to the Lymington River Estuary in southern England (Fig. 6.1). It is characterised as a coastal plain estuarine system with relatively flat/gently sloping surroundings; it comprises of un-vegetated mudflat, cord grass (*Spartina anglica*) marshes and high-level mixed salt marsh located seaward of seawalls. Landward of the seawalls, the site includes three substantial streams, saline lagoons and a belt of fresh and brackish grazing marsh (known in the UK as coastal grazing marsh), which is especially rich in plants and invertebrates and important for breeding wading birds (English Nature 2005). The total area protected from flooding at mean high-water springs is approximately 500 ha, including substantial areas of the town of Lymington and the scattered hamlets of Keyhaven, Woodside and Lower Pennington (Martin 1994). The existence of historic and active landfill sites behind the seawalls adds complexity to the management and planning process of the site as these are a potential source of environmental hazard during flood events.

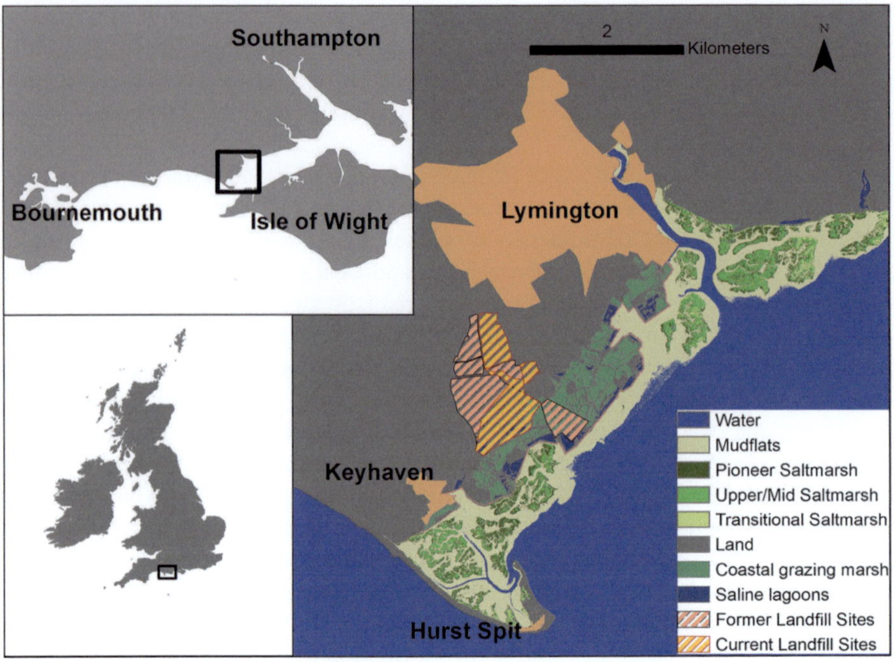

Fig. 6.1 General location of the study site (Hurst to Lymington) including visualised habitat features and key land use/cover classes in 2004 (the baseline year)

The three major intertidal habitat types considered in this investigation are (1) un-vegetated mudflat, (2) salt marsh and (3) coastal grazing marsh (and associated saline lagoons). They are all closely linked to the dynamic spaces provided by tidal variation and seawall protection within the coastal plan form and are intrinsically linked. Salt marsh and mudflats are threatened by coastal squeeze driven by erosion, sediment starvation and rising sea levels, while coastal grazing marsh and saline lagoons are threatened by inundation, rising water tables and realignment of defences. Hence the habitat response to sea-level rise is largely influenced by anthropogenic decisions (Watkinson et al. 2007). Although these habitats are designated separately as a RAMSAR site and SPA under the Birds Directive and as a SAC under the Habitats Directive, they together form an important ecosystem. For example, coastal grazing marsh provides high-water roosts for protected birds, and the mudflats provide feeding sites at low water. However, the links between coastal grazing marsh and salt marsh in terms of ecosystem integrity are unknown, particularly with regard to its resource for protected wading birds, which is often the reason for designation in the first place (Gardiner et al. 2007). The site has experienced substantial loss of salt marsh throughout the twentieth century, exposing the seawall to increased wave loads during storms (Sarah Gardiner Pers. Comm.); these losses are expected to continue. There have been major floods, most notably in December 1989 (Ruocco et al. 2011).

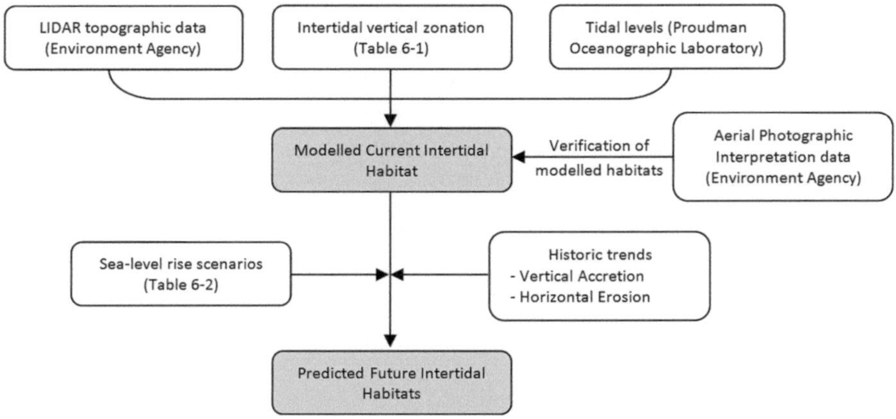

Fig. 6.2 Flow chart illustrating the methodology implemented to predict changes in intertidal habitats

The methodology used here combines the modelling of intertidal habitats with tidal and elevation data (LIDAR) on the current distribution of habitats. By combining data on historic accretion and erosion rates with future sea-level rise data, the ability of these marshes to withstand sea-level rise can then be assessed. Land use can also be used to assess what future management strategies are available in order to facilitate habitat migration and prevent habitat losses; this includes the managed realignment of the coast by retreating sea defences. The flow chart illustrated in Fig. 6.2 describes this method and the sources of data used.

6.2.1.1 Current Intertidal Habitats

Salt marsh plant communities consist of species that colonise different vertical profiles on the shoreline according to inundation frequency and duration and tolerance of salinity (Chapman 1960; Leggett and Dixon 1994). Although there are other factors, the elevation of the land is considered the single most important factor in determining the composition of the vegetation (Burd et al. 1994). The criteria for defining salt marsh types were therefore determined using elevation and position within the tidal frame, with areas below mean high-water neaps (MHWN) but above the lowest astronomical tide (LAT) being characterised as intertidal mudflats (Table 6.1).

At the other extreme, in those areas only covered occasionally by the highest tides and surges, the vegetation is characterised as transitional marsh and is made up of mildly salt-tolerant and non-halophytic plants. In between mean high-water springs and the mean high-water neaps is the main body of salt marsh characterised by halophytic vegetation; this was subdivided into high/mid marsh and pioneer marsh, with the demarcation between upper/mid and pioneer marsh lying at mean high water.

Table 6.1 Tidal criteria for modelling vertical zonation of intertidal areas

Coastal habitats	Criteria for habitat occurrence based on tidal level and elevation
Non-wetland	>highest astronomical tide (HAT)
Transitional marsh	Highest astronomical tide (HAT) – mean high-water springs (MHWS)
Upper/mid salt marsh	Mean high-water springs (MHWS) – mean high water (MHW)
Pioneer salt marsh	Mean high water (MHW) – mean high-water neaps (MHWN)
Mudflats	Mean high-water neaps (MHWN) – lowest astronomical tide (LAT)
Standing water	<Lowest astronomical tide (LAT)

After Blott and Pye (2004), Chapman (1960), Leggett and Dixon (1994) and Pye and French (1993)

LIDAR (Light Detection and Ranging) data were used to create the baseline topography and define the distribution of intertidal habitat (e.g. Blott and Pye 2004; Cope et al. 2008). Combining the baseline topography with local tidal data, provided by the National Oceanographic Centre, and knowledge of the elevation within the tidal frame that intertidal habitats exist allowed a map of current salt marsh and mudflat to be produced. This was then verified using Aerial Photographic Interpretation data.

6.2.1.2 Predicted Future Intertidal Habitats

Here we consider sea-level rise to be the primary driver of coastal change through the twenty-first century. However it is recognised that other drivers, such as storms, waves and precipitation, may be significant at a local level. From the verified baseline data, the position of intertidal habitats under a number of sea-level rise scenarios and time spans was predicted; this simply involved adding the expected sea-level rise onto the tidal parameters (i.e. tidal range does not change). Sea-level rise scenarios were based on global projections by the IPCC (Church et al. 2001) and downscaled by UKCIP (Hulme et al. 2002). Scenarios were developed for four emission scenarios (low, medium-low, medium-high and high) over three time periods (2020s, 2050s and 2080s), relative to a 1990s baseline. These incorporated glacial isostatic adjustment and were adjusted for potential Atlantic deviations from the global-mean rise, as recommended by Hulme et al. (2002). The resulting sea-level rise for the case study site is given in Table 6.2. The low scenario is similar to current trends, while the higher scenarios represent acceleration in the rate of sea-level rise.

Knowledge of sediment supply and historic erosional trends, where available, were included in the analysis to provide more realistic estimates of habitat changes. Accretion was applied only to areas colonised by salt marsh and was modelled using three time steps: current–2020s, 2020s–2050s and 2050s–2080s. Several values for both erosion and accretion were modelled in order to assess a range of future habitat distributions, taking higher and lower values from the literature available for the individual sites (BRANCH Partnership 2007).

Table 6.2 Sea-level rise scenarios (relative to 1990) used in the analysis based on Hulme et al. (2002)

Time Scale	Sea-Level Rise (cm)			
	Low	Medium-Low	Medium-High	High
2020s	4	11	16	23
2050s	7	23	35	57
2080s	9	37	64	108

Note: shaded values are the main focus of this chapter

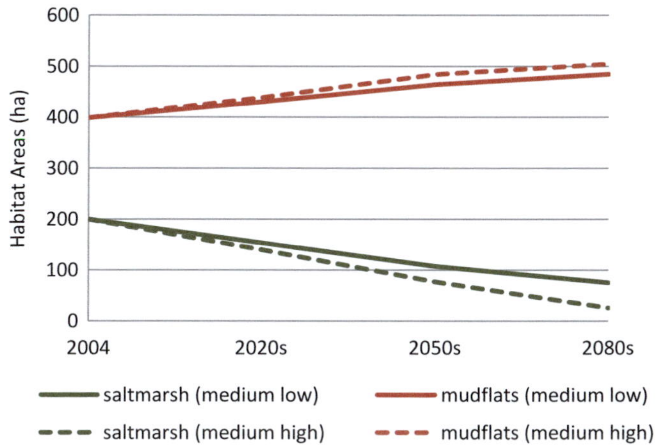

Fig. 6.3 Intertidal areas with the 'hold-the-line' coastal management option (no managed realignment)

Following the model determination of intertidal areas, the impact on other coastal habitats was assessed by evaluating the potential of a habitat to migrate or adapt in line with possible future management/adaptation options. Managed realignment options were also assessed, in order to compensate for habitat losses within the coastal zone.

6.2.1.3 Results and Challenges of Future Management

The selected medium-low and medium-high sea-level rise scenarios combined with 3 mm/year vertical accretion of salt marsh and 3 m/year of horizontal erosion produced a steady decline in the areas of intertidal salt marsh, especially toward the end of the twenty-first century if the current defences are maintained (Fig. 6.3). Under the medium-high sea-level rise scenario, mudflat will almost certainly be the only intertidal habitat present at the site. Furthermore, even with the managed realignment of available habitats, modelling indicates that mudflats will remain the dominant habitat because the elevation of intertidal land behind the defences is too low for the establishment of replacement salt marsh. This means that any managed realignment at the site will also require soft engineering to optimise the levels of tidal water in the realigned areas for enhancing salt marsh creation. This could include regulated tidal exchange and/or elevating lands by beneficial use of dredged materials.

Several management challenges exist at the site and more widely in the region. Development and historic planning decisions have meant that all the available space for intertidal habitats has been reduced, often to near zero. The extension of intertidal areas using deliberate managed realignment is frequently seen as a method of maintaining and enhancing stocks of salt marsh and intertidal mudflats. In the Solent region, two small managed realignment schemes have been conducted to date at Thorness Bay and in Chichester Harbour, with a much larger planned scheme at Medmerry near Selsey Bill (ABPmer 2011; Cope et al. 2008). However, areas of coastal grazing marsh and saline lagoons currently located behind maintained coastal defences are displaced. In common with salt marsh and mudflats, these habitats are often 'designated' under the EC Habitats and Birds Directives and are also required to be maintained. This puts them in direct conflict with intertidal habitats and requires that a choice be made in terms of their relative importance/value. As coastal grazing marsh exists largely due to the presence of the defences, it is probably less economically sustainable to maintain them, with increased maintenance and improvement of the defences being required as the seaward, protective, intertidal areas decline under predicted sea-level rise. Creation of grazing marshes adjacent to rivers offers an alternative replacement strategy, if this is deemed suitable (Gardiner et al. 2007).

6.2.2 Coastal Wetlands and Uncertainty

The preceding analysis takes a deterministic view of future change and does not consider uncertainty. However, there is considerable uncertainty associated with sea-level rise projections and models used for simulating environmental processes (Sutherland 2006; Freckleton et al. 2008). The modelling accuracy is usually limited by our understanding of the systems involved (i.e. both climate system and ecological system). Although empirical models can be used as alternatives as they make fewer assumptions about the underlying processes, they require extensive data sets, and without a mechanistic basis it may be unwise to predict outside the data range. Furthermore, there is often uncertainty over the model structure and parameter values that need to be addressed for better model precision (Sutherland and Freckleton 2012). Consequently we require a modelling approach that takes account of uncertainties in our understanding of habitat processes and can incorporate the updating of future climate and sea-level scenarios. The suggested outcome-driven approach provides spatial distributions and quantitative estimates of possible habitat outcomes. These habitat outcomes are associated with likelihood values that can be updated when information changes. Furthermore the relative simplicity of this approach makes it easier to challenge the assumptions. In the following sections, the implementation of the method will be described, and the results and discussion will demonstrate its potential in accounting for uncertainty and being informative to the decision-making process.

6.2.2.1 Outcome-Driven Modelling: Principles and Implementation

The method is based on the premise that, despite an infinite number of sea-level rise scenarios, future changes in intertidal habitats can be effectively reduced to a limited number of representative outcomes associated with likelihood estimates, which can be easily updated when information (i.e. new climate projections and knowledge) changes. It is described briefly in Sutherland (2006) and also explored in a nonspatial manner by Ridley et al. (2006). In summary, the method uses soft computing to combine all available information from Bayesian analysis, probabilistic reasoning and fuzzy logic in order to address problems associated with imprecision, uncertainty and partial truth.

The aim is to produce a number of habitat outcomes, referred to here as habitat states. Each habitat state is associated with a likelihood value that changes over time due to the effect of sea-level rise and other anthropogenic-related conditions. Table 6.3 illustrates this concept using a hypothetical representation of modelling results. In the baseline year, the likelihood of occurrence of 'habitat state 1' is 100 %, while the likelihood values for the other habitat states are zeros. Under future sea-level rise conditions, these likelihood values will change according to the probabilities associated with sea-level rise scenarios and probability of habitat distribution within the tidal frame. In the 2020s there are three possible habitat outcomes with different likelihood values; the most likely one is 'habitat state 3'. In the 2050s, a different set of habitat outcomes is possible with the most likely one being 'habitat state 4'.

With new sea-level rise projections and/or improved knowledge in environmental processes, the modelling results are not redundant; only the likelihood values need to be updated. This principle makes the outcome-driven approach more relevant to the development of a transparent and flexible policy-driven framework

Table 6.3 Example of habitat outcomes and changing likelihood values over time

	Habitat state 1	Habitat state 2	Habitat state 3	Habitat state 4	Habitat state 5	–	Habitat state n
Baseline year	100 %	0	0	0	0	–	0
2020s	0	35 %	45 %	30 %	0	–	0
2050s	0	5 %	20 %	55 %	12 %	–	0
–	–	–	–	–	–	–	–

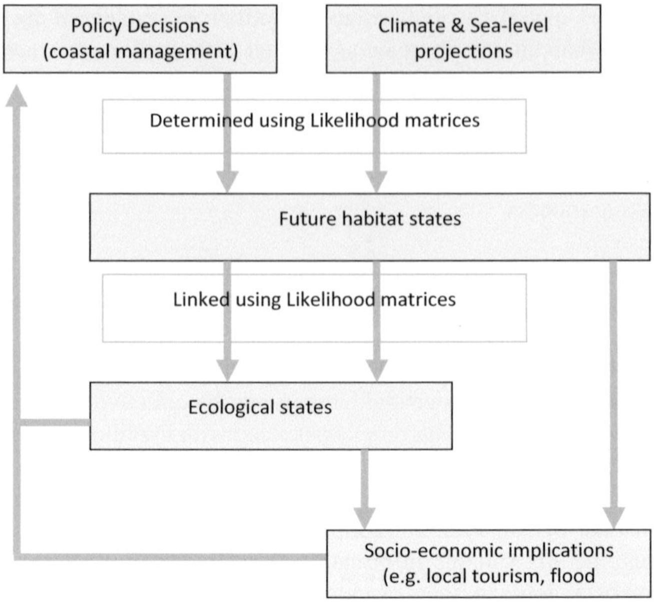

Fig. 6.4 The outcome-driven modelling framework, including climate change drivers and coastal management decisions (adapted from Sutherland 2006; Ridley et al. 2006)

than conventional methods (Fig. 6.4). Within the framework, ecological states with likelihood matrices can be established from the habitat states, and then socio-economic implications can be assessed using indicators that are relevant to ecosystem services, such as local tourism and flood protection buffers. These indicators combined with the predicted ecological states are used to assess different coastal management policies against desired outcomes so that more informed decisions may be made. These ecological or socio-economic analyses persist if the conclusions change (the reworked analysis simply changes the likelihood of each state). This is very different from the standard approach of outputs such as predicted loss of salt marsh or lines on maps which then require a completely new ecological or socioeconomic analysis to the novel conditions.

The methodology is illustrated here for the same Hurst to Lymington study site as discussed in Sect. 6.2.1. It incorporates the results from that model to simulate the full range of habitat outcomes (referred to here as habitat states) that could arise from combining sea-level rise scenarios, coastal management policies and tidal amplitude levels. The investigated management policies include: hold the line, minor realignment, major realignment and no defences. For each management policy, 16 provisional habitat states are identified for a range of sea-level rise values ranging through 0–150 cm in 10 cm increments. The tidal amplitude in this context represents the sea-water level behind seawalls only when managed realignment options and no defences are applied, which can be achieved by implementing regulated tidal exchange. Four levels of tidal amplitude of the observed tidal range are explored: 25 %, 50 %, 75 % and 100 %. Consequently, 208 provisional habitat states are produced. The 208 provisional habitat states were grouped to produce the 17 final habitat states by using cluster analysis. In general, cluster analysis encompasses algorithms that aim to group similar objects into respective categories. Thus, objects can be sorted into groups in a way that the degree of association between objects is maximal if they belong to the same category and minimal otherwise. Given this, cluster analysis is used in this work to group similar provisional habitat states based on examining change in salt marsh and change in mudflats. The most common method of 'Euclidean distance' is used to calculate geometric distances between simulations in a two-dimensional space (Fig. 6.5); the geometric distance is then used to establish the rule of similarity between simulations. The shorter the distances between simulations, the more similar they are. In this case, 17 clusters are apparent in the two-dimensional space of salt marsh and mudflats changes.

Assigning likelihood values for habitat states is central to this method. Estimates of the probability distributions of sea-level rise scenarios are used to account for scenario uncertainty, while expert opinion is used to calculate the probability of habitat distribution within the tidal frame. Here, expert opinion is used to deal with the complex domain in which the relationship between sea-level rise and its effect on the distribution of wetland habitats within the tidal frame is inexact. The likelihoods of provisional habitat states were first calculated from the probability distributions of sea-level rise scenarios and then summed in a way that corresponds to the above-described hierarchical clusters. This gave aggregated likelihoods that

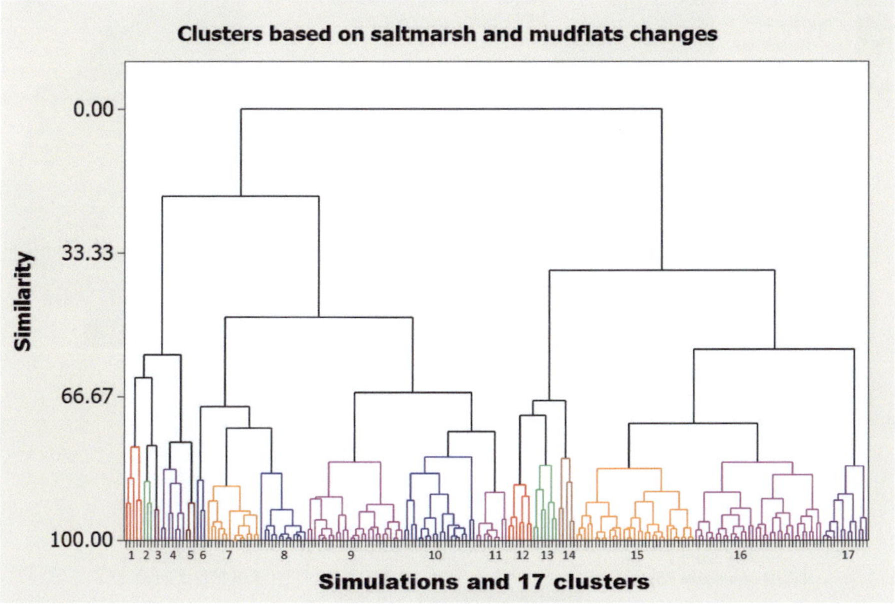

Fig. 6.5 Dendrogram of the cluster analysis using the Euclidean Distance method showing the 17 clusters of habitats and level of similarity in hierarchical structure

correspond to sea-level rise scenarios over time (i.e. the three time slices: 2020s, 2050s and 2080s). For instance, sea-level rise is expected to accelerate in time, and accordingly habitat states representing higher sea-level scenarios will have progressively higher probabilities over time.

6.2.2.2 Results and the Prospects of the Approach

Although the presented outcome-driven approach requires extensive simulation of a wide range of scenarios, it produces a limited set of representative outcomes. In this work, there are 17 habitat states that represent all the possible outcomes that may be produced from a range of climate conditions and policies over the twenty-first century, even allowing for very large rises in sea level. Each habitat state is associated with a likelihood value expressing its probability of occurrence under these conditions. Unless the likelihood value of one habitat state is 100 %, multiple habitat states are possible. The habitat state with the highest likelihood value is the outcome that is more likely. Figure 6.6 shows the two most likely outcomes by 2050s under the medium-high sea-level rise scenario, major managed realignment and 75 % tidal level (i.e. regulated tidal exchange); the likelihood of occurrence is 78 % and 15 %, respectively. This leads to having a range of possible

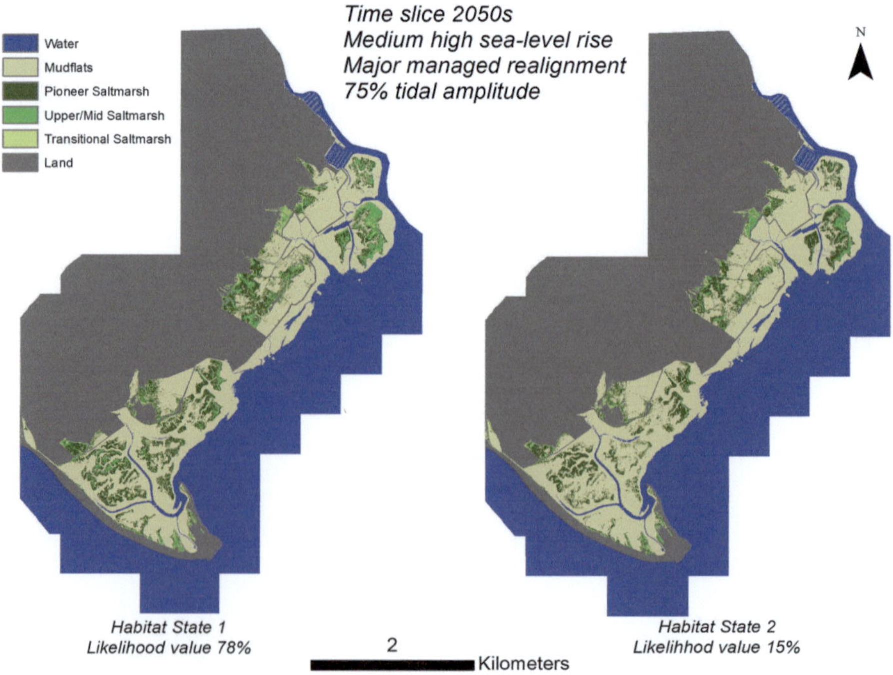

Fig. 6.6 Example of the spatial distribution of habitat states: habitat state 1 represents the most likely habitat outcome with a likelihood value of 78 %

answers to the impact assessment in relation to climate change and management policies. This approach brings benefits but at the same time challenges coastal managers and the decision-making process. It provides more informative results than the conventional, scenario-based approach shown in Sect. 6.2.1; this usually produces one or just a few deterministic views of future habitats without any indication of which is the more likely. The probabilistic approach produces an envelope of future states and estimates their likelihoods. This in turn may encourage coastal managers to consider risk-based decisions in regard to managing adverse impacts.

The probabilistic approach requires a large number of simulations and provides an opportunity to analyse the sensitivity of coastal systems to future sea-level rise and the effectiveness of possible responses. This helps to capture the big picture in relation to understanding the impacts and the range of policies available. For example, the simulation results in this study confirm that 'hold the line' could maintain mudflat areas but at the expense of almost total loss of salt marshes assuming small (<30 cm) rises in sea level (Figs. 6.7 and 6.8). In general, mudflats will dominate the site under all management regimes, but even these will decline after a sea-level rise threshold between 40 and 90 cm has been reached under each management option. On the other hand, the managed realignment options show that up to 90 ha of extra salt marsh can be achieved without any alteration to the land elevations behind the defences. This possible gain can be maximised if managed realignment is applied combined with artificial modification of the land profile to optimise habitat creation (this is not shown in Figs. 6.7 and 6.8). In addition, the differences between habitat states under a range of tidal amplitudes suggest a significant impact of tidal amplitude on habitat creation and that regulated tidal exchange could be considered for the site.

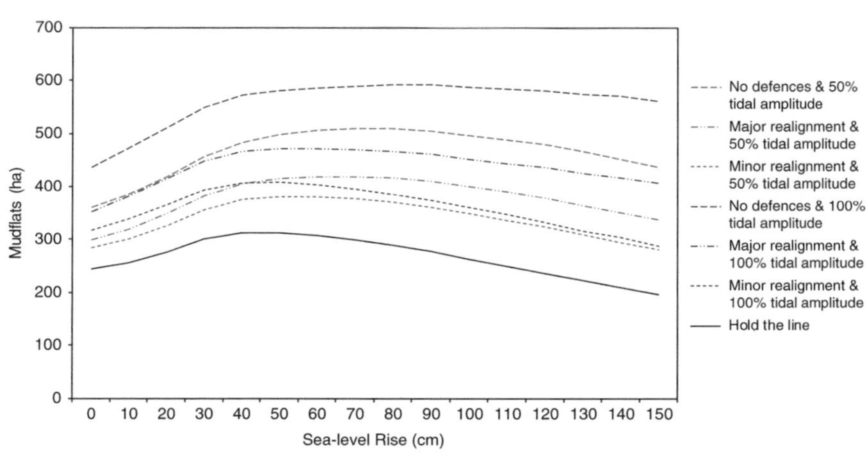

Fig. 6.7 Change patterns in intertidal mudflats under a sea-level rise of up to 150 cm and a range of management policies. Note that each management policy has its own baseline for no sea-level rise

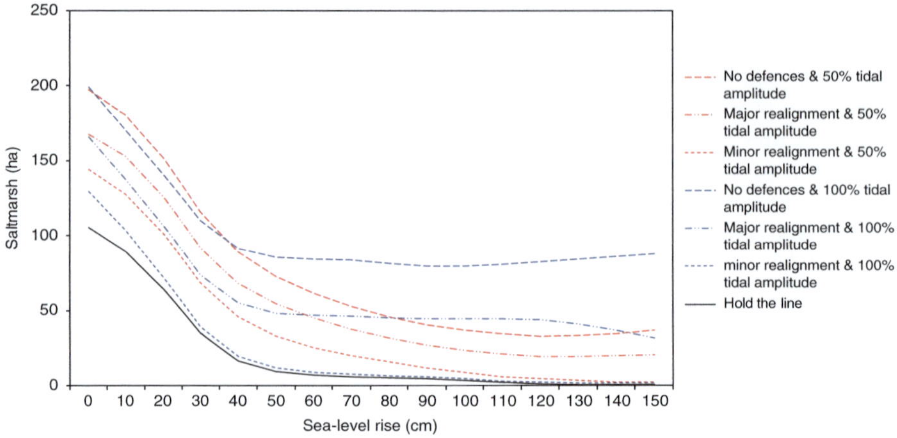

Fig. 6.8 Change patterns in intertidal salt marsh under a sea-level rise of up to 150 cm and a range of management policies. Note that each management policy has its own baseline for no sea-level rise

The habitat states and their likelihoods can be integrated into any GIS platform to allow potential users easy access to the results through an intuitively designed interface (as illustrated in Chap. 11). Thus, for any given combination of time slice, sea-level rise scenario, management policy and tidal amplitude level, there is a probabilistic combination of habitat states. These habitat states represent habitat predictions on the best available scientific understanding to the impact of sea-level rise on coastal wetlands. The interface allows users to alter the probability distributions of sea-level rise scenarios such that the likelihoods of habitat states will be recalculated automatically. As a proof of concept, the system has been developed further as a decision support tool to assess policies based on indicators of key habitats including those located seaward and landward of coastal defences. For example, the decision support system can assess an explored set of management policies and provide a list of policies that may lead to a specific objective, such as the maintenance of coastal gazing marsh with 10 % loss of salt marsh under a specific sea-level rise scenario. Thus, the appropriate policies can be identified, and the most likely distribution can also be accessed and evaluated.

6.3 Conclusions and Next Steps

This research has described two approaches to analyse the future of coastal habitats that might be included in a coastal simulator.

Coastal management decisions will be a major driver of coastal outcomes for coastal habitats, as well as sea-level rise and climate change. Consequently, adaptation to climate change in the coastal zone requires understanding both the future impacts of climate conditions on coastal wetlands and how these interact with management decisions. This requires quantifying habitat loss, change and creation, within the context of spatial planning at broad rather than local scales. In the example we have presented here, managed realignment leads to the loss of coastal grazing marsh, which is of equal importance to the other intertidal habitats according to the Habitats Directive. Planning for adaptation within a wider context provides the possibility of compensating for lost coastal grazing marsh with freshwater grazing marsh as an example (Gardiner et al. 2007).

Deterministic approaches for assessing climate impacts provide more limited information on habitat changes and do not account for uncertainty. This may lead to a misrepresentation and oversimplification of likely future changes and hence poor management decisions; for example, this may underplay the consequences of low-likelihood/high-impact outcomes. Schneider (2001) advocated the use of more probabilistic formulations of potential impacts by the policy communities due to the associated uncertainty. With a range of steps in the analysis (e.g. sea-level rise, resulting flooding risk, resulting changes to habitats, resulting socio-economic changes), these uncertainties multiply with each level.

Outcome-driven modelling provides a balance between traditional single realisations of future states and full probabilistic approaches for assessing climate impacts. It narrows down the possible impacts to a limited number of outcomes with associated likelihood values that are quite easy to understand. The method provides flexibility for any future adjustment to expected driver values by allowing ready adjustment of the likelihood value for each outcome, without repeating the analysis. This also allows for the consideration of low-likelihood/high-impact outcomes. Further investigation of outcome-driven approaches is warranted.

The principles of outcome-driven modelling are generic as the approach can be applied to assess climate impacts and adaptation policies in different places and for many habitat types, as long as the future outcome can be linked to the relevant drivers. In the Tyndall Coastal Simulator, it has been used to explore ecological changes in the North Norfolk region of East Anglia, UK. It is also used to investigate coastal geomorphic changes (see Chap. 5) and to assess coastal habitat changes as described in this chapter. However, coastal geomorphic changes and wetland habitat change are significantly linked, and they can be driven by climate and non-climate changes. Ideally the system should be treated as a whole to establish inter-linked geomorphic, habitat and ecological outcomes with indicators of ecosystem services for a more effective assessment framework. Furthermore, the concept of modelling inter-linked sectors using this approach is already expanding to wider areas. For example, the Theseus project on innovative coastal defence is considering the relationship between coastal engineering, coastal habitats and social systems, including the potential for synergistic effects (see www.theseusproject.eu).

Hence the methods outlined here form an important pillar in the integrated assessment of coastal systems and need to be developed further with other methods outlined in this book for that purpose.

References

ABPmer. (2011). *Online managed realignment guide*. http://www.abpmer.net/omreg/. Cited 10 Dec 2013.

Blott, J., & Pye, K. (2004). Application of Lidar digital terrain modelling to predict intertidal habitat development at a managed retreat site: Abbotts Hall, Essex, UK. *Earth Surface Processes and Landforms, 29*, 893–905.

BRANCH Partnership. (2007). *Planning for biodiversity in a changing climate – BRANCH project final report*. Sheffield: Natural England.

Burd, F., Clifton, J., & Murphy, B. (1994). *Sites of historical sea defence failure. Phase II study* (Report no. Z038-94-F to English Nature). Hull: Institute of Estuarine and Coastal Studies, University of Hull.

Chapman, V. J. (1960). *Saltmarshes and salt deserts of the world*. Aberdeen: University Press.

Church, J. A., Gregory, J. M., Huybrechts, P., Kuhn, M., Lambeck, K., Nhuan, M. T., Qin, D., & Woodworth, P. L. (2001). Changes in sea level. In J. T. Houghton, Y. Ding, D. J. Griggs, M. Noguer, P. J. van der Linden, X. Dai, K. Maskell, & C. A. Johnston (Eds.), *Climate change 2001: The scientific basis. Contribution of working group I to the third assessment report of the Intergovernmental Panel on Climate Change*. Cambridge: Cambridge University Press.

Cope, S. N., Bradbury, A. P., & Gorczynska, M. (2008). *Solent dynamic coast project, main report: A tool for SMP2*. New Forest District Council/Channel Coastal Observatory, UK. Southampton: Channel Coastal Observatory

English Nature. (2005). *Natural areas: Solent and Poole Bay*. Available at http://www.naturalareas.naturalengland.org.uk/Science/natural/NA_Details.asp?NA_ID=109&S=solent&R=0. Cited 10 Dec 2013.

Freckleton, R. P., Sutherland, W. J., Watkinson, A. R., & Stephens, P. A. (2008). Modelling the effects of management on population dynamics: Some lessons from annual weeds. *Journal of Applied Ecology, 45*, 1050–1058.

Gardiner, S., Hanson, S., Nicholls, R. J., Zhang, Z., Jude, S. R., Jones, A., Richards, J. A., Williams, A., Spencer, S., Coggins, P. C. C., Gorczynska, M., Bradbury, A. P., McInnes, R., Ingleby, A., & Dalton, H. (2007). The Habitats Directive, coastal habitats and climate change – Case studies from the south coast of the UK. In *Proceedings of the ICE international conference on coastal management 2007*, Cardiff, pp. 193–202.

Gill, J., & Sutherland, W. (2006). *Towards an integrated coastal simulator of the impact of sea level rise in East Anglia: Part B2- coastal simulator and biodiversity: Models of biodiversity responses to environmental change* (Tyndall Centre technical report 42B2). Norwich: Tyndall Centre.

Hopkinson, C. S., Lugo, A. E., Alber, M., Covich, A. P., & Van Bloem, S. J. (2008). Forecasting effects of sea-level rise and windstorms on coastal and inland ecosystems. *Frontiers in Ecology and Environment, 6*, 255–263.

Hulme, M., Jenkins, G. J., Lu, X., Turnpenny, J. R., Mitchell, T. D., Jones, R. G., Lowe, J., Murphy, J. M., Hassell, D., Boorman, P., Mcdonald, R., & Hill, S. (2002). *Climate-change scenarios for the UK* (The UKCIP02 scientific report). Norwich: Tyndall Centre for Climate Change Research, University of East Anglia.

Jones, L., Angus, S., Cooper, A., Doody, P., Everard, M., Garbutt, A., Gilchrist, P., Hansom, J., Nicholls, R., Pye, K., Ravenscroft, N., Rees, S., Rhind, P., & Whitehouse, A. (2011). Coastal margins. In *The UK National Ecosystem Assessment technical report*. Cambridge: UK National Ecosystem Assessment/UNEP-WCMC.

Lee, M. (2001). Coastal defence and the Habitats Directive: Predictions of habitat change in England and Wales. *The Geographical Journal, 167*, 39–56. doi:10.1111/1475-4959.00004.

Leggett, D. J., & Dixon, M. (1994). Management of the Essex saltmarshes for flood defence. In R. A. Falconer & P. Goodwin (Eds.), *Wetland management*. London: Institution of Civil Engineers.

Martin, D. J. (1994). Impact of conservation issues on a sea defence scheme at Pennington, Hampshire. *Journal of the Institute of Water and Environmental Management, 8*, 567–577.
Mustin, K., Sutherland, W. J., & Gill, J. A. (2008). The complexity of predicting climate-induced ecological impacts. *Climate Research, 35*, 165–175.
Natura 2000 Network. http://ec.europa.eu/environment/nature/natura2000/index_en.htm. Cited 10 Dec 2013.
Nicholls, R. J., Wong, P. P., Burkett, V. R., Woodroffe, C. D., & Hay, J. E. (2008). Climate change and coastal vulnerability assessment: Scenarios for integrated assessment. *Sustainability Science, 3*(1), 89–102.
Nicholls, R. J., Woodroffe, C. D., Burkett, V., Hay, J., Wong, P. P., & Nurse, L. (2012). Scenarios for coastal vulnerability assessment. In M. van den Belt et al. (Eds.), *Treaties on estuarine and coastal science*. Oxford: Elsevier.
Pye, K., & French, P. W. (1993). *Saltmarsh processes and morphology. Erosion and accretion processes on British saltmarshes* (Vol. I). London: Ministry of Agriculture Fisheries and Food (MAFF).
Ridley, J., Gill, J., Watkinson, A., & Sutherland, W. (2006). *Towards an integrated coastal simulator of the impact of sea-level rise in East Anglia, Part B: Coastal simulator and biodiversity, design and structure of the coastal simulator* (Tyndall Centre technical report 42B1). Norwich: Tyndall Centre.
Ruocco, A. C., Nicholls, R. J., Haigh, I. D., & Wadey, M. P. (2011). Reconstructing coastal flood occurrence combining sea level and media sources: A case study of the Solent, UK since 1935. *Natural Hazards, 59*(3), 1773–1796. doi:10.1007/s11069-011-9868-7.
Schneider, S. H. (2001). What is "Dangerous" climate change? *Nature, 411*, 17–19.
Sutherland, W. J. (2006). Predicting the ecological consequences of environmental change: A review of the methods. *Journal of Applied Ecology, 43*, 599–616.
Sutherland, W. J., & Freckleton, R. P. (2012). Making predictive ecology more relevant to policy makers and practitioners. *Philosophical Transactions of the Royal Society B, 367*, 322–330.
Taylor, C. M., & Hasting, A. (2004). Finding optimal control strategies for invasive species: A density-structured model for Spartina alterniflora. *Journal of Applied Ecology, 41*, 1049–1057.
Watkinson, A., Nicholls, R. J., Sear, D., & Ledoux, L. (2007). Environmental impacts of future flood risk. In C. Thorne, E. Evans, & E. Penning-Rowsell (Eds.), *Future flood and coastal erosion risks*. London: Thomas Telford.

Chapter 7
Simulating the Shore and Cliffs of North Norfolk

Mike Walkden, Mark Dickson, James Thomas, and Jim W. Hall

Abstract This chapter describes geomorphic modelling of the main Norfolk study area, the 50 km of the coast between Blakeney Spit and Winterton Ness, specifically the eroding cliffs, shore platforms, beaches, dunes and coast protection structures. The models represent both the natural eroding coastal system and past and future management interventions. Numerical experiments are also used to explore the possible dynamics of this coast if it were in a natural state.

The 'reduced complexity' modelling tool Soft Cliff and Platform Erosion (SCAPE) is core to the analyses. It is described, along with the development of the quasi-three-dimensional Norfolk model. This model links marine hydrodynamic models to those of cliff failure, coastal flooding and economic risk. These links are described, along with the cliff instability module, the representation of beach fluctuation and beach nourishment (both natural and artificial).

Results are shown to illustrate processes operating at a range of scales, including (millennial scale) emergence of the shore profile and beach volumes; (century scale) foreshore lowering, sensitivity to variations in rock strength, sediment transport potential and rate of sea-level rise; and (decadal scale) growth of wave impact forces, emergence of large-scale sand waves, probability density functions of beach volume, probabilistic cliff recession and the effects of beach nourishment.

M. Walkden (✉)
WSP, Keble House, Southernhay Gardens, Exeter EX1 1NT, UK

WSP Group, Exeter, UK
e-mail: mike.walkden@wspgroup.com

M. Dickson
School of Environment, The University of Auckland,
Private Bag, 92019 Auckland, New Zealand
e-mail: m.dickson@auckland.ac.nz

J. Thomas
Experience Prototyping Technician, School of Design, Northumbria University,
Newcastle Upon Tyne NE1 8ST, UK
e-mail: James.E.Thomas@northumbria.ac.uk

J.W. Hall
Environmental Change Institute, Oxford University Centre for the Environment,
University of Oxford, South Parks Road, Oxford OX1 3QY, UK
e-mail: jim.hall@eci.ox.ac.uk

R.J. Nicholls et al. (eds.), *Broad Scale Coastal Simulation*, Advances in Global Change Research 49, DOI 10.1007/978-94-007-5258-0_7

The benefits of such 'reduced complexity' modelling in general are highlighted. These include a deeper understanding of the coastal geomorphology, improved predictability, long-term model stability and ready coupling to other models. The approach is also relatively quick to apply and so can be implemented in practitioner studies, and the results are informing coastal planning in North Norfolk. Important features of the work include its large scale, which the coastal management process greatly needs, and the potential to use probabilistic methods, which allow economic risk to be derived to inform management decision-making.

In terms of the Tyndall Coastal Simulator, the SCAPE model was a key element in linking the offshore conditions to human consequences (flood and erosion risk). It stresses the mediating role of coastal geomorphology in this link. Depending on how the coast is managed, those geomorphic processes can enhance (via erosion) or reduce (via accretion) these risks. The beach volumes were a direct input to the inundation analysis (Chap. 8), which together with the clifftop position informed the risk analysis (Chap. 9).

Keywords Coastal management • Coastal erosion • Climate change • Sea-level rise • Cliff recession

7.1 Introduction

Coastal geomorphic change is a very important element of coastal management in northeast Norfolk. This coast includes a long stretch of fairly rapidly eroding cliff and also extensive areas of land below levels of extreme high tides. These lowlands depend on the (developing) state of the beach fronting them for defence against coastal flooding. There are many communities along this coast, and it is important to project future cliff recession and beach evolution in order to understand the erosion and flood vulnerability of these settlements. Furthermore, projections are needed to understand how this vulnerability may change under increased sea-level rise and under alternative coastal management policies.

This chapter describes model simulations of the 50 km of cliffs, shore platforms and beaches between Blakeney Spit and Winterton Ness, largely in sub-cell 3b (see Sect. 1.4). These simulations were driven with sea-level projections and insights into future surge and wave conditions described in Chap. 2 and also the output of wave modelling (Chap. 3). In turn their outputs influenced 'downstream' models. Predicted clifftop locations drove calculations of erosion risk (Chap. 9), whilst the simulated beach levels were used to determine sea defence structure stability in flood inundation and risk modelling (Chap. 9). The model outputs were also built into an interactive GIS tool (Chap. 11) and computer visualisations (Chap. 10) to aid and assist stakeholder engagement and decision-making.

It is within this element of the Tyndall Coastal Simulator that natural physical processes first meet human systems, in the form of settlements, erosion defences and flood protection structures. It is also here that coastal management and climate change are explicitly represented (through the use of scenarios), allowing their effects to cascade through to the subsequent quantification of erosion and flood risk.

The analysis was performed using the Soft Cliff and Platform Erosion (SCAPE) tool, developed by Walkden and Hall (2005). The challenges included the need to construct models capable of (1) simulating long periods of time (up to one century at least), (2) capture the important internal dynamics of the coastal (geomorphic) system, (3) respond to both weather and climatic drivers and (4) capture response to anthropogenic defence activities (i.e. alternative coastal management policies). SCAPE has been applied to a series of different locations, for the purposes of both research and consultancy. Most recently it has been used to model the shore of Hunstanton (Royal Haskoning 2012) and to quantify the general sensitivity of eroding consolidated shore profiles to accelerated sea-level rise (Walkden and Dickson 2008; Ashton et al. 2011).

The first SCAPE models of the Norfolk Coast were built and run for the Overstrand to Walcott Strategy Study (North Norfolk District Council 2005; Walkden and Hall 2011). These models represented a 35 km of coast, stretching past Sheringham and Sea Palling. They were calibrated against an erosion record 87 years in length and were used to explore coastal change over a future time frame of 50 years, under alternative coastal management options and different rates of sea-level rise.

That original Norfolk model was further developed during the Tyndall Centre research programme to explore questions beyond those addressed in the original strategy study. More refined management options were tested, but also more fundamental questions about the long-term behaviour of soft-cliff coasts and their response to interventions and climate change. The Tyndall model was developed in two stages: the first involved 45 deterministic projections (see Dickson et al. 2007), whilst the second, which was completed in 2008, comprised 5,000 probabilistic simulations. This implementation of SCAPE extended the spatial scale from the root of Blakeney Spit to Winterton Ness.

The following section describes aspects of the coast of particular relevance to this element of the Tyndall Coastal Simulator. The modelling tool used to represent the coast is outlined in Sect. 7.3, along with a description of the model construction and its integration with other elements of the Tyndall Coastal Simulator. This section deals, in particular, with the representation of cliff stability, structural coastal interventions and beach nourishment. The scenarios of climate change and coastal management that were used to make future projections are then described in Sect. 7.4. Various results are given in Sect. 7.5 to illustrate model stability, sensitivities and performance and to show how elements and characteristics of the coast may evolve over the twenty-first century. Section 7.6 includes conclusions and reflections on potential future developments.

7.2 Description of the Coast

Most of coast between Blakeney Spit and Winterton Ness (the area covered by SMP6) is comprised of eroding coastal cliffs up to 50 m in height. Their geology changes between Cromer and Weybourne, from predominantly chalk (towards the north) to predominantly till (towards the south). At Cart Gap, and further south, the land is low lying, with a system of dunes fronting floodplains that extend up to 10 km inland, as far as the Norfolk Broads.

The erosion of the cliff slopes occurs largely through processes of mass movement resulting from over-steepening driven by cliff toe recession. Clearly climatic change may alter the incidence of cliff failure by, for example, increasing the frequency of heavy rainfall events. However, ultimately the primary controls on long-term rates of cliff retreat are the gradient and elevation (relative to sea level) of the shore platform and beach (e.g. Fig. 7.1), as these control the ability of waves to clear fallen debris and then attack and destabilise the cliff toe.

In the north of the study area, the beaches are composed of shingle, but this grades to sand with distance south. At Sheringham the beach is comprised of shingle above a thin sand layer, on the underlying chalk platform. Very little sand is found southeast of Cromer. The net direction of sediment drift along most of this coast is towards the south. However, a drift divide occurs near the town of Sheringham, so that sediment flows northwest at Blakeney Spit. The eroding cliffs and shore platform are the main source of sediments for the beaches to the south and east, including the low-lying coast southeast of Happisburgh. Protection of these cliffs over the nineteenth and twentieth centuries has resulted in falling sediment supply to downdrift beaches.

Fig. 7.1 Exposed clay platform in the bay just south of Happisburgh after a major storm surge in November 2007

Coast protection measures have been employed along this coast for many years. A timeline of these interventions, which was ascertained from reports, historical photographs, maps and local knowledge, is shown schematically in Fig. 7.2.

As is indicated in Fig. 7.2, the major settlements of Sheringham, Cromer and Mundesley have been protected with groynes and seawalls since the latter part of the nineteenth century. Overstrand was also protected but did not have a seawall until the early twentieth century. Most major engineering works south of Mundesley occurred between 1950 and 2000 following the 1953 storm surge, with construction mainly comprising groynes and palisades.

For more detailed descriptions of this coast, including the bathymetry, history of sea-level rise and tidal and wave conditions, the reader is referred to Chap. 1 and Dickson et al. (2007).

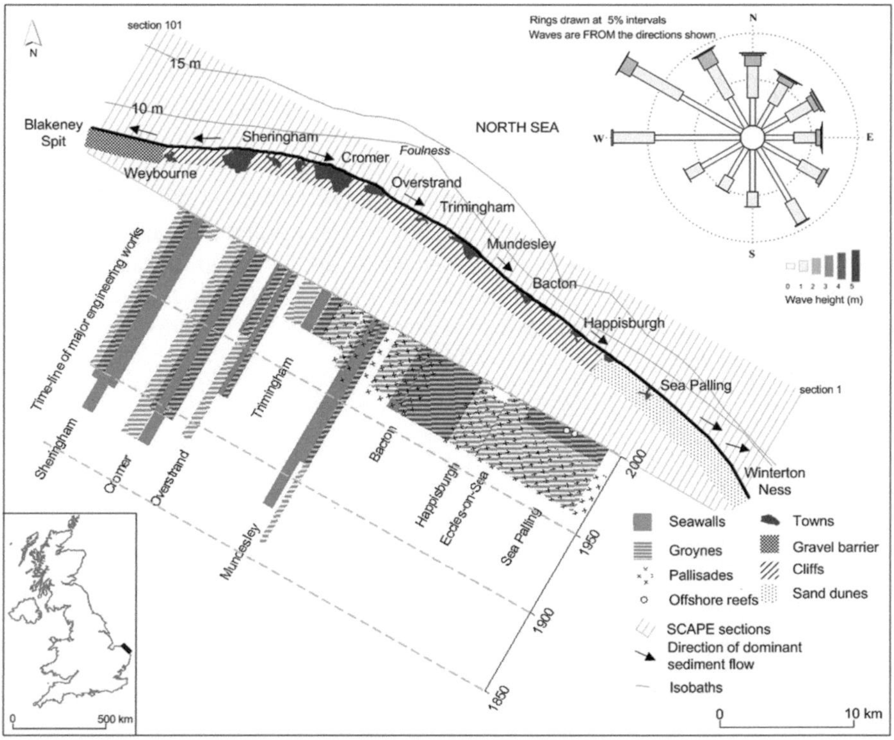

Fig. 7.2 The study area, model grid, offshore wave rose, generalised geomorphology and a chronology of major engineering interventions. The spokes of the wave rose show the proportion of waves (indicated by line length) of different heights (indicated by thickness) (Adapted from Dickson et al. 2007)

7.3 Model Construction

7.3.1 Soft Cliff and Platform Erosion (SCAPE)

The construction of the Norfolk SCAPE model used in this study is described in this section, with an overview of the modelling tool. More extensive descriptions of SCAPE, including its numerical basis, may be found in Walkden and Hall (2005) and Walkden and Dickson (2008). In addition, Walkden and Hall (2011) describe the earlier version of Norfolk model, which was constructed for the Overstrand to Walcott Strategy Study (North Norfolk District Council 2005).

SCAPE describes physical processes but does so in abstract terms. This allows the representation of a 'whole' system and reduces run times. The model contains several linked modules, as illustrated in Fig. 7.3.

Offshore wave transformation is resolved in a separate model (described below and in Chap. 3) which propagates waves inshore to a water depth of approximately 15 m. SCAPE then refracts and shoals these waves to the point at which they begin to break, using linear wave theory. At this stage, profile erosion is calculated (see below) and alongshore transportation of beach sediments is derived using the CERC equation (Shore Protection Manual 1984). The 'one-line' approach (Pelnard-Considere 1956) is used to model the beach morphology. In contrast to most applications of the one-line approach, the lower surface of the beach (the top of the shore platform) is also calculated. We assume that the upper surface of the beach follows a Bruun curve (Bruun 1954) and that its capability to protect the shore platform below it varies with its thickness. Other important parameters in the calculation of erosion include wave height and period, (locally discretised) shore profile slope and a calibrated material resistance term (Walkden and Hall 2005). Talus development

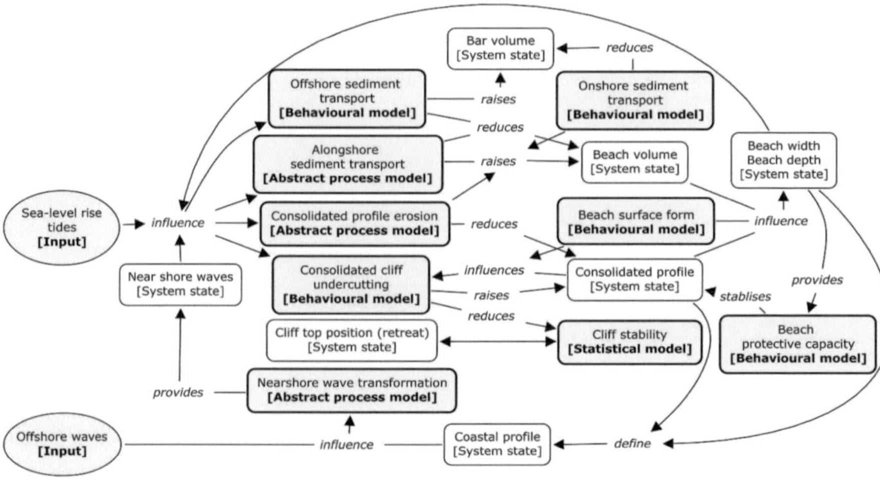

Fig. 7.3 Conceptual flowchart of modules within SCAPE

and removal, offshore sediment transport and offshore sediment storage are described by other modules (Walkden and Hall 2011).

The model integrates processes over individual tidal cycles, and this fixes the model timestep at 12 h, 25 min. At each timestep input conditions were extracted from time series of wave and tidal records. The wave record was hindcast from 23 years of wind data, which was then extended to 1,000 years. A long time series was necessary to allow the model to be run long enough for dynamic equilibrium forms to emerge. A 13-year tide record was also extended to 1,000 years and input to the model, with extreme values resampled from a distribution taken from Dixon and Tawn (1997).

A series of nearshore wave transfer functions were used to account for the effect of offshore topography on waves. These transfer functions were calculated by Kuang and Stansby (2004) (see Chap. 3) using the TOMOWAC model (part of the EDF TELEMAC suite). The wave model domain extended 30 km from the coast with an alongshore extent of 50–100 km, to account for wave propagation from all significant directions (see also Hall et al. 2005). The SCAPE model was run with these transfer functions until dynamic equilibrium profile forms and beach volumes were re-established.

The starting condition of each model section was arbitrarily chosen to be a vertical cliff in deep water. Profiles gradually emerged (see, e.g. Fig. 7.6) through simulation of many thousands of years of erosion. Their shapes ultimately depended on tidal characteristics, the cross-shore distribution of erosion, the size of beach (if present), material strength, rate of sea-level rise and the incident wave conditions. Erosion is controlled by feedback through the form of the beach and platform. This stage of modelling was intended to represent the shore in a natural, un-engineered, state (for fuller descriptions of these stages of model preparation, see Walkden and Hall 2011). The effects of major historical engineering interventions, from the mid-nineteenth century to 2000, were then represented.

Engineering interventions are implemented in SCAPE in ways that perturb the dynamic equilibrium of the model. Seawalls stop cliff toe retreat but do not prevent shore platform lowering, palisades reduce the height of waves passing through them by half, and groynes reduce longshore sediment transport (except when beaches are wider than the structures are long). After simulating the effect of engineering construction, model behaviour was compared with historical recession data measured from Ordnance Survey maps. This chapter includes results associated with two processes of validation (which are described in Dickson et al. 2007; Walkden and Hall 2011). Due to space constraints, only the latter is represented graphically here, in Fig. 7.4. The model performs well over the region of cliffed coast (solid black line), but there is more uncertainty in the boundary regions where unconsolidated coasts are present. Here, rather than cliff toe recession rate, the rate of change of the model still water level (SLW) is shown. The reduction in the quality of fit is seen most strongly in the southern limit of the final epoch; the model shows recession, whereas in reality the coast has been relatively stable. The shoreline stability is due, in large part, to beach nourishment, which has been a management practice here since 1994. This influence on beach position is not represented in this hindcast SCAPE simulation.

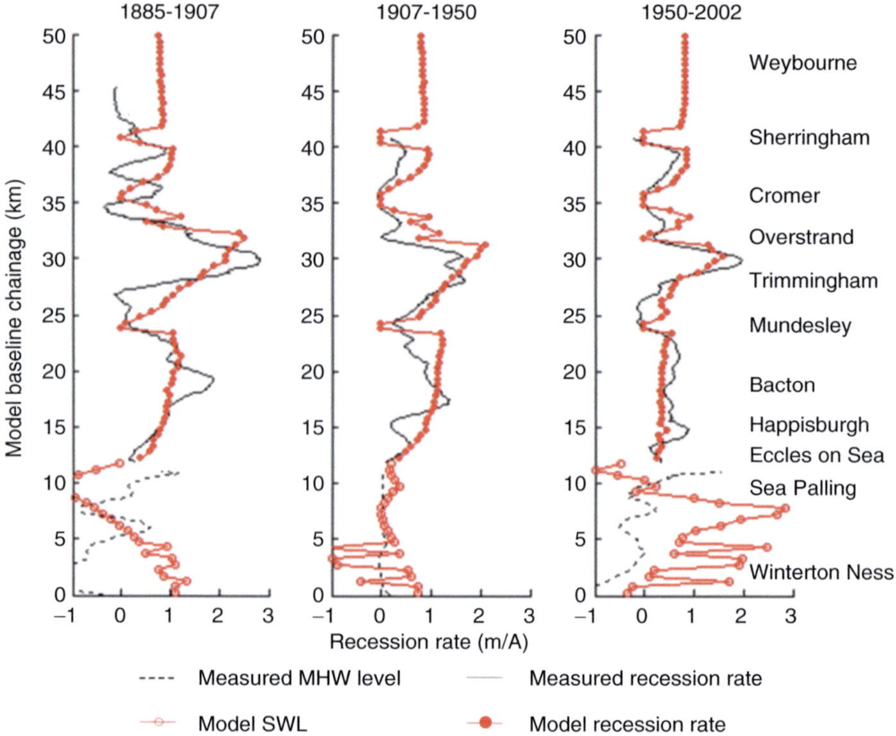

Fig. 7.4 Comparison of SCAPE model recession rates with those measured from historical maps over three eras, 1885–1907, 1907–1950 and 1950–2002 (Adapted from Dickson et al. 2007)

7.3.2 Cliff Stability

The SCAPE simulations predict (amongst other things) the position of the cliff toe. For economic appraisal, predictions are required of when individual clifftop assets will be lost due to cliff failure.

Coastal cliff landsliding occurs as a consequence of both cliff toe recession and geotechnical processes within the cliff slope. Landsliding of these cliffs is preceded by marine removal of material from the cliff toe, resulting in coastal slope steepening. Eventually a landslide occurs that delivers debris to the beach and reduces the coastal slope. The time at which this occurs depends on the rate of material removal and geotechnical processes and cannot be predicted precisely. However, the rate of shoreline retreat from SCAPE can be combined with a geotechnical assessment to generate an approximate probability distribution of the possible clifftop location after failure (Hall et al. 2000). The approach used here is based on the concept of a Cliff Behavioural Unit (CBU), which is a frontage of cliff line which behaves in a broadly uniform way. Within a CBU, the cliff face may be expected to fail when it is steepened to an average angle α_f and will fail to an angle α_s. Neither α_f nor α_s can

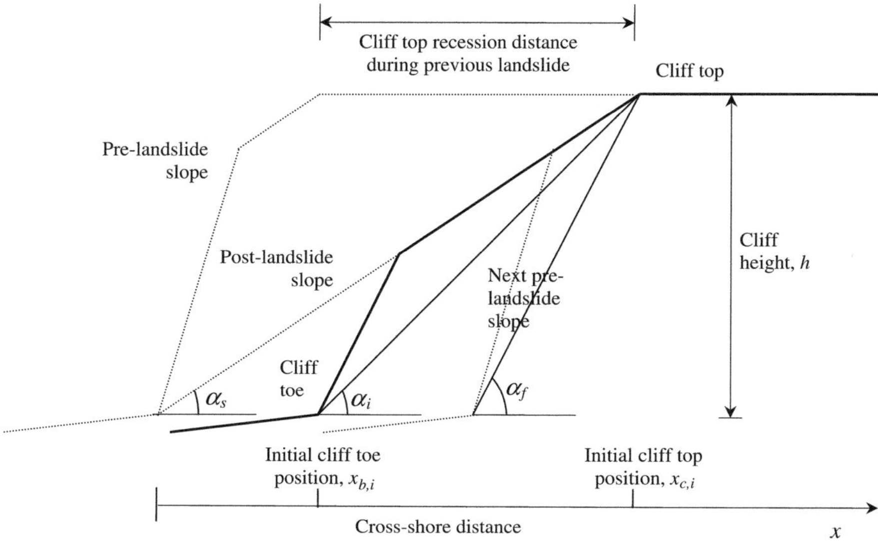

Fig. 7.5 Coastal landsliding model (Modified from Dawson et al. 2009)

be known precisely but will vary with local variations in cliff strength and composition and temporal variations in pore pressure. This uncertainty has been included in the analysis by representing α_f and α_s as normally distributed random variables, with means and variances obtained from a geomorphological assessment of each CBU. This is illustrated diagrammatically in Fig. 7.5.

The initial cliff angle is also represented as a normally distributed random variable, with mean and variance based on observations within each CBU. Further details of how the parameters were derived and the model run can be found in Dawson et al. 2009.

7.3.3 Beach Level

One of the key outputs of the Regional Coastal Simulator is the representation of coastal flood risk across the Norfolk Broads. This depends strongly on whether the seawalls south of Happisburgh remain stable, which, in turn, depends on the levels of the beaches fronting them (see Chap. 9). Although SCAPE simulates the beach surface (see Walkden and Hall 2011) as well as volume, the approach used was felt to be too crude to inform seawall stability assessment. It was therefore improved in the area of the model south of Happisburgh.

A function was required to predict sensitivity of the beach surface to changes in both incident wave conditions and longer-term changes in beach volume. Following work by Dawson et al. (2009), the Vellinga profile (Vellinga 1983) was selected for this purpose. The Vellinga approach describes beach profile shape as a function of offshore wave height. This profile was calculated every timestep and then translated horizontally at each SCAPE section until it encompassed the known beach volume. The point of intersection of the Vellinga profile and the seawall then identified the beach elevation for subsequent seawall stability calculations. Following this approach, low beach volumes do not necessarily lead to depressed beach elevations at the seawall, so long as wave conditions are not too aggressive. However, they do make lower beaches more likely.

The beach-level information was passed to the flood element of the Tyndall Coastal Simulator project (Chap. 9), where it was used for seawall stability assessment. The information was also used, within the SCAPE model, to simulate the management practice of beach nourishment.

7.3.4 Beach Nourishment

Coastal flood risk in the Sea Palling area is managed through beach nourishment, as well as the maintenance of seawalls. Nourishment takes place every few years and is initiated by observation of low beach levels along the seawalls. These are assessed against 'trigger' thresholds at which it has been calculated that seawalls may become unstable. Around 140,000 m^3 is added to the beaches per year, on average.

In representing this process in the SCAPE model, it was not appropriate to simply increase beach volumes at the known historic rates because nourishment practice is expected to adapt in the future. It was therefore necessary to simulate the nourishment *decision process*. The following approach was adopted: (1) At every model section, the average monthly beach level was calculated. (2) If this was found to be lower than the 'trigger' thresholds in this area, then (3) 10 cubic metres was added to that beach section, per metre length, per tide during the following month. This rate was chosen to represent a feasible rate of nourishment. Because nourishment may be triggered during a series of months, the *annual* rate of nourishment depends on beach evolution. This approach simulated the nourishment decision-making process and so allowed this element of the model to respond to changes in the climate and cliff management practice.

7.4 Future Climate Change and Coastal Management

7.4.1 Coastal Management

Five scenarios of coastal management were represented to explore how human interventions may change future coastal evolution and flood risk and also influence the system's sensitivity to climate change. These are described in Chap. 1 and summarised below:

1. Removal of all coast protection structures (i.e. those intended to prevent erosion) in the year 2000
2. Implementation of the policies recommended by the North Norfolk Shoreline Management Plan (North Norfolk District Council 2010)
3. Maintenance of all existing structures
4. Construction of seawalls along the entire coast
5. Implementation of Scenario 2, without beach nourishment in the Sea Palling area

Scenario 2 is the most realistic of the set. Scenario 5 was intended to reveal the value of beach nourishment in reducing flood risk. Scenario 3 was included with the intention of informing public debate around the consequences of maintaining historic defence policies. Scenarios 1 and 4 were included to explore gross changes in coastal management.

7.4.2 Climate Change

The following three aspects of coastal climate change were considered for representation: wave activity, surge levels and rates of relative sea-level rise. The Tyndall Coastal Simulator project progressed in conjunction with the derivation of the UK Climate Impacts Programme's 2009 projections (UKCP09, see Chap. 2). The UKCP09 projections correspond to specific SRES (Special Report on Emissions Scenarios) scenarios provided by the Intergovernmental Panel on Climate Change.

Results generated within the Tyndall project showed negligible for future (climate-driven) change in surge magnitude or wave height off the Norfolk coast (see Chap. 2). Consequently these elements of climate were assumed to be stationary (i.e. to neither grow nor decay). UKCP09 (Lowe et al. 2009) gives four projections of sea-level rise, representing conditions under alternative scenarios of greenhouse gas emissions and climate sensitivity. These are termed (in order of increasing magnitude) 'low' (SRES scenario B1), 'medium' (A1B), 'high' (A1F1) and 'H++'. The last of these was interpreted from the IPCC fourth Assessment Report in response to lack of scientific understanding of the behaviour of ice sheet failure. Each is given as a range. The ranges of the lower three represent, nominally, the 5th and 95th percentile, which are equidistant from a central projection. The lower limit of the H++ range is based on the maximum global mean sea-level rise value given by the IPCC Fourth Assessment Report. The top of the H++ range is based on indirect observations of sea-level rise during the last interglacial. This upper limit is believed to be highly unlikely and was provided to inform contingency planning.

All four projections (shown in Fig. 7.6) were used to drive the Tyndall Coastal Simulator. To do this, it was necessary to make two assumptions, which were as follows: (1) all projections are normally distributed and (2) the H++ trend is similar in form to the other three projections. The first of these assumptions seems reasonable given the central position of the average estimates of sea-level rise within UKCP09.

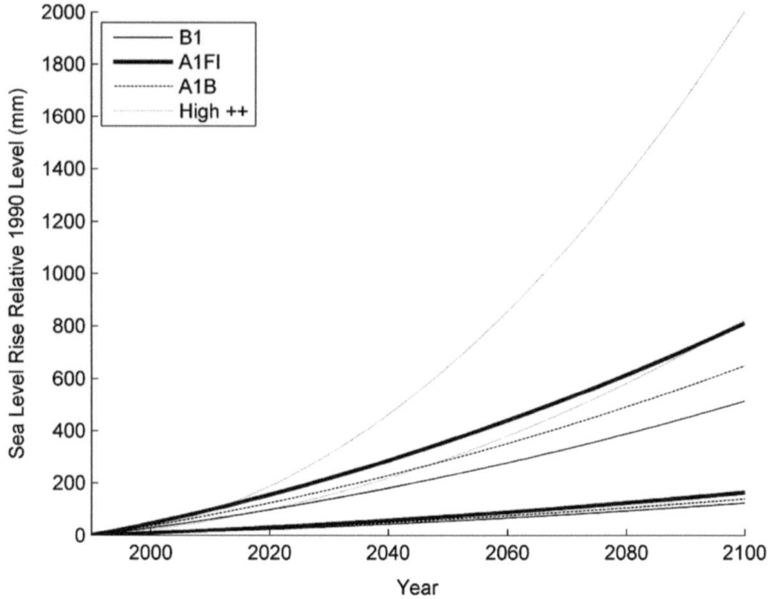

Fig. 7.6 Projected sea-level rise around North Norfolk, following UKCP09 projections (Lowe et al. 2009)

7.4.3 Monte Carlo Simulations

The five management scenarios were combined with the four climate change (essentially sea-level rise) projections to provide a set of twenty general scenarios. These were modelled using the Monte Carlo approach, involving 250 simulations per condition. A total of 5,000 SCAPE simulations were therefore run. In each simulation, the following three datasets were sampled from their assumed distributions: (1) wave sequence, (2) sea-level rise trajectory and (3) year of failure of each structure.

7.5 Results

The 'downstream' elements of the Regional Coastal Simulator required quantitative projections of clifftop position (for erosion risk assessment) and beach levels to allow seawall stability calculations for inundation modelling. Practitioners estimate such quantitative information after first building qualitative knowledge of coastal behaviour. Qualitative understanding is normally brought to a project through expert interpretation, in which insights gained at other places, or from historic records, are extrapolated to the problem in hand. This may be supported by quantitative methods, but these tend to be very broad-brush in nature, such as sediment budgeting or crenulate bay theory.

The development of quantitative and qualitative information was much more entwined in the Norfolk SCAPE simulations. This is important, not least because it allows deeper model validation. Model confidence is built not only on its ability to replicate changes in shoreline position (as in Fig. 7.4) but also to mimic a broader set of coastal behaviours. The results shown below demonstrate some of these behaviours, as well as examples of the final quantitative projections required for flood and erosion risk quantification.

At the largest scales, the model captured the emergence of shore profile form and shore planshape. The profile shape was formed in the model from an arbitrary starting condition (a vertical cliff in deep water) and was reshaped every tide over thousands of years, until it converged on a form in dynamic equilibrium (see Fig. 7.7). The initial planshape could not be arbitrary. Instead this was assumed to be similar to a smoothed (and displaced) version of the current coastal planform. Given this form the model was left to run for many hundreds of years to stabilise beach volumes, shore profile shapes and alongshore planshape. These three are intimately related and had to be allowed to stabilise their interrelationships.

The general planshape evolution mimicked that seen at Norfolk; the cliffs eroded, and the sediment produced accumulated in the Happisburgh to Winterton area, generally straightening the coast (for more information on this model behaviour, see Figure 4.1 of Walkden and Rossington 2009). The shore platform profile became

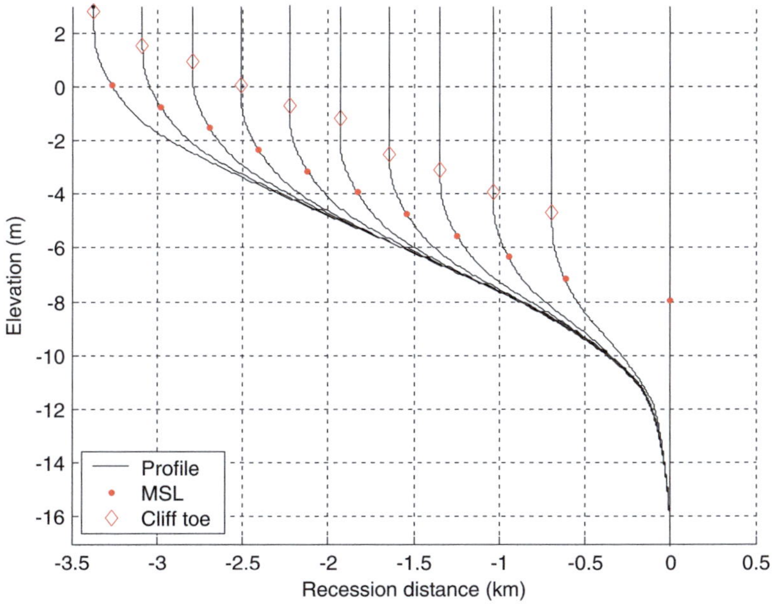

Fig. 7.7 Development of a shore profile over 4,000 years (assuming sea-level rise at 2 mm/year)

deeper in response to the beach (see Figure 15 of Walkden and Hall 2011). The beach volumes initially became very high and then decayed to dynamic stability, as illustrated in the 1,000-year records shown in Fig. 7.8.

This long-term stability is a key property of this model. It reflects dynamic balance between the key elements of the geomorphic system: (retreat of) the cliff toe, (release of) cliff material (transport of) beach sediments and (relative lowering/raising of) shore platform level. Having simulated this general large-scale balance, it was then possible to explore how it might be affected by changes.

Interesting insights came from introducing artificial internal imbalance, through sensitivity testing of the two calibration coefficients: the rock strength and longshore transport scale (*R* and *K* respectively; see Fig. 7.9, which is reproduced in modified form from Walkden and Hall 2011). Two distinct modes of behaviour were found. In the northern end of the model, the shore recession was limited by rock strength and governed by fluctuations in profile shape. In the south of the model, shore recession was governed by the ability of the model to lower beach levels by transporting sediment away. These different modes of behaviour meant that coastal erosion was quite insensitive to large changes in rock strength in the south (see the left side of the upper panel of Fig. 7.9) and relatively insensitive to changes in sediment transport rate in the north (indicated in the right side of the lower panel of the same figure).

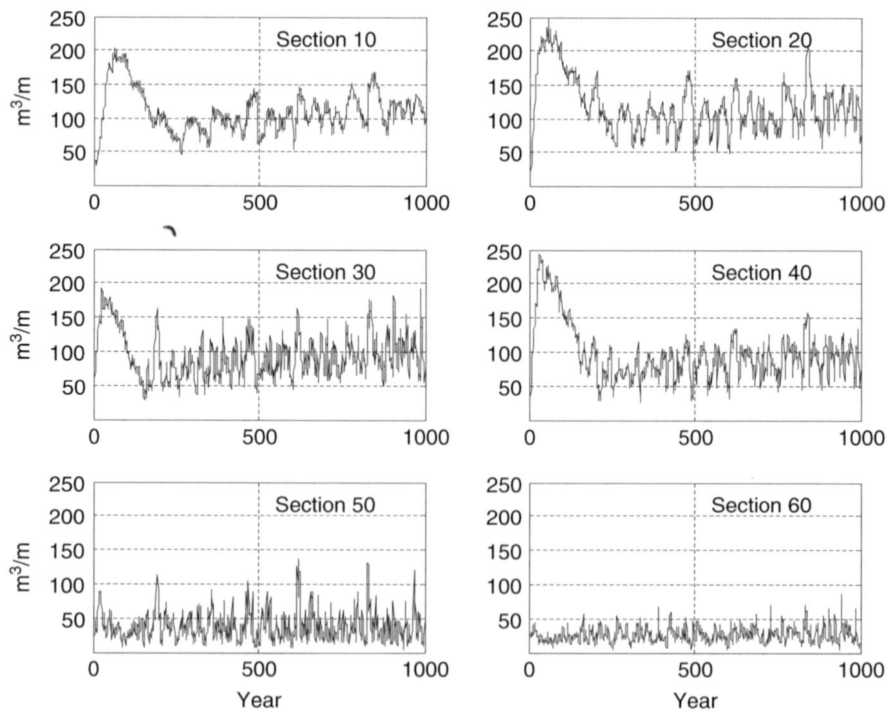

Fig. 7.8 Yearly averaged beach volumes at 6 sections over 1,000 years

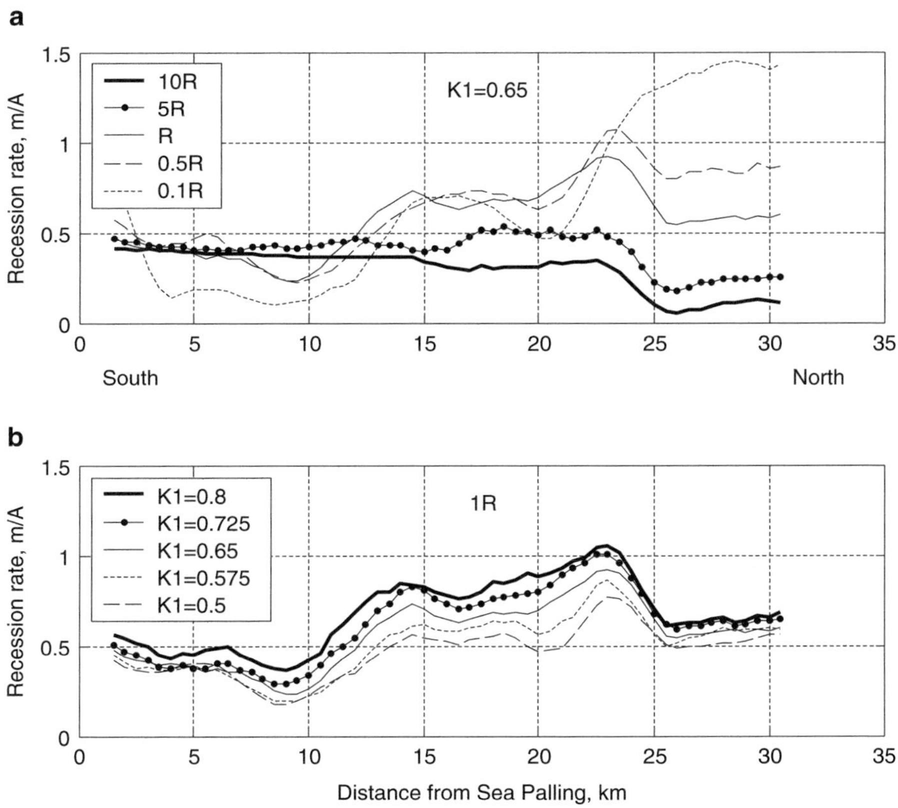

Fig. 7.9 Sensitivity of erosion rates to: (**a**) variations in the sediment transport coefficient (K1) and (**b**) rock strength coefficient R (Adapted from Walkden and Hal 2011)

Closely related behaviour appears to offer natural protection against accelerated sea-level rise for the most southern part of this coast. Figure 7.10 shows shoreline behaviour under two different rates of sea-level rise under the (unreal) scenario of the complete absence of structures. In the north of the model, higher sea-level rise leads to higher recession. In the south, this relationship becomes less clear and even reverses. This behaviour results from a high influx of sediment from the more rapidly eroding cliffs in the north, which is transported south where it increases beach volume. The larger beaches provide more protection.

Any management intervention directly influences at least one of the elements of the coastal system; and in doing so it indirectly influences the others. For example, the local effect of building a seawall is illustrated with SCAPE model output in Fig. 7.11. This shows an example of the development of the upper profile (below the beach) during a period within which a seawall is constructed. The data represents the Mundesley area, where a seawall has existed since 1910. The shore profile is pushed into a state of imbalance in which the junction between the platform and cliff is artificially lowered at the structure. This direct effect also influences (1) sediment release and therefore beach volume, (2) the elevation of the underside of the beach

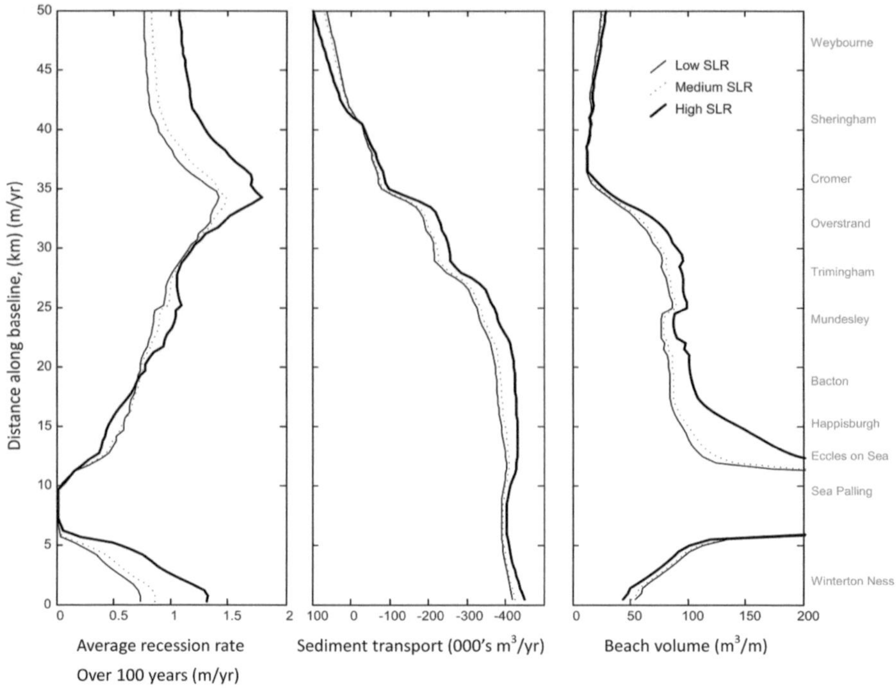

Fig. 7.10 Predicted sensitivity of an undeveloped coast to accelerated sea-level rise (Adapted from Dickson et al. 2007)

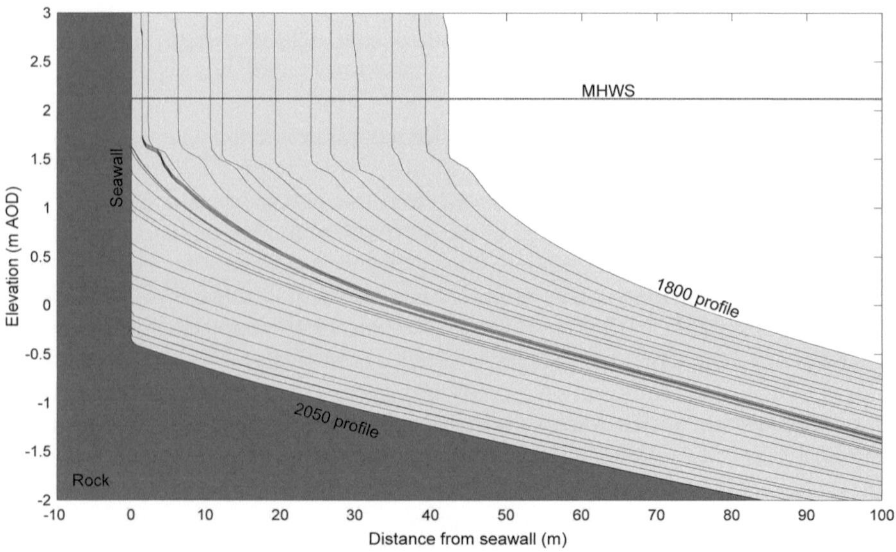

Fig. 7.11 Example evolution of the model cliff toe and upper platform profile at Mundesley (1800–2050). The seawall was constructed in 1910 (Adapted from Walkden and Hall 2011)

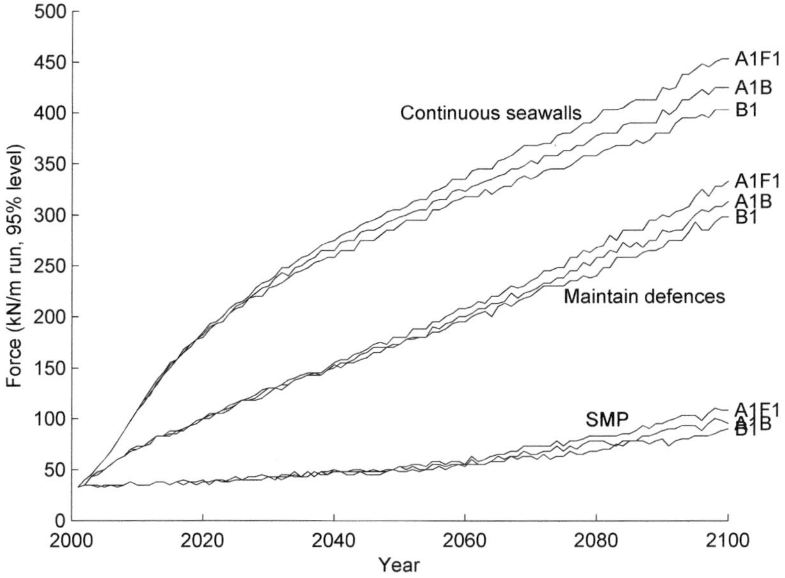

Fig. 7.12 Impact forces on the Cromer seawall throughout the twenty-first century, under three management scenarios and three UKCP09 sea-level rise projections (upper 95 % levels derived from 2,250 simulations)

and therefore beach surface level and (3) cliff planshape (assuming neighbouring areas are not protected) and also, therefore, alongshore sediment transport.

Such change also, of course, leads to structure undermining and growth of overtopping and wave loading. The last phase of model simulations included calculations of wave impact loading, using the method of Blackmore and Hewson (1984), and example (probabilistic) results are shown in Fig. 7.12.

This figure shows projected impact forces (and by implication uprush and vovertopping) growing significantly and being much more sensitive to management scenario than to sea-level rise. Although future impacts almost double under the SMP management scenario, maintaining more structures worsens the situation by up to an order of magnitude. The low sensitivity to sea-level scenario is largely because beaches tend to rise with the sea level. Maintaining or increasing the number of defences means that there is less sand available to support this mitigating process. Reduced sand volumes also allow more platform downwearing, which further lowers the beach, increasing the impact forces. Such changes imply a shortening of structure lifetime and a necessary increase in the size of their replacements. In addition the lowering of the beach and platform tends to narrow the intertidal zone, reducing the beach aesthetic and amenity value. Such trends are common along UK coasts and represent a growing problem.

The structures along the cliffed coast have a strong influence on the release and transport of beach sediments and therefore the volumes and levels of beaches along the flood-vulnerable frontages between Ostend and Winterton. Figure 7.13 illustrates

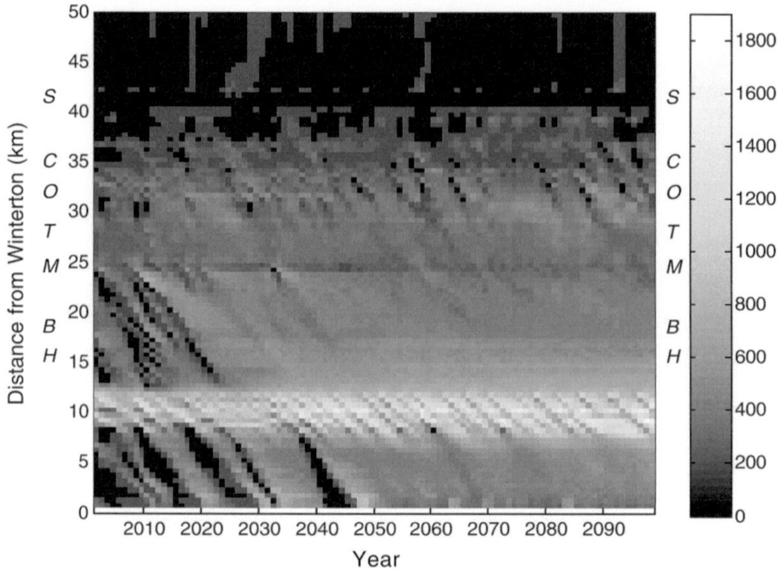

Fig. 7.13 Beach volumes over the twenty-first century, showing crests/troughs of waves in sand volume propagating southward from the Mundesley area (indicated by the letter '*M*'); the influence of the Sea Palling offshore reefs can be seen as a horizontal (i.e. stationary) band of high beach volume (*light shading*); units are m^3/m (Adapted from Dawson et al. 2009)

the evolution of beach volumes throughout the model under a single simulation, over the twenty-first century.

There is a general increase in beach volume with increasing closeness to Winterton. Volumes are particularly large around the shore-parallel reefs of Sea Palling. Diagonal bands of low/high volume can be seen south of Mundesley, particularly before the year 2050. These show southward movement of sand waves at an approximate rate of 0.8 km/year. These are promoted, in the model, by the presence of engineering structures. This means that the spatiotemporal distribution of beach volume (and therefore economic risk) is more complex under scenarios with more engineering interventions. There is anecdotal evidence of the existence of sand waves of similar speed and scale travelling south from Mundesley, and this provides some confidence that these features are not merely artefacts of the model. They have not been deeply explored, and so this model behaviour should be viewed with caution, but it is worth noting that these sand waves are coupled to cycles of erosion in a manner similar to that observed by Pringle (1985) at the Holderness coast.

These large-scale alongshore fluctuations in beach level have an important influence on coastal flood risk between Cart Gap and Winterton Ness. They are also, of course, superimposed on higher-frequency fluctuations associated with cross-shore processes. Storm events move beach material seaward, temporarily reducing the stability of seawalls and increasing the probability of their failure. The flood risk calcula-

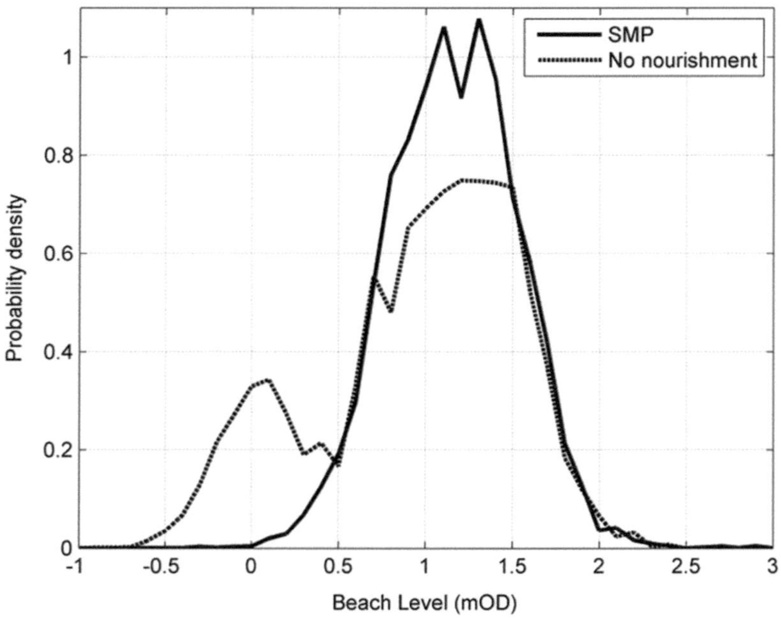

Fig. 7.14 Example probability density functions representing beach level at one model section, in a particular year, under management scenarios 2 ('SMP') and 5 ('SMP without nourishment')

tions within the Tyndall Coastal Simulator were performed on an annual basis and were linked to the morphological simulations via structure failure functions. These described the likelihood of sea defence overturning as a function of beach level. Consequently the morphological simulations were set to output beach levels as annual probability density functions, and two examples are shown in Fig. 7.14.

Figure 7.14 illustrates the influence of beach nourishment, without which a distinct peak in probability emerges at lower beach levels, associated with post-storm conditions. An equivalent pdf was calculated for every model section within the flood-vulnerable area for every year, under each management scenario. The annual probability of failure of each sea defence was then found by multiplying these probability density functions to the sea defence failure functions (see Chap. 9).

Cliff erosion was also calculated in probabilistic terms. Each of the SCAPE simulations projected the future recession of the bases of the cliffs, and for each of these, a set of 50 simulations of cliff failure were performed. This provided a set of 12,500 projections of clifftop position for each location, under each management and climate change scenario. These were used to derive probability density functions (pdf) of clifftop position. Fig. 7.15 shows two example results from these pdf. Only the range of clifftop positions is shown (from the 5 to 95 % level), relative to the cliff toe in the year 2000.

No recession is shown in areas less than around 12 km of Winterton Ness (the left hand side of the figures) because here there is no cliff. It can be seen that under the SMP2 management policy, the clifftop is close to the year 2000 cliff base position

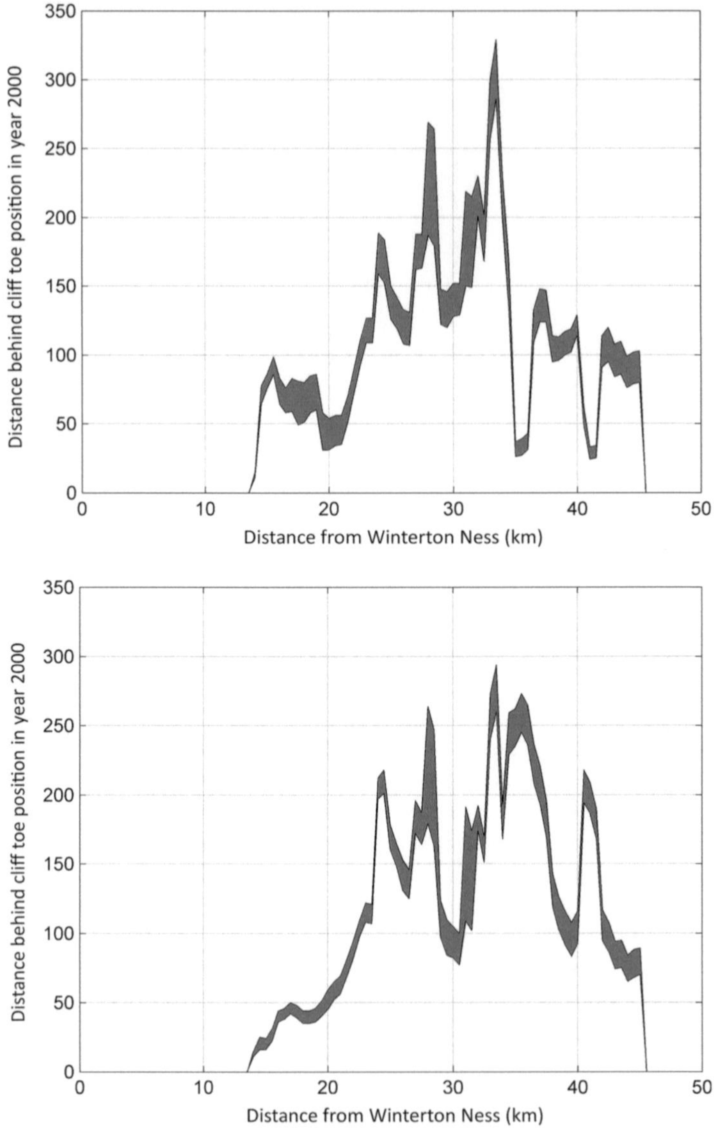

Fig. 7.15 Example projections of clifftop position (relative to the cliff toe) during the 2,190 s, under management scenarios 'SMP2' (*top*) and 'defences removed' (*bottom*). The bandwidth represents the central 90 % of projections

at locations around Cromer (36 km) and Sheringham (41 km). In these locations, under this scenario, cliff protection structures are maintained, preventing cliff toe retreat and minimising cliff destabilisation. The clifftop position is not, however, certain (i.e. a range of possible positions is shown) because the cliff face has some potential to fail, even though its base is not retreating. The highest recession in the two figures is seen under the SMP2 scenario, at around 33 km. This is immediately to the southeast of Cromer and 'downdrift' of it. The high recession results from disturbance of the beach sediment supply and transport caused by the 'updrift' defences. The removal of all structures results in generally more even clifftop recession across the frontage. There are still large variations, however, as can be seen at Sheringham (42 km), Cromer (36 km) and Mundesley (28 km), where defences are allowed to fail, and these local areas exhibit accelerated, though transient, recession rates. These localised higher recession rates bring some benefit to downdrift areas. The retreat distances from around 12 to 20 km of Winterton Ness are all significantly lower and more certain under the 'defences removed' scenario. The benefits of sediment release are particularly strong for the low-lying and flood-vulnerable land close to Winterton Ness. Here artificial nourishment is used to manage the flood risk. The greater release of sediment with fewer defences reduces the need for artificial nourishment.

Figures 7.14 and 7.15 show projections of nourishment rates for the flood-vulnerable frontage south of Happisburgh, under the lowest and highest emission scenarios. These are average values from the Monte Carlo probabilistic simulations. The most similar scenario to the current climate and management policy is 'low emissions' 'maintain defences', the results of which are in Fig. 7.16. It can be seen that the projected nourishment rate is quite steady at around 140,000 m^3/year. This compares very well to the current understanding of the nourishment requirement, which the Environment Agency (2002) reports as 150,000 m^3/year.

As expected, projected nourishment rates are highly sensitive to coast protection policy. The removal of all defences results in the least nourishment, almost nothing

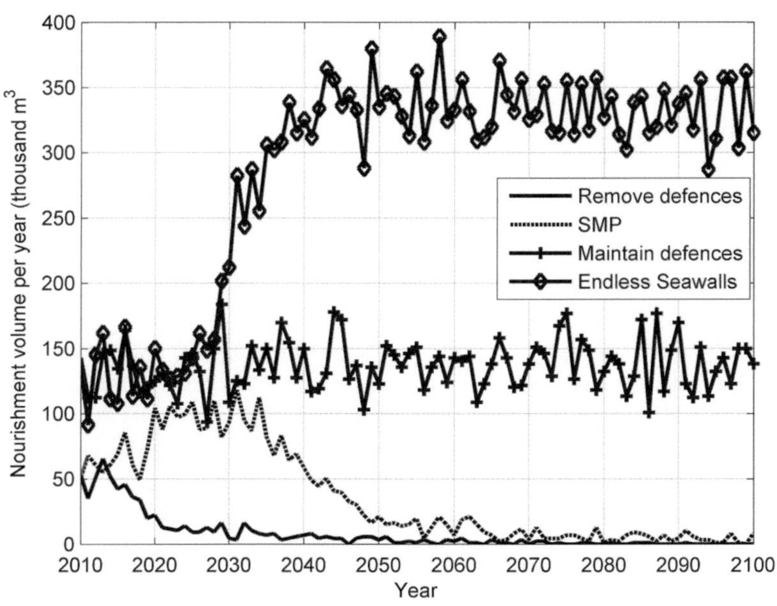

Fig. 7.16 Projected beach nourishment volumes per year, emission scenario 1 ('low')

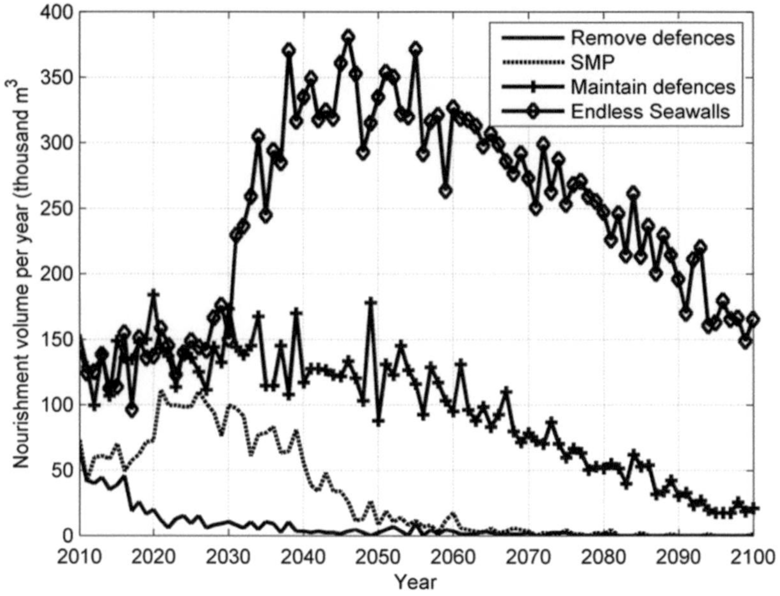

Fig. 7.17 Projected beach nourishment volumes per year, emissions scenario 4 ('high ++')

after 2050. At the other extreme, the construction of continuous seawalls necessitates the highest rates of up to around 330,000 m^3/year. The necessary nourishment stabilises near this level because here it is replacing the entire natural flux of beach sediment from the cliff.

The nourishment projections were relatively insensitive to sea-level rise. However, as can be seen in Fig. 7.17, the highest rate of sea-level rise influences the average nourishment rates after around 2050. The effect of higher water levels is, perhaps surprisingly, to reduce the required rate, presumably due to beach levels rising with the mean sea level.

7.6 Conclusion and Next Steps

The simulations of the shore and cliffs of North Norfolk gave new insights into the behaviour of this coast and others like it, and their projections continue to be used to inform the development of coastal management policies in the region.

The SCAPE model simulated more than a century of known historic coastal recession and also reproduced unexpected behaviours, such as the formation and migration of large-scale littoral waves in beach volume. There is some observational evidence to support this prediction. Other processes, such as transient accelerations in recession following structure failure, were simulated plausibly but could not be tested due to lack of observational data.

Much of the more interesting model behaviour was tied to large-scale processes. This regional-scale simulation capacity was possible because the model was constructed to represent the interactions of the shore platform, cliff and beach. Importantly these allowed quantification of the alongshore benefits/costs of coastal retreat/protection. These included reductions in recession rate due to updrift erosion and, at a larger scale, quantification of the trade-off between erosion risk and coastal flood risk (see Chaps. 8 and 9).

The sensitivity of coastal erosion rate to increases in sea-level rise was found to be less than predictions made using the (modified) Bruun rule (Dickson et al. 2007). This lower sensitivity arises from the model's ability to represent changes in the profile form as the rate of sea-level rise changes and the associated hysteresis that this transition introduces. In contrast the Bruun methodology includes assumptions of profile conservation and instantaneous response. Further encouraging indications into the response of the Norfolk Coast (and similar areas) to accelerated sea-level rise arose from large-scale alongshore interactions. Simulations indicated that, under certain management policies, higher sea-level rise rates may increase beach volumes in downdrift areas and in doing so reduce artificial nourishment rates, coastal erosion, seawall failure probability and flood risk.

SCAPE used wave conditions from the TOMOWAC models (Chap. 3) and provided beach volumes and clifftop positions for use in the flood and erosion risk analysis (Chaps. 8 and 9). Coupling SCAPE and the cliff stability analysis was straightforward because the former operated at a higher resolution than the latter, which only required an annual cliff toe position. The probabilistic output of the cliff stability analysis passed readily into the subsequent erosion risk analysis (see Chap. 9). Providing output appropriate for the coastal flood risk simulations was more challenging, requiring tailored development of the response of the beach profile to wave action. In addition the beach level at the sea defence structures had to be quantified probabilistically. This was necessary to drive the structure fragility functions that formed the interface with the flood risk element of the Tyndall Coastal Simulator (see Chap. 9).

This work could be extended in several alternative directions, which include (1) coupling a regional SCAPE model with a model of nearshore and offshore sediment transport; (2) exploration of the influence of varying geology on the development of the form the shore; (3) exploring the cliff-platform-beach interactions of other coastline forms, such as embayments, or the roots of spits and barrier beaches; and (4) nesting higher-resolution SCAPE models within the regional model to explore more detailed response to management changes. Elements of (1), (2) and (3) are being taken forward in the iCOASST Project which is considering Suffolk and Liverpool Bay (Nicholls et al. 2012). Perhaps the most important scientific message from this element of the Tyndall Coastal Simulator is, however, that modelling focussed on the representation of a 'whole' system, accepting physical descriptions of reduced detail, can provide compelling and useful long-term projections of coastal geomorphology.

From the perspective of the coastal practitioner, the most striking elements of this modelling are its scale and the capture of the coast's response to management interventions and sea-level rise. The Norfolk SCAPE models have already proved themselves useful in contributing to real coastal management. The first iteration of the model was a successful element of a coastal strategy study (and was built under typical practitioner programme and financial constraints). Its results also informed the region's subsequent Shoreline Management Plan review. This work continues to be applied to the practicalities of coastal management in Norfolk. It has, for example, been used to quantify cliff recession and sediment flux between Cromer and Winterton Ness as part of a recent Coastal Management Study (North Norfolk District Council 2013).

References

Ashton, A., Walkden, M., & Dickson, M. (2011). Equilibrium responses of cliffed coasts to changes in the rate of sea-level rise. *Marine Geology, 284*, 217–229.

Blackmore, P., & Hewson, P. (1984). Experiments on full scale wave impact pressures. *Coastal Engineering, 8*, 331–346.

Bruun, P. (1954). *Coast erosion and development of beach profiles*. In: Technical Memorandum No. 44, Beach Erosion Board, US Army Corps of Engineers USA.

Dawson, R., Dickson, M., Nicholls, R., Hall, J., Walkden, M., Stansby, P., Mokrech, M., Richards, J., Zhou, J., Milligan, J., Jordan, A., Pearson, S., Rees, J., Bates, P., Koukoulas, S., & Watkinson, A. (2009). Integrated analysis of risks of coastal flooding and cliff erosion under scenarios of long term change. *Climatic Change, 95*, 249–288. doi:10.1007/s10584-008-9532-8.

Dickson, M. E., Walkden, M. J. A., & Hall, J. W. (2007). Systemic impacts of climate change on an eroding coastal region over the twenty-first century. *Climatic Change, 84*(2), 141–166.

Dixon, M. J., & Tawn, J. A. (1997). Estimates of extreme sea conditions. *Proudman Oceanographic Laboratory Report, 112*, 217.

Environment Agency. (2002). *Strategy review: Happisburgh to Winterton sea defences: Stage three*. Peterborough: Environment Agency Anglian Region.

Hall, J. W., Lee, E. M., & Meadowcroft, I. C. (2000). Risk-based benefit assessment of coastal cliff protection. *Proceedings Institution of Civil Engineers Water Maritime and Energy, 142*, 127–139.

Hall, J. W., Dawson, R. J., Walkden, M. J. A., Dickson, M. E., Stansby, P. S., Zhou, J., Brown, I., Watkinson, A., & Nicholls, R. J. (2005). Broad-scale analysis of morphological and climate impacts on coastal flood risk. In *Proceedings of international conference on coastal dynamics*, Barcelona.

Kuang, C. P., & Stansby, P. S. (2004). Modelling directional random wave propagation inshore. *Proceedings Institution of Civil Engineers Water Maritime and Energy, 157*, 123–131.

Lowe, J. A., Howard, T. P., Pardaens, A., Tinker, J., Holt, J., Wakelin, S., Milne, G., Leake, J., Wolf, J., Horsburgh, K., Reeder, T., Jenkins, G., Ridley, J., Dye, S., & Bradley, S. (2009). *UK Climate Projections science report: Marine and coastal projections*. Exeter: Met Office Hadley Centre.

Nicholls, R., Bradbury, A., Burningham, H., Dix, J., Ellis, M., French, J., Hall, J., Karunarathna, H., Lawn, J., Pan, S., Reeve, D., Rogers, B., Souza, A., Stansby, P., Sutherland, J., Tarrant, O., Walkden, M., & Whitehouse, R. (2012). iCOASST – Integrating coastal sediment systems. In P. Lynett, J. McKee Smith (Eds.), *Proceedings of the 33rd international conference on coastal engineering 2012*, Reston, US. Coastal Engineering Proceedings, *1*(33):sediment.100. doi:10.9753/icce.v33.sediment.100.

North Norfolk District Council. (2005). *Overstrand to Walcott strategy study*, report EX 4692 (100 pages), Cromer, UK.

North Norfolk District Council. (2010). *Kelling Hard to Lowestoft Ness Shoreline Management Plan Review*.

North Norfolk District Council. (2013). *Cromer to Winterton Ness coastal management study*, study report, July, Cromer, UK. http://www.northnorfolk.org/files/C2WS_Cromer_to_Winterton_Ness_Study_Report.pdf

Pelnard-Considere, R. (1956). Essai de theorie de l'evolution des forms de ravage en plage de sable et de galets. *4th Journees de l'Hydraulique, Les Energies de la Mer Question, III*, 792–808.

Pringle, A. W. (1985). Holderness coast erosion and the significance of ords. *Earth Surface Processes and Landforms, 10*(2), 107–124.

Royal Haskoning. (2012). *Wash East coastal management strategy*. Prepared for the Environment Agency, Peterborough, UK

Shore Protection Manual. (1984). *U.S. Army Engineer Waterways Experiment Station* (4th ed., Vol. 2). Washington, DC: US Government Printing Office.

Vellinga, P. (1983). *Verification of predictive computational model for beach and dune erosion during storm surges*. Delft: Delft Hydraulics Laboratory.

Walkden, M., & Dickson, M. (2008). Equilibrium erosion of soft rock shores with a shallow or absent beach under increased sea-level rise. *Marine Geology, 251*(1–2), 75–84.

Walkden, M. J. A., & Hall, J. W. (2005). A predictive mesoscale model of the erosion and profile development of soft rock shores. *Coastal Engineering, 52*(6), 535–563.

Walkden, M., & Hall, J. (2011). A mesoscale predictive model of the evolution and management of a soft-rock coast. *Journal of Coastal Research, 27*(3), 529–543.

Walkden, M., & Rossington, K. (2009). *Characterisation and prediction of large scale, long-term change of coastal geomorphological behaviours: Proof of concept modelling Joint Defra/Environment Agency Flood and Coastal erosion Risk Management R&D Programme*. Science report: SC060074/PR1. Bristol: Horizon House. ISBN: 978-1-84911-088-4.

Chapter 8
Broadscale Coastal Inundation Modelling

Xing Zheng Wu, Jim W. Hall, Qiuhua Liang, and Richard J. Dawson

Abstract To understand the implications of changes to marine climates and the impact on people, analysis is required of the spatial extent and depths of flooding in urban and rural areas. Over extended timescales, the uncertainties related to possible sea-level rise and changes in storminess increase significantly. As such, there is no certainty regarding when and if a major storm or breach will occur. The increasing availability of some details of changing wave climate (described in Chaps. 2 and 3) and full range of beach evolutionary behaviours (given in Chap. 7) offer an opportunity to improve the implications of these uncertainties on flood extents.

Uncertainties in beach level in front of flood defences, future sea level, extremes of wave height and water level and breach location(s) require that a wide range of current and potential future conditions are considered to fully understand flood risk. It is therefore crucial to capture and model a wide range of possible future inundation events. We simulated shallow flow hydrodynamics and validated this with a comparison of the recorded inundation extent from a flood event in 1938. Subsequently, with 1,344 simulations for each breaching scenario, we analyzed the full range of hydraulic loadings, beach morphologies, coastal defence structural responses (overtopping and breaching) and coastal management policy in order to understand the full range of uncertainties that influence long-term changes in flood impacts.

X.Z. Wu (✉)
Department of Applied Mathematics, School of Applied Science, University of Science and Technology Beijing, 30 Xueyuan Road, Haidan District, Beijing 100083, P. R. China
e-mail: xingzhengwu@gmail.com

J.W. Hall
Environmental Change Institute, Oxford University Centre for the Environment, University of Oxford, South Parks Road, Oxford OX1 3QY, UK
e-mail: jim.hall@eci.ox.ac.uk

Q. Liang
School of Civil Engineering and Geosciences, University of Newcastle, Room 3.03: Cassie Building, Newcastle upon Tyne NE1 7RU, UK
e-mail: qiuhua.liang@ncl.ac.uk

R.J. Dawson
School of Civil Engineering and Geosciences, Tyndall Centre for Climate Change Research, Newcastle University, Newcastle upon Tyne NE1 7RU, UK
e-mail: richard.dawson@newcastle.ac.uk

R.J. Nicholls et al. (eds.), *Broad Scale Coastal Simulation*, Advances in Global Change Research 49, DOI 10.1007/978-94-007-5258-0_8

Initially, LISFLOOD-FP was used to simulate inundation over the full extent of the floodplain. Later in the case study development, we benefitted from increased desktop computing power and were able to use the more advanced NewChan model for the more detailed analysis of flooding in the Norfolk Broads. Flood depth and extent in the Norfolk Broads are dominated by extreme water levels and to a lesser extent wave heights. Rising sea levels will notably increase the area at risk of flooding, posing substantial challenges for coastal managers in this region. The results of these simulations are integrated with information on damages to quantify flood risk in Chap. 9.

Keywords Inundation modelling • Flood depth and extent • Overtopping • Defence breach • Sea-level rise

8.1 Introduction

In the event of wave overtopping or structural failure of a flood defence, computationally demanding inundation modelling is necessary to obtain realistic estimates of flood damages and other impacts. The size of these flood impacts will be sensitive to the location of a flood defence breach and the magnitude of the storm surge and wave heights. The sensitivity of flood depths and extent, and hence flood damages, to a range of variables requires that a large number of inundation simulations be performed in order to fully understand the profile of flood impacts.

This chapter highlights the importance of inundation modelling for investigating flood defence breach and overtopping flows and the subsequent flood routing within the floodplain. Flooding is considered from two mechanisms: overtopping of defences or where a breach in the defences forms due to drawdown of the beach in front of the defences. The discharge through the breach is estimated taking into account detailed information on wave, water level and beach level condition. Discharge due to overtopping can be added into breach events, though the volume of overtopping is typically negligible compared to the volume flowing through the breach. Over extended timescales, the sea level, extreme events, beach morphology and other driving conditions vary considerably. To evaluate longer-term trends in inundation probability (and ultimately flood risk, see Chap. 9) requires analysis of a large number of flood simulations. Thus, part of this chapter considers the implications of different variables on flooding.

This chapter starts with consideration of approaches to coastal inundation modelling, with a focus on the two approaches used in the Tyndall Coastal Simulator. This is followed by a description of boundary conditions for inundation models and further details on how the model(s) was set up for the North Norfolk case study site. A number of results are subsequently presented for a range of breach and loading scenarios. These highlight the need for considering a wide range of inundation conditions and therefore an efficient inundation algorithm and flood risk calculation as described in Chap. 9.

8.2 Approaches to Coastal Inundation Modelling

Simulation of inundation over low-gradient floodplains with dike structures requires at least a two-dimensional hydrodynamic modelling approach with relatively high spatial resolution to represent the complex geometry of the floodplain. For example, a 'bath-tub' model has been shown to be extremely inaccurate for such a low-lying and shallow floodplain (Bates et al. 2005).

The Norfolk application of the Tyndall Coastal Simulator used a two-dimensional diffusive wave model (or storage cell model) (Horritt and Bates 2001) and a shallow water equation model (Causon et al. 2000; Liang et al. 2004). These simulations require the identification of boundary conditions (surge, wave heights, breach location, etc.), subsequent calculation of inflow discharges and inundation simulation. Flood wave propagation is simulated by solving the continuity and momentum equations discretized over a grid. A diffusive wave model may be discretized using a finite difference technique and an explicit scheme for the time dependence, which allows the model to represent 2D dynamic flow fields on the floodplain. The resulting flow between two grids is simply a function of the free surface height difference between those grids by definition of their volumetric flow rates, as reported by Bates et al. (2005). Alternatively, the differential equation may be recast in integral form with the solution variables stored at cell centres as integral averages over each cell area. The spatial discretization then involves flux balances across each cell interface. Such finite volume approaches calculate shallow water flows by integrating the simplicity of finite difference methods with the geometric flexibility of finite element methods (Toro 1997; Causon et al. 2000; Liang 2008).

Two inundation models were used in this analysis. LISFLOOD-FP (Bates et al. 2005) was used in the first phase of the work to provide inundation simulations at 250 m resolution between Lowestoft and Sea Palling (see, e.g. Fig. 8.10). During the second stage of the analysis, which focused on the smaller area of the Norfolk Broads, we were also able to benefit from increased desktop computing power and data so the NewChan model (Liang et al. 2004) was used to provide more accurate representation of the flooding process.

8.2.1 LISFLOOD-FP Raster Inundation Model

Floodplain flows are described in terms of continuity and momentum equations, discretized over a grid of square cells, which allows the model to represent 2D dynamic flow fields on the floodplain. We assume that the flow between two cells is simply a function of the free surface height difference between those cells (Estrela and Quintas 1994):

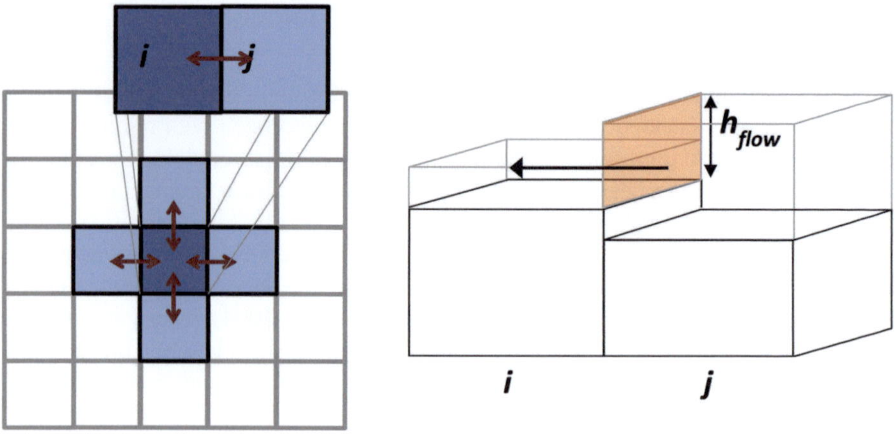

Fig. 8.1 Representation of flow between floodplain cells in LISFLOOD-FP

[AU1]

$$\frac{\mathrm{d}h^{i,j}}{\mathrm{d}t} = \frac{Q_x^{i-1,j} - Q_x^{i,j} + Q_y^{i,j-1} - Q_y^{i,j}}{\Delta x \Delta y} \tag{8.1}$$

$$Q_x^{i,j} = \frac{h_{\mathrm{flow}}^{5/3}}{n_{\mathrm{f}}} \left(\frac{h^{i-1,j} - h^{i,j}}{\Delta x} \right)^{1/2} \Delta y \tag{8.2}$$

Here, $h_{i,j}$ is the water free surface height at the node (i, j), Δx and Δy are the cell dimensions, n_{f} is the effective grid scale Manning friction coefficient for the floodplain, and Q_x and Q_y describe the volumetric flow rates between floodplain cells. Q_y is defined analogously to Eq. 8.2. The flow depth, h_{flow}, represents the depth through which water can flow between two cells and is defined as the difference between the highest water free surface in the two cells and the highest bed elevation (Fig. 8.1). While this approach does not accurately represent diffusive wave propagation on the floodplain, due to the decoupling of the x- and y-components of the flow, it is computationally simple and has been shown to give very similar results to a more accurate finite difference discretization of the diffusion wave equation (Horritt and Bates 2001). However, the position of the performance maximum was shifted by the switch to a diffusion wave approximation for floodplain flows, and the model became more sensitive to changes in floodplain friction.

8.2.2 Shallow Water Equation and NewChan

When the horizontal dimension of a floodplain is much larger than the water depth, the three-dimensional Navier-Stokes equations are usually integrated over the depth, and the two-dimensional St. Venant equations, or shallow water equations, are used. In matrix form, the hyperbolic conservation law formed by the two-dimensional nonlinear shallow water equations by neglecting the Coriolis, wind and viscous forces, can be written as:

$$\frac{\partial\{u\}}{\partial t}+\frac{\partial\{fu\}}{\partial x}+\frac{\partial\{gu\}}{\partial y}=\{Su\} \tag{8.3}$$

Here, t denotes time, x and y are the Cartesian coordinates, and $\{u\}$, $\{fu\}$, $\{gu\}$ and $\{Su\}$ are the vectors representing the fluxes in x- and y-direction and the source terms, respectively:

$$\{u\}=\begin{Bmatrix}\varepsilon\\ uh\\ vh\end{Bmatrix},\{fu\}=\begin{Bmatrix}uh\\ u^2h+0.5g\left(\varepsilon^2-2\varepsilon z_b\right)\\ uvh\end{Bmatrix},$$

$$\{gu\}=\begin{Bmatrix}vh\\ uvh\\ v^2h+0.5g\left(\varepsilon^2-2\varepsilon z_b\right)\end{Bmatrix},\{Su\}=\begin{Bmatrix}0\\ -\dfrac{\tau_{bx}}{\rho}-g\varepsilon\dfrac{\partial z_b}{\partial x}\\ -\dfrac{\tau_{by}}{\rho}-g\varepsilon\dfrac{\partial z_b}{\partial y}\end{Bmatrix} \tag{8.4}$$

where ε is the water surface elevation above the datum; z_b is the bed elevation above the datum, water depth $h=\varepsilon-z_b$; u and v are the depth-averaged velocity components in the two Cartesian directions; g is the gravitational acceleration; τ_{bx} and τ_{by} are the bed friction stresses; ρ is the density of water; and $\frac{\partial z_b}{\partial x}$ and $\frac{\partial z_b}{\partial y}$ are the bed slopes in the x- and y-directions, respectively. The bed stress terms τ_{bx} and τ_{by} represent the effect of bed roughness on the flow and may be estimated using the empirical formulae $\tau_{bx}=\rho C_f u\sqrt{u^2+v^2}$ and $\tau_{by}=\rho C_f v\sqrt{u^2+v^2}$. The bed roughness coefficient C_f can be defined using $C_f=gn_f^2/h^{1/3}$, where n_f is the Manning coefficient.

This formulation of the shallow water equations can be solved by a finite volume Godunov-type scheme (Causon et al. 1999; Hu et al. 1998) to capture different types of shallow flow hydrodynamics over uneven bed bathymetry. The model, NewChan, developed by Liang (2008) is used here. NewChan automatically preserves still water level over a complex domain without any need for specialized numerical treatment of the source terms. Furthermore, the model makes some effort to achieve overall higher-order accuracy by combining a second-order Runge-Kutta time integration method with the MUSCL (Monotone Upstream Scheme for Conservation Laws) linear reconstruction (see Liang 2008 and Liang and Marche 2009 for further details). The model has been extensively validated elsewhere against idealized benchmarks (Liang 2008; Liang et al. 2008; Liang and

Marche 2009). Given appropriate inflow boundary conditions and topographical data, NewChan can provide a range of relevant outputs, including water depth, flow velocity and inundation duration.

8.2.3 *Model Boundary Conditions*

Wave overtopping and defence breaching are frequent causes of major flood events, and our analysis considers both these mechanisms.

(a) *Non-breaching Overtopping Inflow*

When the crest level of the dike or embankment seawall is lower than the highest wave run-up level, the defence will likely be overtopped. The estimated discharge under all kinds of combinations of wave height and water level can be fed directly into an inundation model. Owing to the stochastic nature of wave breaking and wave run-up, empirical formulae derived from physical experiments are the main mechanism for calculating overtopping rates. Here, wave overtopping discharge is calculated for a sloping structure using EurOtop (2007) guidance, which lists empirical relationships for overtopping rates for a number of flood defence structures based upon wave height and period, freeboard, as well as geometrical characteristics of the structure. Overtopping is usually expressed in terms of a mean discharge per linear metre of width, q , in $m^3/s/m$.

Defence crest freeboard, $R_c = (H_c - W)$, can be defined as the vertical difference between the still water level W and the crest height of a structure H_c . Wave run-up height exceeded by 2 % of the incoming waves $R_{u2\%}$ is usually influenced by the geometry of the dike or seawall and the wave properties and includes the uncertainty of the prediction. Considering the prediction has significant scatter, a formula indicating 5 % upper exceedance limit of the 2 % measured wave is given by (EurOtop 2007):

$$R_{u2\%} = \left(c_1 - c_2 / \sqrt{\gamma_b}\right) H_{m0} \gamma_f \gamma_\beta \tag{8.5}$$

Here, c_1 and c_2 are empirical coefficients (4.3 and 1.6, respectively, for all structures in the study); H_{m0} is the significant wave height at the toe of the structure based on the zero moment of a wave spectrum, and this value is called H_s if the wave height is determined by measurements; and γ_b is the breaker parameter or surf similarity parameter that relates slope steepness, $\tan\alpha$, to the wave steepness, $s_b = H_s / L_w$ where $\gamma_b = \tan\alpha / \sqrt{s_b}$, and is often used to distinguish different breaker types. The defence slope, α, is provided by defence survey information. L_w is the deep water wave length, $L_w = gT^2 / 2\pi$, and T is the wave period. γ_f describes structure roughness, for example, for concrete or natural block revetment systems which are along some of the shoreline, $\gamma_f = 1.0$. γ_β captures the influence of wave direction; it is equal to one when waves are parallel to the defence axis (angle of wave attack $\beta = 0$). A schematic of a typical defence and shore cross section, defining key parameters, is shown in Fig. 8.2.

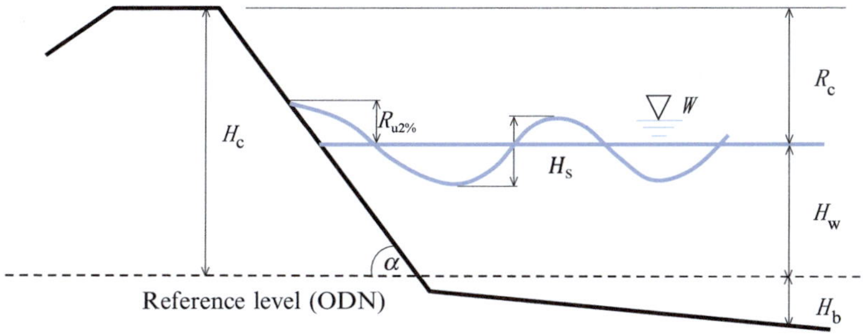

Fig. 8.2 A typical beach cross section relating wave run-up, $R_{u2\%}$, to storm surge level, W, with flood defence of height, H_c (in most sections, this comprises dunes and structures); freeboard, R_c; and beach level $\boldsymbol{H}_b$

If water level is lower than the crest of the defence, i.e. R_c is nonnegative, TAW (2002) recommends a formula for deterministic design and safety assessment of dike or embankment seawall with simple slopes in deep foreshore (breaker parameter γ_b <5):

$$q_1 = 0.2\exp\left(-2.3\frac{R_c}{H_s}\right)\sqrt{gH_s^{\ 3}} \tag{8.6}$$

For shallow foreshore with very heavy breaking and long waves to avoid a larger underestimation of wave overtopping (breaker parameter γ_b >7):

$$q_2 = 0.21\exp\left(-\frac{R_c}{H_s\left(0.33+0.022\gamma_b\right)}\right)\sqrt{gH_s^{\ 3}} \tag{8.7}$$

A linear interpolation can determine discharge q_{12} for breaker parameter $5 \le \gamma_b \le 7$. If the water level is not lower than the crest of the dike or embankment seawall (i.e. R_c <0), the amount of water flowing to the landward side of the structure is composed by a part which can be attributed to overflow q_{ow} and a part which can be attributed to overtopping q_{ot0} at zero freeboard. The effect of combined wave run-up and wave overtopping is given by the superposition of overflow and wave overtopping as a rough approximation:

$$q_3 = q_{ow} + q_{ot0} = 0.6\sqrt{g\left|-R_c^{\ 3}\right|} + 0.0537\gamma_b\sqrt{gH_s^{\ 3}} \tag{8.8}$$

(b) *Breached Inflow*

Once a beach level is lower than a critical level or threshold b_t, a collapse of the flood defence is considered to have occurred (c.f. Thomas and Hall 1992). Breaching width, b_d, is computed from the simple empirical expression:

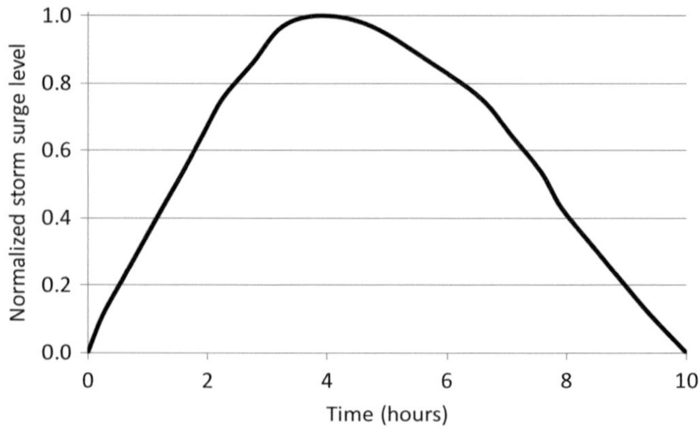

Fig. 8.3 Normalized profile of water level during a typical (as observed in 1953) storm surge event

$$b_d = \min(10R_c c_a, L_s) \tag{8.9}$$

Here, R_c is the crest freeboard, L_s is the length of a dike section, and c_a is the empirical coefficient, which varies from 3, for cohesive materials (HR Wallingford 2004), to 15 for non-cohesive materials (Visser 1998); 12 is taken in this study. Once a breach occurs in a defence section, its crest height in the model is reduced to the beach level, and the freeboard is recalculated as:

$$R_c = (H_b - W) \tag{8.10}$$

This is used to recalculate a breached discharge, q_b, using Eq. 8.7.

Discharges through the defence system are translated into a hydrograph using a dynamic tidal profile corresponding in shape to an extreme event, such as the well-documented 1953 surge (Rossiter 1954) shown in Fig. 8.3. The peak discharge is therefore determined by mean high water level, relative sea-level rise, wave height and beach level.

8.2.4 *Adapting Boundary Conditions for Climate and Morphological Change*

The boundary conditions are adjusted to assess a range of current and potential future climate by analyzing a range of defence breach locations; storm surge levels, *W*; wave heights, H_s; and antecedent beach levels, H_b, which reflect natural variability and the range of beach management scenarios (Appendix A).

8.3 Application to North Norfolk

8.3.1 Site Description

The case study site is fully introduced in Chap. 1, but points particularly salient to this analysis are now highlighted. Much of the land is reclaimed marshland and is below the level of mean high water spring tides. Extensive flooding, often the result of defence breaching (Fig. 8.4), has occurred a number of times over the last 100 years. In the 1953 event, 17 people died in this area (Kelman 2003). Within the LISFLOOD-FP model domain (Fig. 8.12), there are almost 20,000 properties at risk of flooding in an area approximately 340 km^2. Over 3,000 of these properties are nonresidential (including commercial and industrial activities). Almost 50 % of the floodplain is premium agricultural land, with only 5 % less productive agricultural land. Some 35 % is registered as being environmentally sensitive (with salt marsh, national nature reserve destinations).

The overall wave climate is complex; however, the dominant wave direction is from the northeast. Longshore sediment transport from Sheringham (43.1 km north-northwest of Norwich) is southeastward but considerably variable (Vincent 1979; Wang and Reeve 2010). Counted from Winterton to Happisburgh, the coastline is subdivided into $N_s = 30$ subsections. LiDAR calculated crest elevations for these flood defences were cross-checked against the values in the UK's National Flood and Coastal Defence Database (DEFRA 2006) to estimate defence crest levels (Fig. 8.5). The dunes southeast of Happisburgh are protected by concrete seawalls with a combined height of over 10 m high (Environment Agency 2006). An extensive system of dunes has been reinforced by construction of seawalls and groynes along this coastline with a standard of protection of approximately 1 in 200 years. In the 1990s, the progressive lowering of the beaches in front of the structures near Sea Palling, and consequent increase in inundation probability, was addressed by construction of nine detached breakwaters and extensive beach nourishment (Thomalla and Vincent 2003). Further south at Winterton, the beaches are wider with timber groynes, and there is a more extensive dune system, which may act to restrict inland flooding.

Extreme water levels were estimated using the method of Dixon and Tawn (1997). Based upon observations up to 2010, the 1 in 1,000 storm water level is

Fig. 8.4 Sea Palling after the 1953 storm surge (Image courtesy of Norfolk County Council Library and Information Service)

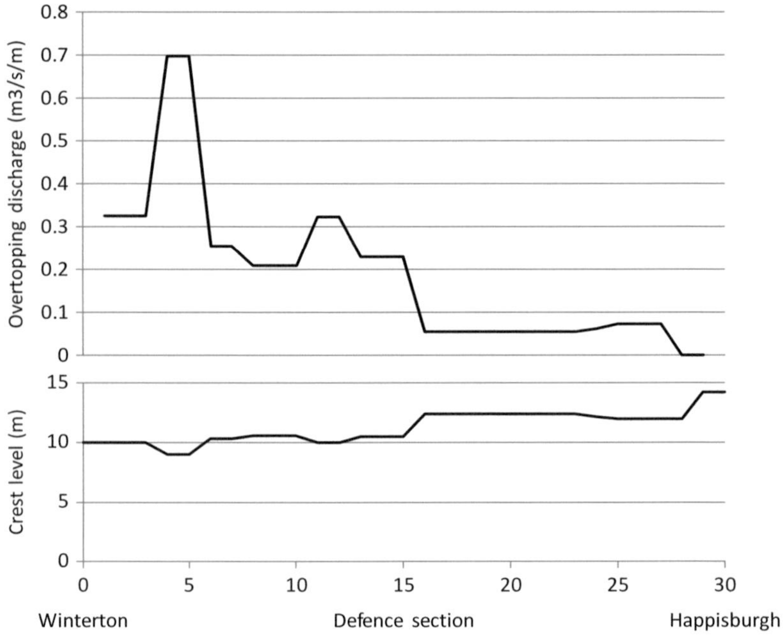

Fig. 8.5 Crest level of defence sections from Winterton (Section 0) to Happisburgh (Section 30) and their associated overtopping discharges for $W=7$ m, $H_s=3$ m

4.40 m ODN, and other extreme water levels are shown in Table 8.1. Accounting for the sea-level rise scenarios reported in Chap. 2 and Appendix A, by 2100, the 1 in 1,000 year event could lead to extreme water levels as high as 6.12 m. We analyzed the combinations in Table 8.2 to capture uncertainties in future surge levels, extreme wave heights (of up to 3 m as described in Dickson et al. 2007) and beach level at the instance of breaching (Chap. 7).

A 250 m (some 161 × 168 grid cells) spatial resolution DEM was derived from airborne laser altimetry (LiDAR) data. The elevation of large area of land behind the coastal defences is below 5 m in many locations (Fig. 8.6). The NewChan and LISFLOOD-FP models were calibrated against the recorded inundation extent from the major flood event in 1938 with a Manning coefficient of 0.025 s/m$^{1/3}$ (Bates et al. 2005), as shown in Fig. 8.6. To simulate the breach, the DEM is modified at Winterton to include the breach dimensions given by Mosby (1938). Unfortunately, no dynamic time series of water levels over the 1938 event is available, and the only boundary condition information consisted of the maximum recorded water level at Winterton. The 1953 storm profile (Fig. 8.3) was therefore used as a representative surge event. For the calibration simulation, a peak water level of 3.0 m and a beach level of 0.8 m at the time of breach were taken. Inflows in conjunction with storm-induced overtopping or breaching were calculated over the 10 h duration shown in Fig. 8.3, which was used for all the other simulations. Both the LISFLOOD-FP and NewChan model replicate the flood extent reasonably well, with differences resulting from different

Table 8.1 Return period of extreme water levels (m ODN) at Sea Palling in 2010

Return period	Extreme water level (m)
10	3.35
25	3.59
50	3.69
100	3.90
250	4.11
1,000	4.40

Table 8.2 Range of loading conditions simulated

Variable	Symbol	Value(s)
Wave period (sec)	T	6.5 (if no value is available)
Structure roughness	γ_f	1.0
Wave attack direction	γ_β	1.0
Manning coefficient	n_f	0.025
Breached beach level (m)	H_b	0.8; 0.4; 0; –0.4; –0.8; 1.2; –1.6; –2
Still water level (m)	W	0.5; 1.0; 1.5; 2.0; 2.5; 3.0; 3.5; 4.0; 4.5; 5.0; 5.5; 6.0; 6.5; 7.0
Wave height (m)	H_s	0; 0.25; 0.5; 0.75; 1.0; 1.25 ; 1.5; 1.75; 2.0; 2.5; 3.0; 3.5

approaches to calculating inflow discharges. In the following sections, the sensitivity of the model to varying hydrodynamic conditions is explored.

8.3.2 Inundation Scenarios Due to Overtopping Under Varying Wave Heights

Figure 8.5 shows the overtopping discharge for $H_s=3$ m and $W=7$ m along the coastline from Winterton to Happisburgh. The lower crest levels of Sections 4–5 lead to the greatest overtopping flow. Total discharge, computed maximum flood depths and inundated area for a range of wave heights are shown in Table 8.3. Inundation remains relatively close to the coastline for $H_s<1$ m, but Fig. 8.7 shows that water can penetrate deep into the Norfolk Broads for larger wave heights.

8.3.3 Inundation Scenarios Due to Single Breaching Under Varying Water Levels

If the defence system state is characterized by the combination of failed and non-failed defence sections, the inflow of a breached system state will be a function of breached beach level, water level and wave height near the defence(s). The flood

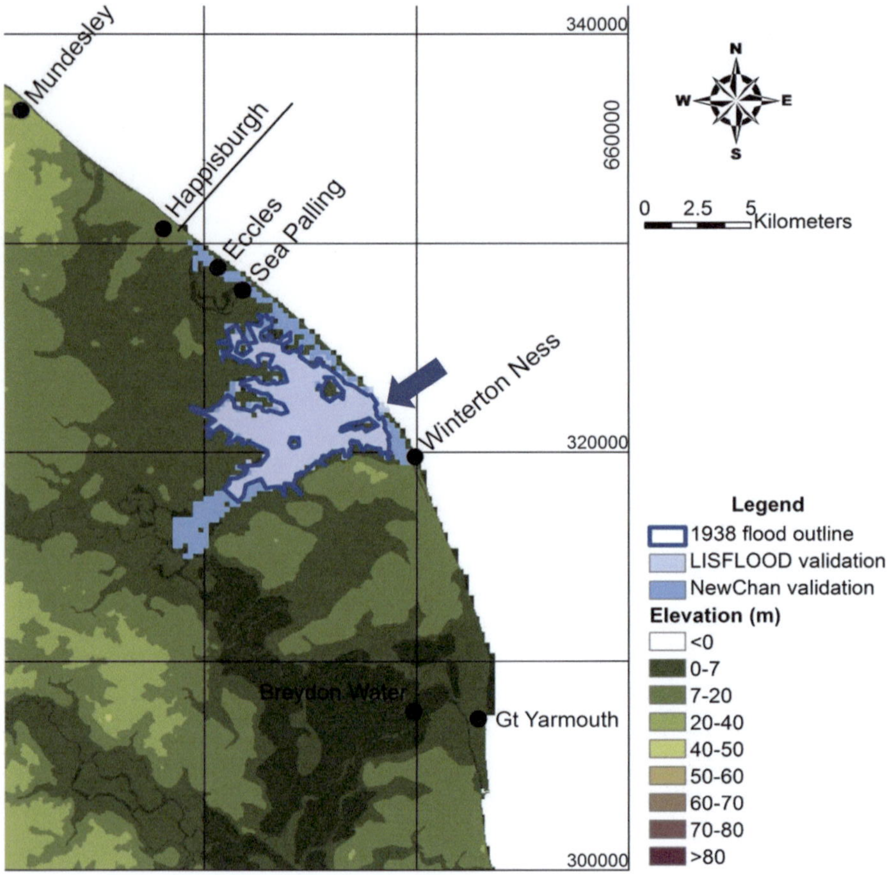

Fig. 8.6 Comparison of observed inundation extent for the 1938 Horsey flood with that predicted by the LISFLOOD-FP model (Bates et al. 2005) and the NewChan model

extent as a result of various water levels and beach levels when Section 3 is breached is now considered. Figure 8.8 shows the flood extent for a range of extreme water levels, with ingress increasing with the magnitude of the event. A relatively small storm surge, $W=3$ m, exceeds the largest overtopping inundation event shown in Fig. 8.7. The computed maximum flood depths and corresponding inundation area for a range of extreme water levels are listed in Table 8.4. These results show that sea-level rise could greatly increase the area at risk of flooding and the impacts of flooding as a result of increased flood depths. Figure 8.9 shows the flood extent 2, 4 and 10 h after the breaching event for $W=7$ m and $H_s=0$ m, with the maximum calculated flood depth for $W=7$ m shown in Fig. 8.10. For comparison, Fig. 8.11 shows the flood extent calculated using the LISFLOOD-FP model over the full spatial extent of the region for $W=3$ m assuming that all defences are removed.

Table 8.3 Computed maximum water depths under various wave heights

H_s (m)	Discharge ($m^3/s/m$)	Inundated area (km^2)	Maximum water depth (m)
1	0.04	0.31	0.70
2	1.17	12.31	1.50
3	5.57	39.57	2.28

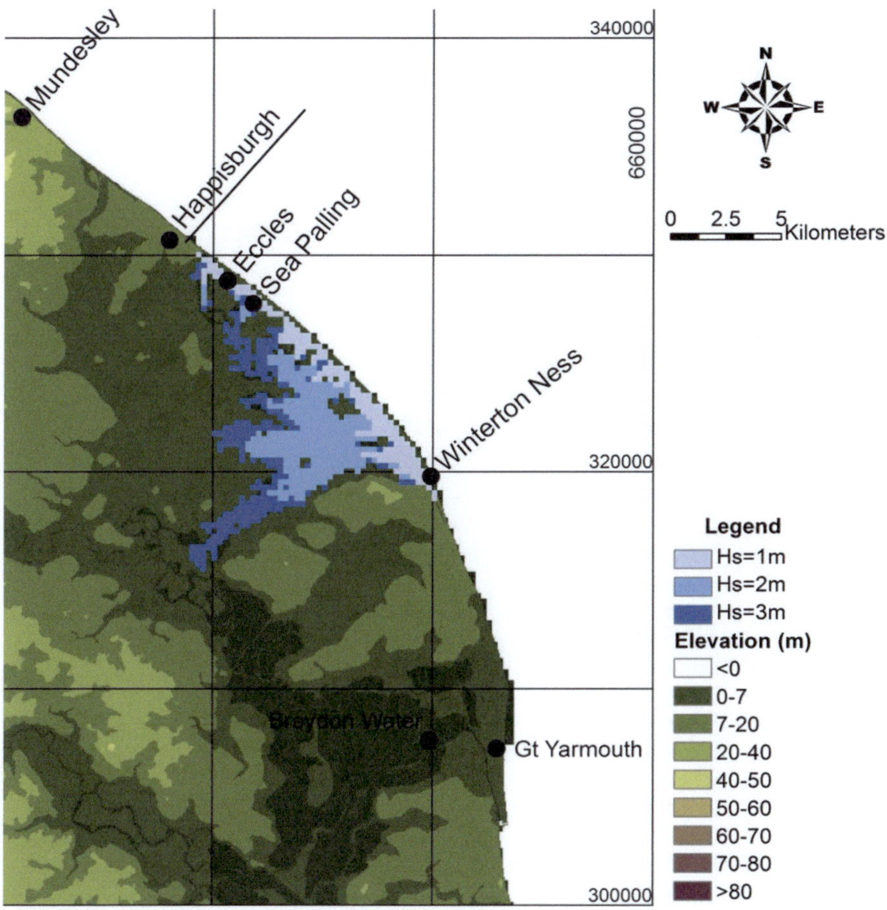

Fig. 8.7 Maximum flood extents for $W=7$ m and for $H_s=1$ m, 2 m and 3 m with no flood defence breach

8.3.4 Inundation Scenarios Due to Single Breaching Under Varying Breached Locations

Defence breach could occur at any section, or combination of sections, along the coastline. Figure 8.12 shows the different flood extents for $W=7$ m, $H_s=0$ m for a breach of Sections 3, 5 and 16. In some cases (*c.f.* Dawson and Hall 2006), the

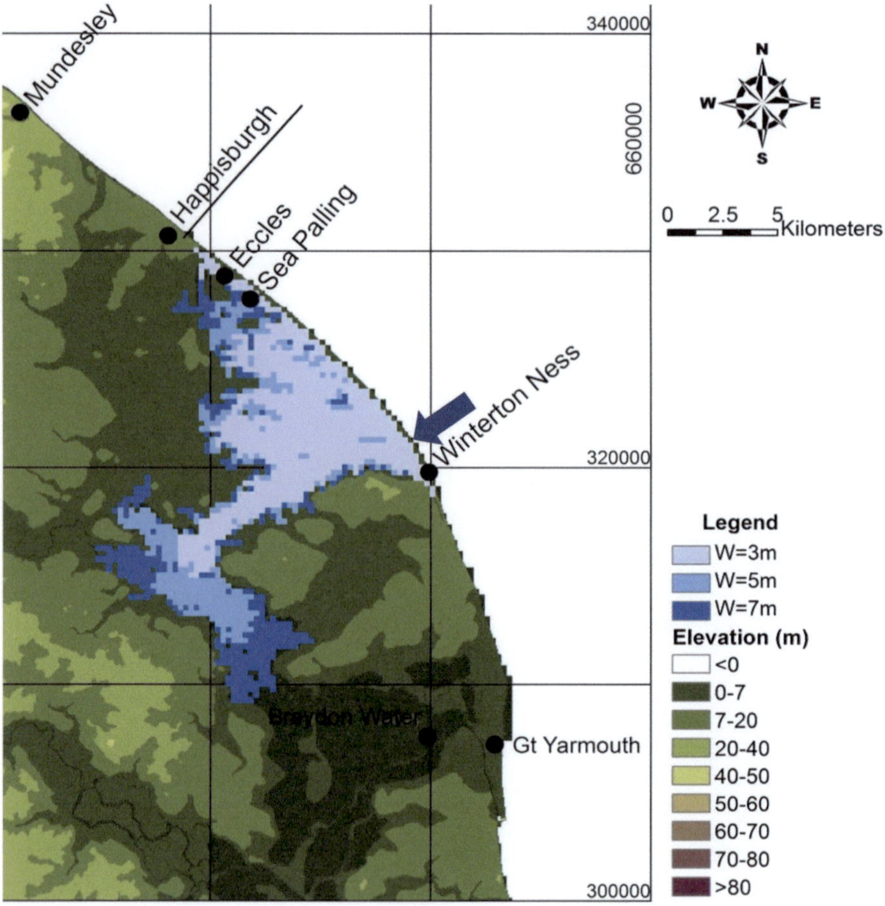

Fig. 8.8 Maximum flood extents for storm surge levels of $W=3$ m, 5 m and 7 m when Section 3 breaches and $H_s=0$ m, $H_b=0.8$ m

Table 8.4 Computed maximum water depths under various water levels

Water level (m)	Discharge ($m^3/s/m$)	Inundated area (km^2)	Maximum water depth (m)
1	0.02	0.13	0.98
2	1.45	15.19	1.72
3	6.13	39.06	2.44
4	10.76	51.88	2.90
5	16.18	63.38	3.46
6	22.28	76.31	3.93
7	29.01	89.69	4.35

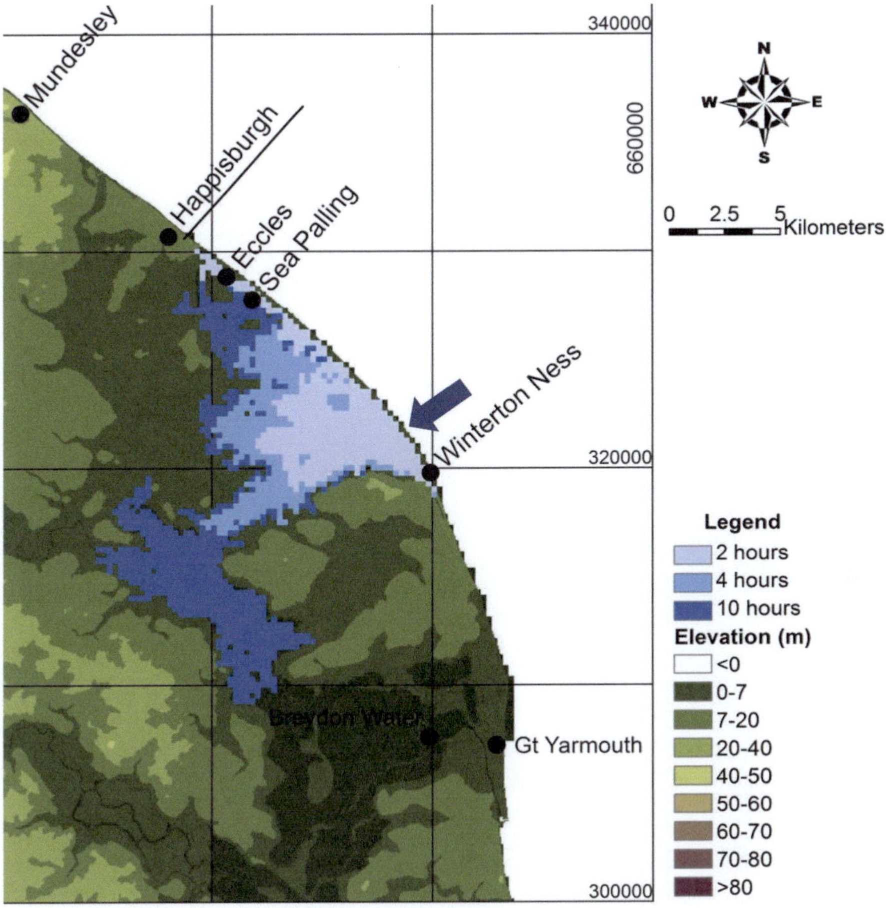

Fig. 8.9 Flood extents after 2, 4 and 10 h propagation when $W=7$ m, $H_s=0$ m, $H_b=0.8$ m

breach location has an enormous influence on flood extent, but here, the difference in flooded area is less than 4 %. However, the location does have a bearing on the timing and movement of the flood wave which has crucial implications for flood warning and evacuation strategies.

8.3.5 Inundation Scenarios Due to Multi-breaching Under Varying Combinations

During an extreme event, it is possible that multiple defence sections could fail. Figure 8.13 compares flood extents for several combinations of Sections 3, 5 and 16 failing. Discharges of these combinations contributed by breaching are 29.01, 49.52 and 70.03 $m^3/s/m$, respectively.

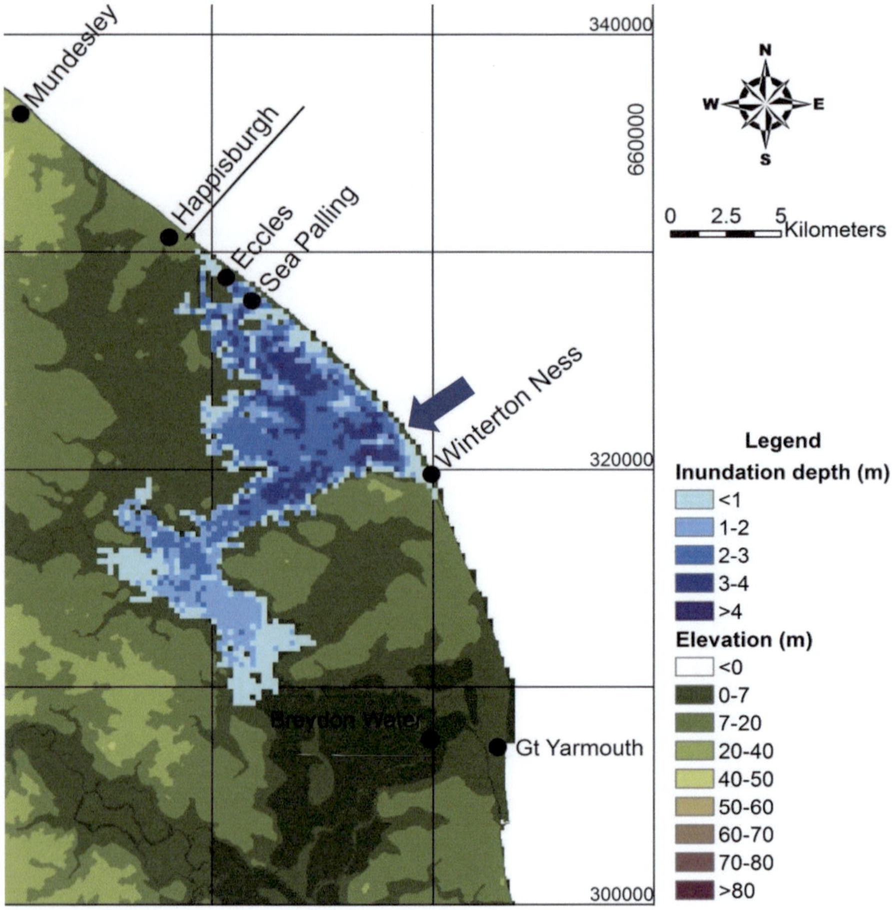

Fig. 8.10 Maximum computed flood depth for event of $W=7$ m, $H_s=0$ m, $H_b=0.8$ m

8.4 Conclusions and Next Steps

Coastal management requires analysis of the current and future flood impacts from overtopping and breaching. This chapter has demonstrated the need to analyze a wide range of boundary conditions to understand flood impacts on dynamic and spatially heterogeneous coastlines. Uncertainties in beach level in front of flood defences, future sea level, extremes of wave height and water level and breach location(s) require that a wide range of current and potential future conditions are considered to fully understand flood risk. It is therefore crucial to capture and model a wide range of events.

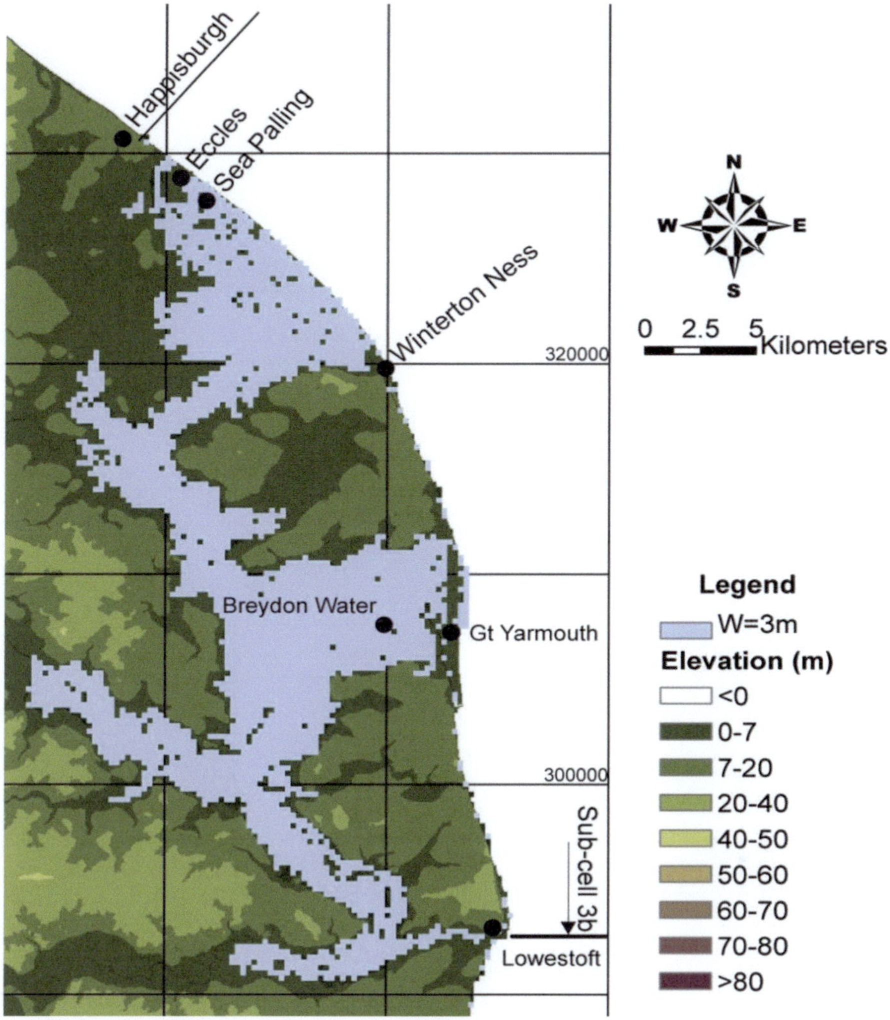

Fig. 8.11 Flood extent, estimated using LISFLOOD-FP, with $W = 3$ m and all defences assumed to have breached thereby enabling floodplain ingress at Lowestoft, Great Yarmouth, and along the Norfolk Broads

Flood depth and extent in the Norfolk Broads are dominated by extreme water levels and to a lesser extent wave heights. Flood impacts are also sensitive to breach locations but much less than other locations due to its topography. This shallow topography also means that a simple 'bath-tub' model overestimates flood extents in this location, and so a hydraulic model is required. The accuracy of the process representation must be balanced with the computational demands of simulating the range of uncertainties associated with loading conditions and defence breaches.

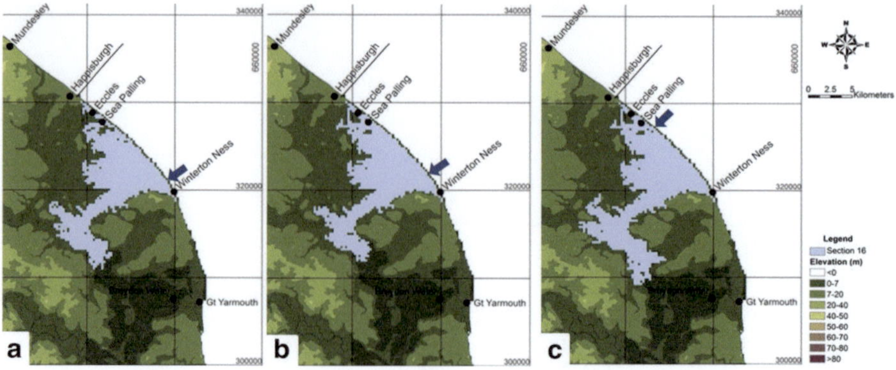

Fig. 8.12 Maximum flood extents for three different flood defence breach locations, Sections (**a**) 3, (**b**) 5 and (**c**) 16, between Sea Palling and Winterton-on-Sea

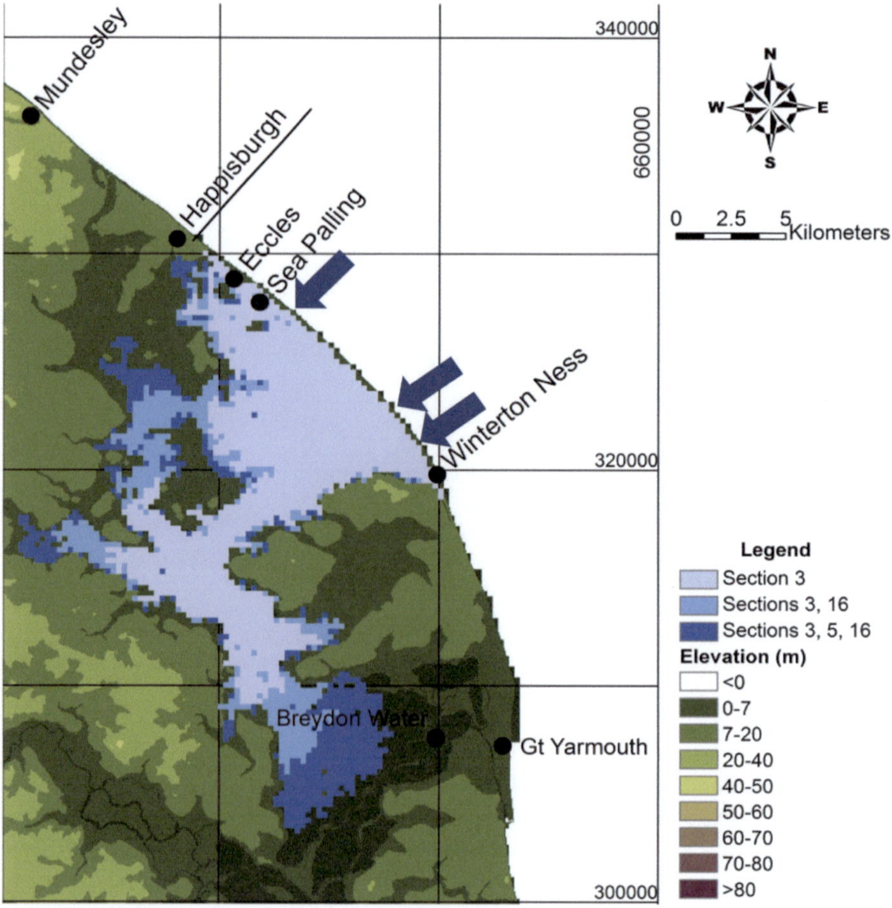

Fig. 8.13 Maximum flood extents resulting from multiple defences failing in combination (**a**) Section 3 (section furthest south), (**b**) Sections 3 and 16 (section furthest north), (**c**) Sections 3, 5 and 16 for a flood event of $W=7$ m, $H_s=0$ m, $H_b=0.8$ m

Initially, LISFLOOD-FP was used to simulate inundation over the full extent of the floodplain. Later in the case study development, we benefitted from increased desktop computing power and were able to use the more advanced NewChan model for the more detailed analysis of flooding in the Norfolk Broads. The results of these simulations are integrated with information on damages to quantify flood risk in Chap. 9.

References

Bates, P. D., Dawson, R. J., Hall, J. W., Horritt, M. S., Nicholls, R. J., Wicks, J., & Hassan, M. A. A. M. (2005). Simplified two-dimensional numerical modelling of coastal flooding and example applications. *Coastal Engineering, 52*, 793–810.

Causon, D. M., Mingham, C. G., & Ingram, D. M. (1999). Advances in calculation methods for supercritical flow in spillway channels. *Journal of Hydraulic Engineering, 125*(10), 1039–1050.

Causon, D. M., Ingram, D. M., Mingham, C. G., Yang, G., & Pearson, R. V. (2000). Calculation of shallow water flows using a Cartesian cut cell approach. *Advances in Water Resources, 23*, 545–562.

Dawson, R. J., & Hall, J. W. (2006). Adaptive importance sampling for risk analysis of complex infrastructure systems. *Proceedings- Royal Society of London A, 462*(2075), 3343–3362.

DEFRA. (2006). *Shoreline management plan guidance – Volume 1: Aims and requirements Volume 2: Procedures*. London: Department for Environment, Food and Rural Affairs.

Dickson, M. E., Walkden, M. J. A., & Hall, J. W. (2007). Systemic impacts of climate change on an eroding coastal region over the twenty-first century. *Climatic Change, 84*, 141–166.

Dixon, M. J., & Tawn, J. A. (1997). *Estimates of extreme sea level – final report. Spatial analyses for the UK coast* (Internal document, no 112). Liverpool: Proundman Oceanographic Laboratory, 217 pp.

Environment Agency. (2006). *Happisburgh to Winterton Sea defences stage 3B works, Environmental Statement*. Peterborough: Environment Agency.

Estrela, T., & Quintas, L. (1994). Use of a GIS in the modelling of flows on floodplains. In J. W. R. White & J. Watts (Eds.), *River flood hydraulics* (pp. 177–189). Chichester: Wiley.

EurOtop. (2007). *Wave overtopping of sea defences and related structures: Assessment manual* (T. Pullen, N. W. H. Allsop, T. Bruce, A. Kortenhaus, H. Schüttrumpf, & J. W. van der Meer, Eds.). Available online: www.overtopping-manual.com. Accessed 21 Apr 2012.

Horritt, M. S., & Bates, P. D. (2001). Predicting floodplain inundation: Raster-based modelling versus the finite element approach. *Hydrological Processes, 15*, 825–842.

HR Wallingford. (2004). *Investigation of extreme flood processes & uncertainty*. Risk and uncertainty (WP5) – Technical report. HR Wallingford, UK.

Hu, K., Mingham, C. G., & Causon, D. M. (1998). A bore-capturing finite volume method for open-channel flows. *International Journal for Numerical Methods in Fluids, 28*, 1241–1261.

Kelman, I. (2003). CURBE fact sheet 3: UK deaths from the 1953 Storm Surge. Version 4.

Liang, Q. (2008). Simulation of shallow flows in nonuniform open channels. *Journal of Fluids Engineering, 130*(1), 011205. doi:10.1115/1.2829593.

Liang, Q., & Marche, F. (2009). Numerical resolution of well-balanced shallow water equations with complex source terms. *Advances in Water Resources, 32*(6), 873–884.

Liang, Q., Borthwick, A. G. L., & Stelling, G. (2004). Simulation of dam- and dyke-break hydrodynamics on dynamically adaptive quadtree grids. *International Journal for Numerical Methods in Fluids, 46*(2), 127–162.

Liang, Q., Du, G., Hall, J. W., & Borthwick, A. G. L. (2008). Flood inundation modeling with an adaptive quadtree grid shallow water equation solver. *Journal of Hydraulic Engineering, 134*(11), 1603–1610.
Mosby, J. E. G. (1938). The Horsey flood, 1938: An example of storm effect on a low coast. *Geographical Journal, 93*(5), 413–418.
Rossiter, J. R. (1954). The North Sea storm surge of 31st January and 1st February 1953. *Philosophical Transactions of the Royal Society of London, 246*, 915371–915400. doi:10.1098/rsta.1954.0002.
TAW. (2002). *Technical Report – wave run-up and wave overtopping at dikes*. Technical Advisory Committee for Flood Defence in the Netherlands (TAW), Delft.
Thomalla, F., & Vincent, C. E. (2003). Beach response to shore-parallel breakwaters at Sea Palling, Norfolk, UK. *Estuarine, Coastal and Shelf Science, 56*(2), 203–212.
Thomas, R. S., & Hall, B. (1992). *Seawall design*. London: CIRIA.
Toro, E. F. (1997). *Riemann solvers and numerical methods for fluid dynamics: A practical introduction*. Berlin/New York: Springer.
Vincent, C. (1979). Longshore sand transport rates – A simple model for the East Anglia coastline. *Coastal Engineering, 3*, 113–136.
Visser, P. J. (1998). *Breach growth in sand defences, Communication on hydraulic and geotechnical engineering*. TU Delft report no. 88–91. The Netherlands.
Wang, B., & Reeve, D. (2010). Probabilistic modelling of long-term beach evolution near segmented shore-parallel breakwaters. *Coastal Engineering, 57*, 732–744.

Chapter 9
Analysing Flood and Erosion Risks and Coastal Management Strategies on the Norfolk Coast

Jim W. Hall, Richard J. Dawson, and Xing Zheng Wu

Abstract Coastal systems are characterised by variability and interdependencies at a range of scales. On the Norfolk coast from Weybourne to Winterton in SMP6, variations in sediment supply, from cliff erosion and beach nourishment, have a profound influence upon the probability of failure of the flood defences that protect the large area of coastal lowlands, including the Norfolk Broads. The risk of flooding is therefore influenced by large-scale and long-term changes in sediment supply as well as by short-term fluctuations which are dominated by the arrival of extreme storms. The reliability of the flood defences also plays a crucial mediating role on the probability of flooding. When flooding does occur, the extent and severity of damage is influenced by patterns of inundation and the human and economic vulnerability of the communities that are flooded.

This chapter brings together science from several of the preceding chapters in order to analyse how the risks of flooding and erosion will change in the future. The analysis makes use of a Monte Carlo simulation methodology, which takes multiple simulations of beach levels (from the SCAPE model described in Chap. 7) and combines this with reliability analysis of the flood defence system, quantification of the inundation depth and extent (from Chap. 8) and changes in coastal vulnerability and land use (from Chap. 4). The evolution of future flood risk is sensitive to the quantity of beach sediment, from eroding cliffs or artificial beach nourishment, underlining the importance of strategic coastal management.

J.W. Hall (✉)
Environmental Change Institute, Oxford University Centre for the Environment,
University of Oxford, South Parks Road, Oxford OX1 3QY, UK
e-mail: jim.hall@eci.ox.ac.uk

R.J. Dawson
School of Civil Engineering and Geosciences, Tyndall Centre for Climate Change Research,
Newcastle University, Newcastle upon Tyne NE1 7RU, UK
e-mail: richard.dawson@newcastle.ac.uk

X.Z. Wu
Department of Applied Mathematics, School of Applied Science, University of Science and Technology Beijing, 30 Xueyuan Road, Haidan District, Beijing 100083, P.R. China
e-mail: xingzhengwu@gmail.com

R.J. Nicholls et al. (eds.), *Broad Scale Coastal Simulation*, Advances in Global Change Research 49, DOI 10.1007/978-94-007-5258-0_9

In the Tyndall Coastal Simulator, the work constitutes a key result and builds on the paper by Dawson et al. (Proc R Soc Lond A 462(2075):3343–3362, 2009). It demonstrates that erosion and flood risk are strongly linked. Alternatively, nourishment can substitute for the sources from cliff erosion. Many of the results from this work are taken forward in the visualisation work (Chap. 10), the interface development (Chap. 11) and the stakeholder interaction (Chap. 12).

Keywords Flooding • Cliff erosion • Risk analysis • Coastal management • Beach nourishment

9.1 Introduction

Climate change with its associated rising sea levels and possible increases in storminess and changes in wave climate can be expected to increase the risks of coastal erosion and flooding in most coastal locations (Nicholls et al. 2007). As described previously in Chap. 7, changes in one length of coastline can influence changes in other morphologically connected areas (Dickson et al. 2007). Sediment movement along beaches and in shallow coastal waters provides a mechanism of interaction between erosion processes and flood risk because the behaviour and properties of beaches are a key determinant of flood risk. Beach morphology has a critical role in dissipating incident wave energy. Moreover, on natural coasts, the beach is the 'defence', whilst on artificially defended coasts, the reliability of structures usually depends on beach volume as the level of the adjacent beach influences the likelihood of toe scour, a critical failure mechanism for coastal structures. Sediments released from erosion may serve to increase beach volume and reduce the probability of structural failure on neighbouring coasts. Conversely, the existence of structures that disrupt sediment movement may starve down-drift beaches reducing their protective function. In situations such as this, coastal managers should no longer consider erosion and flood hazards in isolation – as discussed in Chap. 1. Rather analysis of climate impacts and long-term coastal management must be implemented at an appropriate (broad) scale to account for this morphological interdependence. Thus, the approach to risk analysis takes a systematic and integrative approach to analyse the impact of long-term change on both coastal flood risk and coastal erosion.

This chapter draws together aspects of the analysis described in several proceeding chapters of this volume in order to estimate changing flood risk on the Norfolk coast.

- Chapter 2 introduced a range of sea-level rise scenarios, where uncertainty is attributable both to uncertain greenhouse gas emissions scenarios and epistemic uncertainties in future sea-level rise projections. Here we make use of these emissions scenarios, including a Monte Carlo sample of sea-level rise projections to incorporate these uncertainties in flood risk analysis.
- Chapter 3 explained how the future wave climates were derived for the use in the coastal simulations and analysis of the probability of flood defence failure.

We make use of these hydraulic model results both directly in estimating the probability of flooding and indirectly (from Chap. 7) via their effect on sediment supply and beach levels.

- The analysis of land use change in Chap. 4 has been used to understand the potential effects of socio-economic change on vulnerability to flood risk.
- The SCAPE coastal dynamics simulation model, described in Chap. 7, has been used to generate simulations of beach level for each emissions scenario and management scenario. This provides the primary input to the reliability analysis of flood defences. The beach management strategies (including various strategies for beach nourishment) were also simulated in the SCAPE model.
- The coastal inundation modelling described in Chap. 8 has been used to compute the extent and depth of flooding for a range of possible breach scenarios.

These various inputs are brought together in a probabilistic risk analysis framework which we introduce briefly in the following before summarising the results.

The risk analysis framework (Fig. 9.1) is developed through the project, starting with individual realisations of future conditions, and a focus on the trade-offs between erosion of the cliff and flooding. Subsequent simulations took a Monte

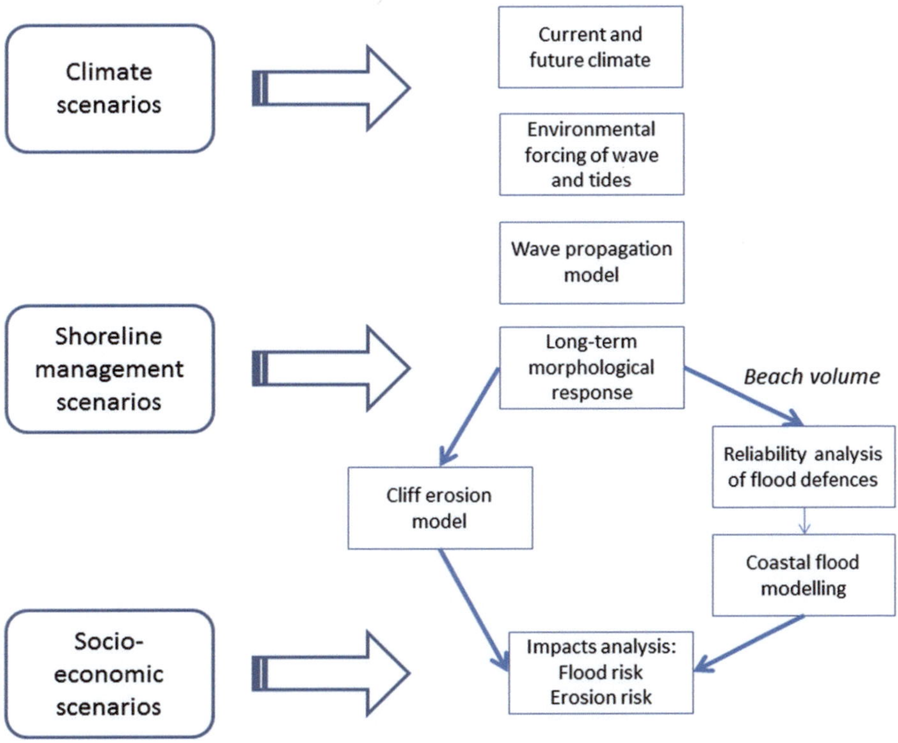

Fig. 9.1 Overview of the risk analysis calculation (From Dawson et al. 2009, and reproduced in Appendix A)

Carlo approach to better characterise the uncertainties and also added beach nourishment as a response option. More details are explained below.

9.2 Risk Analysis Framework

The occurrence of damaging erosion or flooding on the coast is determined by highly variable marine processes (surge tides, extreme waves) and the (uncertain) responses of the coastal system. It is therefore impossible to predict coastal flooding and erosion in deterministic terms. Instead, a risk-based approach is adopted in which coastal behaviour is predicted, through time, in probabilistic terms and combined with assessment of impacts. The primary metric for comparison of scenarios and management options here is economic risk, but this can be supplemented by indicators of potential social or environmental impact if available.

The low-lying coastline Norfolk coastline is naturally protected by a system of dunes that have been artificially reinforced by the construction of solid walls (Fig. 9.2). The dunes are naturally high enough to prevent significant overtopping of sea water into the floodplain. The primary mechanism by which flooding can occur is due to erosion of the dunes leading to a breach and a major inflow of water. Breaching can occur when the beach level is low and the artificial defences are undermined by toe erosion. Thus the analysis of flood risk depends upon the

Fig. 9.2 The beach at Winterton with concrete sea wall in front of sand dune (Richard Dawson)

probability of beach levels lowering below a critical level. As both the beach level, b_i, and the critical toe level, c_i, of the defence vary along the coast, we divide the coast into n sections $i = 1, \ldots, n$.

9.3 Broad-Scale Risk Analysis of Coastal Subcell

The calculations of risk, based upon this causal sequence, are described in full by Dawson et al. (2009, and reproduced in Appendix A) with the fundamental equations reproduced in the following subsections.

9.3.1 *Broad-Scale Erosion Risk Calculation*

The morphological modelling described in Chap. 7 is used to predict shoreline (*i.e.* cliff toe) location, $x_s(y,t)$ at distances y along a shore-parallel baseline and time t years in the future. Given a particular shoreline location, x_s, we then generate a probability density function (p.d.f.), $f_c(x|x_s,y,t)$ of the cliff top location, x, at a distance y along the coast and in year t using the method described by Hall et al. (2000). This p.d.f. expresses the uncertainty in the location of the cliff top, relative to a given shoreline. The expected annual damage due to erosion, r_c, is therefore given by

$$r_c(t) = \iint f_c(x|x_s, y, t) D_c(x,y,t) \mathrm{d}x\mathrm{d}y \tag{9.1}$$

where $D_c(x,y,t)$ is a spatially variable damage function constructed from market values of property and agricultural land, which will depend upon the socio-economic scenario (*cf.* Thorne et al. 2007).

9.3.2 *Broad-Scale Flood Risk Calculation*

In calculating flood risk, we consider the following causal sequence (after Dawson and Hall 2006):

1. Variability in marine conditions lowers or raises beach levels in front of flood defence structures in section i.
2. If the arrival of an extreme storm with wave height, H, and water level, W, causes lowering of the beach in front of the defence such that it drops below a critical level, breaching occurs.
3. Water that passes through a breach in a combination of flood defence sections C_i causes inundation of the coastal floodplain.
4. Damage, D, is caused to coastal people and properties who are flooded, the severity of which increases with flood depth.

The flood risk is therefore calculated according to the following expression:

$$r_f(t) = \iint \sum_{i=1}^{2^n} P(S_i \mid H, W, x_s) f_f(H, W \mid t) D_f(H, W, C_i, t) \mathrm{d}H \mathrm{d}W \tag{9.2}$$

$D_f(H,W,C_i,t)$ is a damage function, t reflects the fact that the frequency of waves and extreme water levels (due to sea-level rise) is expected to change with climate change, whilst socio-economic drivers will alter vulnerability and damages.

The combinations of sea-level rise, wave height and wave direction scenarios and the broad-scale management options that were tested in the coupled analysis are summarised in Appendix A. Further consideration of management scenarios using beach nourishment in front of the Norfolk Broads and incorporation climate model uncertainties is reported later in this chapter.

The results of the risk analysis are presented in 2003 prices. Figure 9.3 illustrates separately the erosion risk, r_c, and the flood risk, r_f. The flood risk at the site is at least an order of magnitude greater than the erosion risk in the base year and potentially increases at a greater rate. The flood risks grow exponentially during the latter half of the twenty-first century, whilst the erosion risk is predicted to remain relatively constant through the twenty-first century, assuming the baseline socio-economic scenario. The difference between the two risks is further illustrated in Fig. 9.3, which shows how the erosion risk is concentrated in a narrow band near the coast, whereas the flood risk zone extends far inland. The differences between the risk functions associated with the two hazards arise because:

- Properties can be flooded on multiple occasions, whereas they can only be eroded from a cliff top once. Taking this to its most extreme: if an entire coastal settlement is eroded over the course of the twenty-first century, the risk will decrease to near zero once the settlement has disappeared and all that remains is the agricultural hinterland.
- There are fewer properties at risk of erosion over the twenty-first century than at risk of flooding which reflects that the flood risk zone is larger and more populous than the erosion risk zone.
- For a given return period, sea-level rise increases the area at risk of flooding and depth of inundation over the entire floodplain, whilst its influence on erosion is more complex as accelerated cliff recession generates additional sediment which can reduce recession rates elsewhere along the coastline (Dickson et al. 2007). In both cases, the relationship between sea-level rise and risk is not linear.

Figure 9.3 also illustrates the trade-off associated with different cliff management options. Whilst cliff protection reduces erosion risks (at a cost), it also reduces sediment supply to the flood-prone down-drift coast, leading to the potential for rapidly increasing flood risks over the twenty-first century (assuming no improvement in flood defence infrastructure). It is particularly interesting to note that the erosion risk for the 100 % protection is never zero, because there is always a finite probability of a failure: despite the presence of cliff toe protection, landslides can still occur (*cf.* Frew and Guest 1997), and so there is still an erosion risk. However,

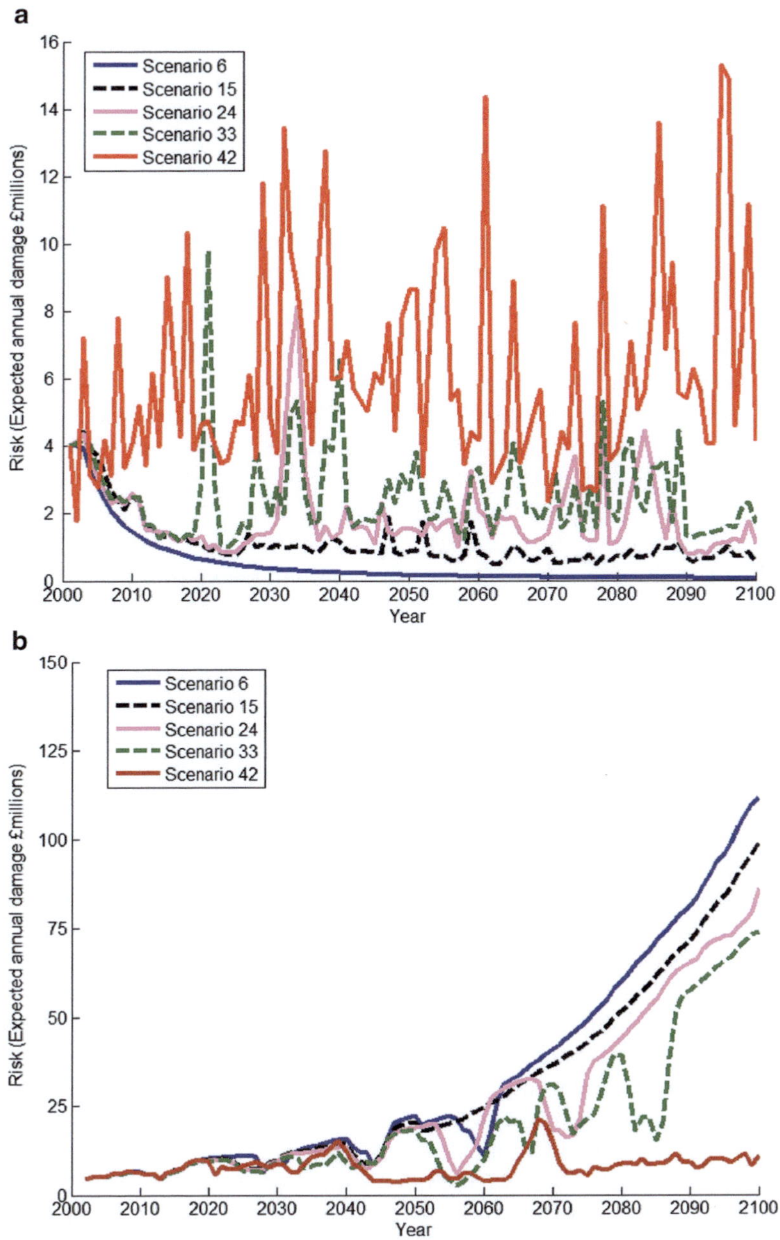

Fig. 9.3 Comparison of the evolution of (**a**) erosion and (**b**) flood risks over the twenty-first century under a high climate change scenario and assuming no change in socio-economic vulnerability: Scenarios 6, 15, 24, 33 and 42 (Note: (**a**) and (**b**) are plotted to different scales) represent decreasing levels of cliff protection (See Dawson et al. 2009)

over the twenty-first century, the erosion risk associated with this scenario decreases due to the slope of the cliff tending towards an increasingly stable angle. Alternatively, additional geotechnical stabilisation measures could be implemented, but these are costly and save relatively few houses. Removal of cliff protection results in increasingly severe economic losses on the cliffed coastline according to the amount of protection lost. On the other hand, whilst the removal of cliff protection will result in significant economic losses on the cliffed coastline, the gains are usually of an order of magnitude greater, in terms of risk reduction, on the flood-prone coast.

Other results, discussed in greater depth by Dawson et al. (2009) and reproduced in Appendix A show that:

- The effects of increases in offshore wave heights had little impact on erosion rates and consequently flood risk.
- Changes in offshore wave direction by +/– 10° proved more significant in terms of erosion; the consequent changes in available sediments were not sufficient to result in a noticeable impact on the flood risk.
- Current observed rates of sea-level rise are reflected in more or less steady risk levels over the twenty-first century.
- A tripling of the rate of sea-level rise from current rates to 4.5 mm/year over the twenty-first century results in only an approximate doubling of the flood risk. The reduction is because rising sea levels will result in increasing release of sediments from the up-drift cliffed coast which provides protection and reduced flood risk in the Norfolk Broads. However, this feedback is insufficient to compensate for the effect of larger changes in climate: a 1.2 m sea-level rise over the twenty-first century results in greatly increased levels of flood risk.
- As shown in Fig. 9.3, protection of the soft cliffs reduces erosion risks, but the benefits these eroded sediments provide in terms of flood risk reduction are an order of magnitude greater.
- The most marked differences between scenarios become noticeable over the second half of the twenty-first century; before that time, the difference between scenarios does not exceed the range of variability in the output from the coastal models. This has important implications for the use of cost benefit analysis based upon net present value calculations of risk – because the risk profiles only diverge later in the twenty-first century the net present risks are similar.
- Socio-economic growth in the region increases both flood and erosion risks. Given the assumptions made in this study, this increase is a similar order of magnitude when compared to a medium sea-level rise scenario but relatively insignificant when compared to the highest potential sea-level rise scenarios.
- Doubling of the number of properties (as occurs in the world markets scenarios) does not imply a doubling of flood risk because the location of development is crucial in determining the change in risk: development in floodplain locations with lower inundation probability contributes less towards flood risk when compared to development in areas of the floodplain that experience more frequent inundation.

9.4 Detailed Analysis of Beach Nourishment in North Norfolk

The preceding results analyse flood risks over the entire domain shown in Fig. 9.4. The modelling tools used to calculate these risks are suited to analysis over such a large spatial domain and made use of the best available climate model outputs at the time.

Although the results provided a number of important insights, it is clear from Fig. 9.3 that there is a large amount of interannual variability in the risk. This is because each time series of risk is the outcome of one morphological simulation – which is in turn driven by one time series of extreme wave and water levels. Alternative time series of storms, which are equally consistent with climate variables and storm surge statistics, would show different interannual fluctuations. However, the advent of probabilistic climate information in 2009 enabled this variability to be explored in greater depth by driving the risk analysis with an ensemble of 250 time series of extreme wave and water levels. Moreover, this enabled a more useful analysis of beach nourishment strategies (following current practise), and their mean nourishment volumes, than is feasible with a single realisation.

This more detailed flood risk analysis focused on the region between Happisburgh and Winterton where flood risk is most sensitive to morphological changes. The region comprising the first 30 beach sections in the SCAPE model described in

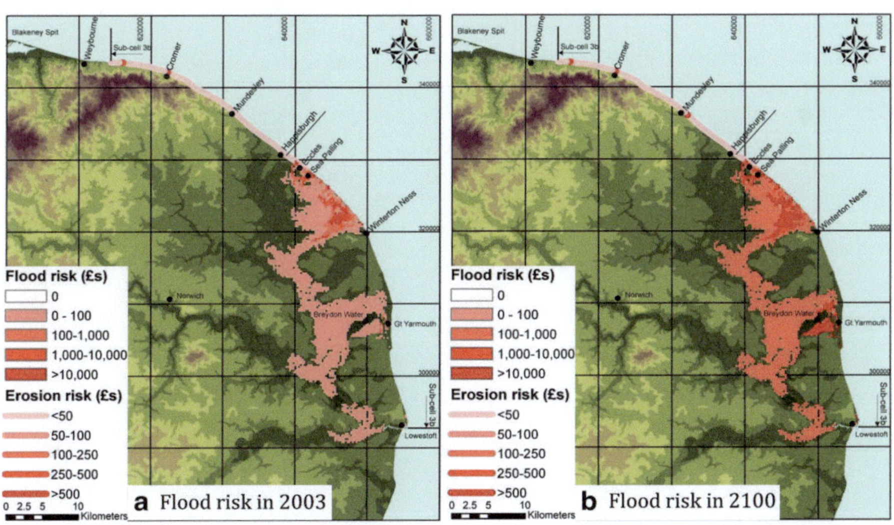

Fig. 9.4 Map of flood and erosion risks, expressed in terms of expected annual damages (**a**) at present and (**b**) in 2100 for Scenario 15 with constant socio-economic conditions. The cliff erosion risk has been scaled to the area of a flood cell (250 m × 250 m). Darker shades imply a greater risk (See Dawson et al. 2009)

Chap. 7 (see Fig. 9.5) also indicates that the major settlements of Horsey corner, Sea Palling and Eccles-on-Sea have been additionally protected with groynes and seawalls since the latter part of the nineteenth century. More recently, a system of offshore reefs has been constructed between Eccles-on-Sea and Sea Palling, and the beaches have been repeatedly nourished with dredged sand: nourishment is triggered when beach levels fall below a critical level. Conversely, nourishment is not needed if enough sediment is supplied to maintain beach levels.

As discussed in Chap. 7, the beach levels on this section of coast are sensitive to the coastal management strategy adopted for this coast and for the up-drift eroding cliffs between Sheringham and Happisburgh. The same management strategies from Appendix A, and described in Chap. 7, are analysed here with respect to their effect of flood risk. Each of these management scenarios, with the exception of a ‘no

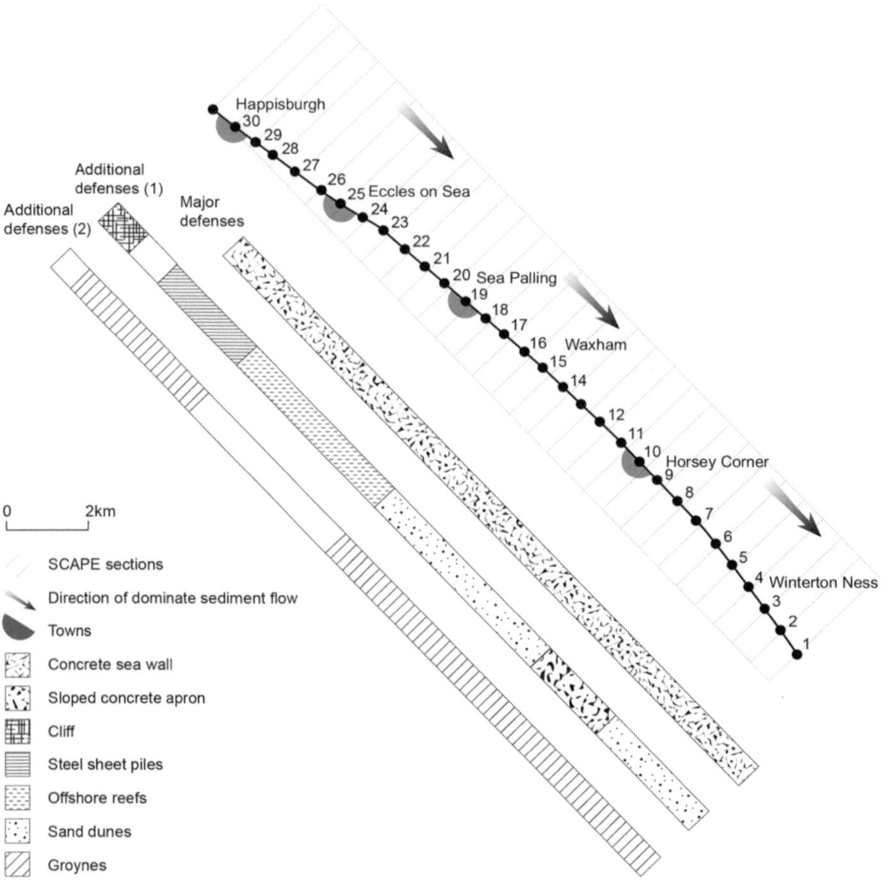

Fig. 9.5 Coastline included in the flood risk analysis, highlighting key settlements and flood defence infrastructure

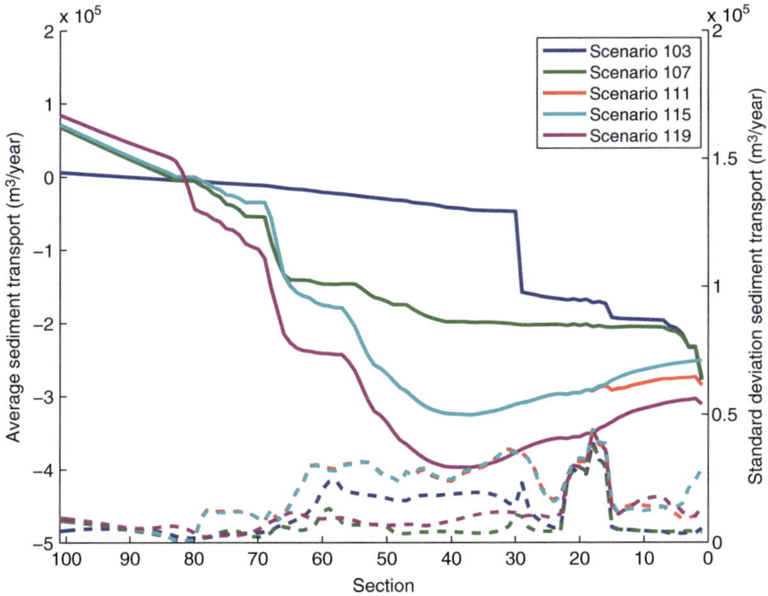

Fig. 9.6 Annual average (*continuous*) (and standard deviation (*dashed*) of 250 projections) sediment transport for a range of management scenarios, all for high rates of sea-level rise, at each model section along the coastline

nourishment' scenario, is supplemented with beach nourishment in order to maintain the beach levels following the current management approach. The effect of these strategies on the mean and variance of sediment transport along the coast is illustrated in Fig. 9.6. Of particular interest is the net volume of sediment being transported into the Happisburgh to Winterton site for different management scenarios. Thus, a combination of net southerly beach drift and periodic nourishment, if needed, results in a gradual increase in the beach level, more or less keeping up with sea-level rise (Fig. 9.7). The beach level resulting from a 'High++' sea-level rise scenario will increase from 1.1 to 2.5 m by the end of twenty-first century.

9.4.1 *Incorporation of Climate Uncertainties into Risk Analysis*

Here we extend the analysis to incorporate the effect of uncertainty in sea-level rise. In common with the approach adopted in Chap. 7, we deal with the uncertainty in sea-level rise by exploring a Monte Carlo sample of 250 sea-level rise simulations s_{ke}: $k = 1, \ldots, 250$ for each emissions scenario e. Each of these simulations is based upon a different random sampling of day-to-day and extreme wave conditions, so as

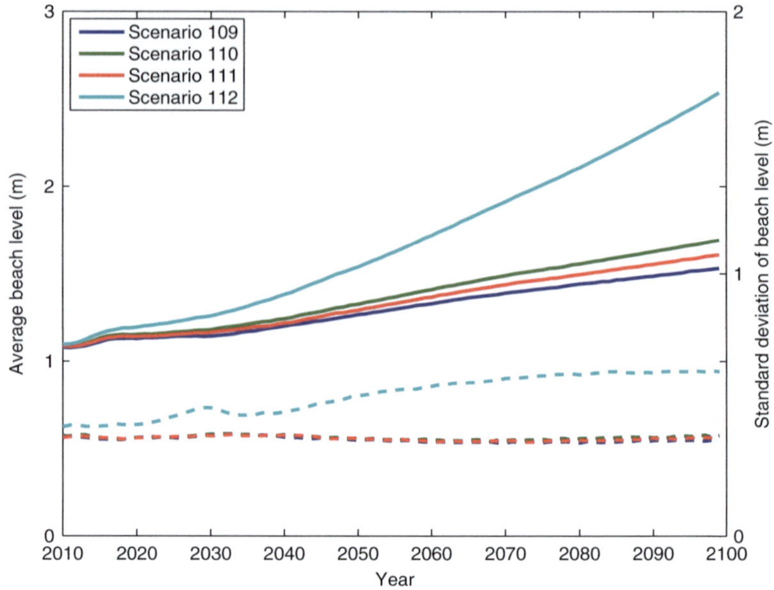

Fig. 9.7 Average level (*continuous line*) of the beach between sections 2 and 27 (and standard deviation of 250 projections (*dashed line*)) each year between 2010 and 2100 under a range of sea-level rise scenarios, for shoreline management strategy M4

well as representing the uncertainty in long-term sea-level rise, the simulations also sample the variability in wave conditions, coastal erosion and beach levels. Whilst 250 is a reasonable number of samples of this variability, we are particularly concerned about the probability of situations in which the beach levels are particularly low, which may not be extensively sampled in our Monte Carlo simulations. We therefore fit a probability density function, $f_b(b|t, s_{ke})$, describing the joint distribution of beach levels $b=(b_1,\ldots, b_n)$ in year t for each sea-level rise simulation s_{ke}. The dependence structure of this joint density function is described by a copula function (Nelson 2006). Attention is focussed upon the joint probability distribution of extremely low beach levels, using a peaks *under* threshold method. Further details of the approach are provided below.

Each element i of the flood defence system may be a 'failed' or 'non-failed' state. There are therefore 2^n combinations of failure and non-failure, which are labelled C_j: $j=1,\ldots, 2^n$. The probability of being in any one of these system states is described by a fragility function $P(C_j|b, c)$ and conditional upon the vector of beach levels b in front of the defences whose critical beach level is characterised by $c=(c_1, \ldots, c_n)$. Given the variability in beach level described by $f_b(b|t, s_{ke})$, the probability of being in a defence system state C_j during year t for sea-level simulation s_{ke} is given by the multidimensional integral:

$$P\left(C_j|t,s_{ke}\right)=\int P\left(C_j|b,c\right)f_b\left(b|t,s_{ke}\right)\mathrm{d}b \tag{9.3}$$

The severity of the damage that occurs during a breach depends upon the incident water level z when the breach occurs. We cannot predict precisely what that water level will be, but we can estimate the likely water level, given that a particular breach scenario has occurred. We write this as the conditional density function $f(z|C_j)$. Given the water level z and the beach level b the flow into the flood zone can be calculated for a given breach scenario as the function $q(z, b, C_j)$. Chapter 8 explained how the flood depths in the flood zone can be computed for a given inflow q. The flood damage can be calculated at each location in the flood zone given the maximum flood depth. All of this computation is condensed into a damage function $D(q)$. The expected annual flood damage $E(D)$ for a given sea-level rise scenario at year t is therefore estimated as

$$E_{s_{ke},t}(D) = \sum_{j=1}^{2^n} \int f_b\left(b \mid t, s_{k,e}\right) P\left(C_j \mid b, c\right) f\left(z \mid C_j\right) D\left(q\left(z, b, C_j\right)\right) \mathrm{d}b\mathrm{d}z \qquad (9.4)$$

9.4.2 Nonstationary Analysis of Extreme Beach Lowering

When beach levels are high, relative to the toe level of the coastal defences, the probability of breaching is negligibly small. The critical situations with respect to flood risk are when beach levels are extremely low, in which case the probability of breaches occurring is high. If the distribution of beach levels was stationary through time, then a conventional extreme value analysis could be applied to the beach statistics. However, beach levels are changing on a range of timescales and will continue to change in future. We therefore apply a nonstationary analysis to simulations of future beach level in order to estimate how the probability of extremely low beach levels will change.

The first step in analysing the probability of extreme beach lowering is to remove the long-term trend from the beach level simulations. To illustrate this approach, Fig. 9.8 shows the simulated beach level under one of 250 projections for the years 2010–2099 for section 3 near Winterton-on-Sea. The 5th percentile of this distribution is identified as the threshold of extreme values, and a quadratic regression curve is fitted through time. The probability distribution of the excesses below the threshold was modelled using a generalised Pareto distribution (GPD) (Coles 2001).

9.4.3 Analysis of Flood Defence Failure Probabilities

Initial analysis of the probability of failure of the flood defence system identified that sections 3, 5 and 16 are by far the most prone to failure. All other sections make negligible contribution to the probability of flooding. The analysis therefore proceeds by just considering the probability of failure of these sections, either

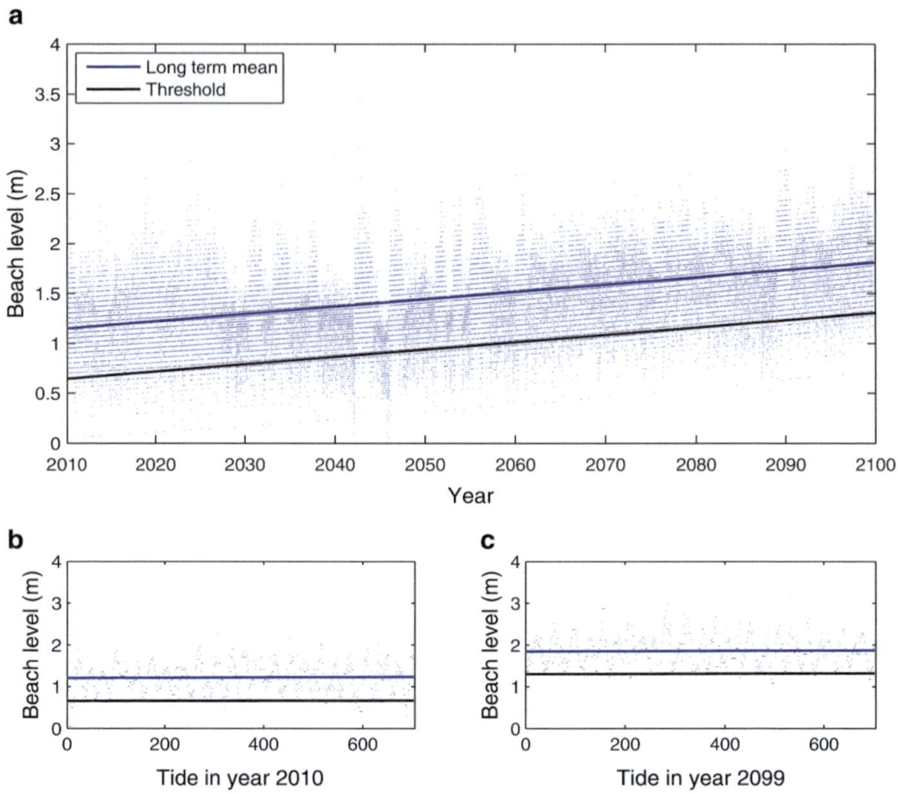

Fig. 9.8 Simulated beach levels at beach section 3 (near Winterton) for Scenario 111: (**a**) 2010–2099 beach level time series; (**b**) beach level time series for time slice 2010; (**c**) beach level time series for time slice 2099

individually or in combination. In this case, there are 2^3 possible combinations, *C*, including 2^3 -1 failed states, as summarised in Table 9.1.

The fragility of these defence sections with respect to breaching has been established by the Environment Agency (Environment Agency 2006; Halcrow 2002) and is described by a normal distribution with mean 0.8 m and standard deviation 0.35 m. The probabilities of failure to possible combined defence system states are calculated from 2010 to 2099 for these system states in Scenario 111 are shown in Fig. 9.9. The probability curve labelled 'Annual' is the sum of the probability of the various failure states. The contribution of each defence combination to the annual failure probability can be distinguished clearly. The annual probabilities are usually dominated by one or two significant failure combinations. The failure combinations C4 and C7 have extreme low probabilities. The failure probabilities of each combination evolve through time with respect to the varying beach conditions.

Table 9.1 Summary of failed defence combinations, referred to as system state

No	Description
C_0	No breach
C_1	Breach in section 3
C_2	Breach in section 5
C_3	Breach in section 16
C_4	Breach in sections 3 and 5
C_5	Breach in sections 3 and 16
C_6	Breach in sections 5 and 16
C_7	Breach in sections 3 and 5 and 16

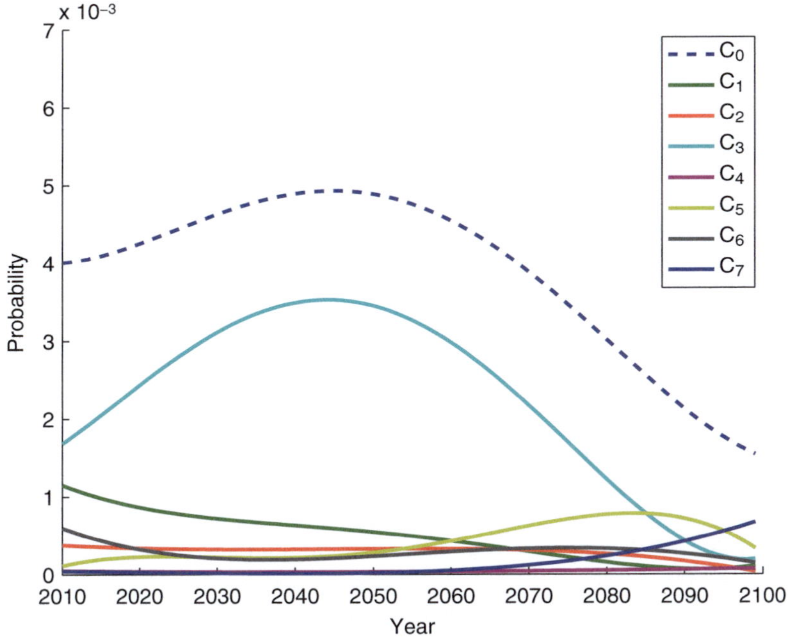

Fig. 9.9 Probability of various flood defence system states, described in Table 9.1, over the twenty-first century for a single climate projection for Scenario 111

It is interesting to note that the probability of failure goes up then down, whilst the beach levels are monotonically rising in Fig. 9.7. The failure probability of failure depends not only on the beach condition but also on the incident tide and wave conditions. Whilst the beach levels may be raising in absolute terms, what is significant is their level relative to sea level and extreme storm conditions.

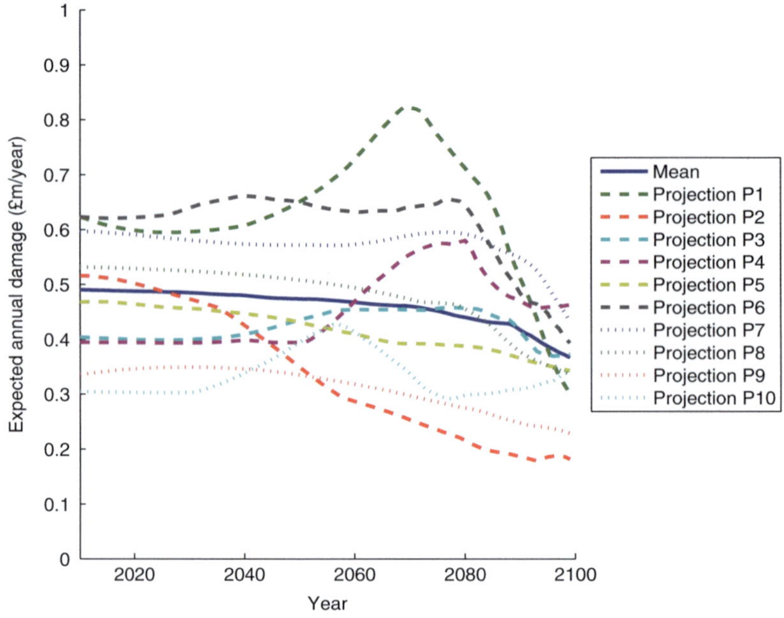

Fig. 9.10 Risk over the twenty-first century for 10 (of a total 250 climate projections) under Scenario 111

9.4.4 Changes and Uncertainties in Flood Risk

Figure 9.10 illustrates how economic flood risk (or expected annual damages, EAD) fluctuates through time for 10 of 250 projections for Scenario 111. The magnitude of the risk is significantly less than the earlier simulations (see also Dawson et al. 2009) because of the effect that the beach nourishment strategy, being implemented in these simulations, has on the probability of flooding. Another difference is that the area of floodplain considered here is rather less than that considered by Dawson et al. (2009) (reproduced in Fig. 9.3). The flood risk is shown to be a complex function of a number of factors: the system state, its probability of occurrence and associated damages subject to loadings. The temporal distribution of EAD follows the trend in flooding probabilities.

Figure 9.11 illustrates spatial distribution of the average and standard deviation of EADs for Scenario 111 in 2099. The patterns of spatial change in these values are similar, but the magnitudes of standard deviations are smaller.

Figure 9.12 illustrates the changing flood risk for different emissions scenarios, averaged over all 250 simulations for each emission scenario. The risk is not particularly sensitive to sea-level rise scenario in the first half of the twenty-first century. Because of the effect of increased sediment supply, from the up-drift coast and due to beach nourishment, the flood risk is actually projected to decrease with the 'High++' scenario in the latter part of the twenty-first century. This demonstrates how sustained beach nourishment can be employed to respond even to high sea-level rise scenarios.

Figure 9.13 shows the influence of beach management strategies on the evolution of flood risk. In particular it illustrates rapid increase in risk if no beach nourishment

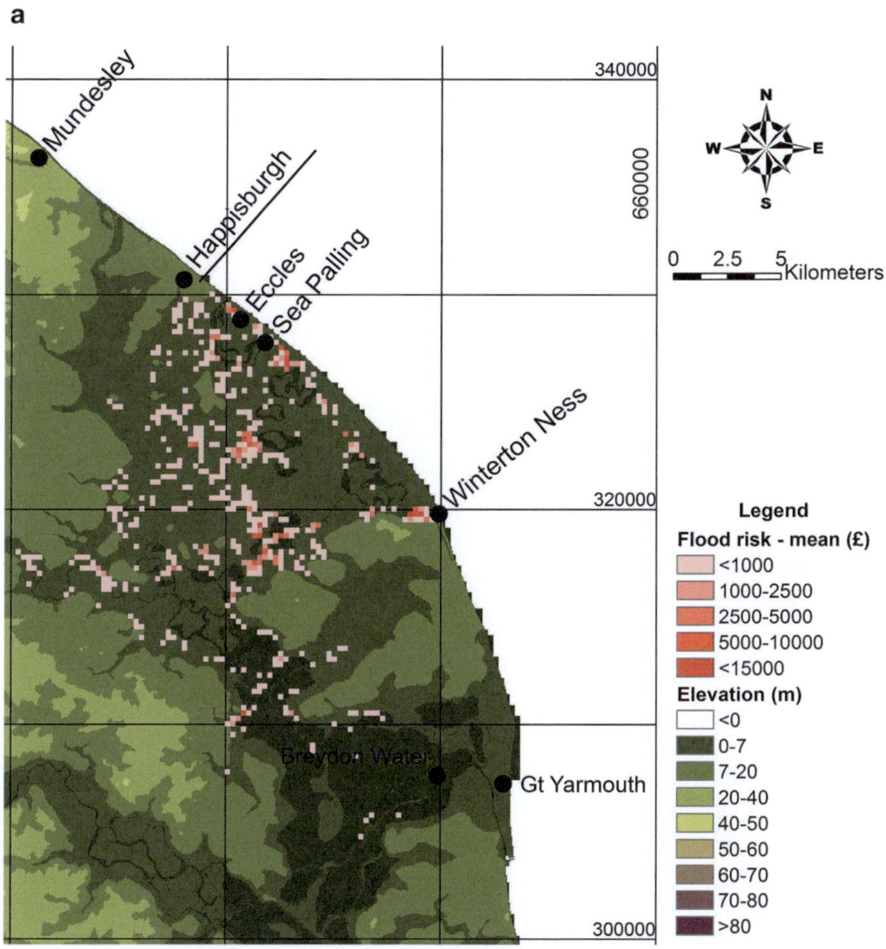

Fig. 9.11 Spatial distribution of risk in 2099, calculated from 250 projections, for Scenario 111: (**a**) mean of the expected annual damages (EAD); (**b**) coefficient of variance (CoV) of the EAD

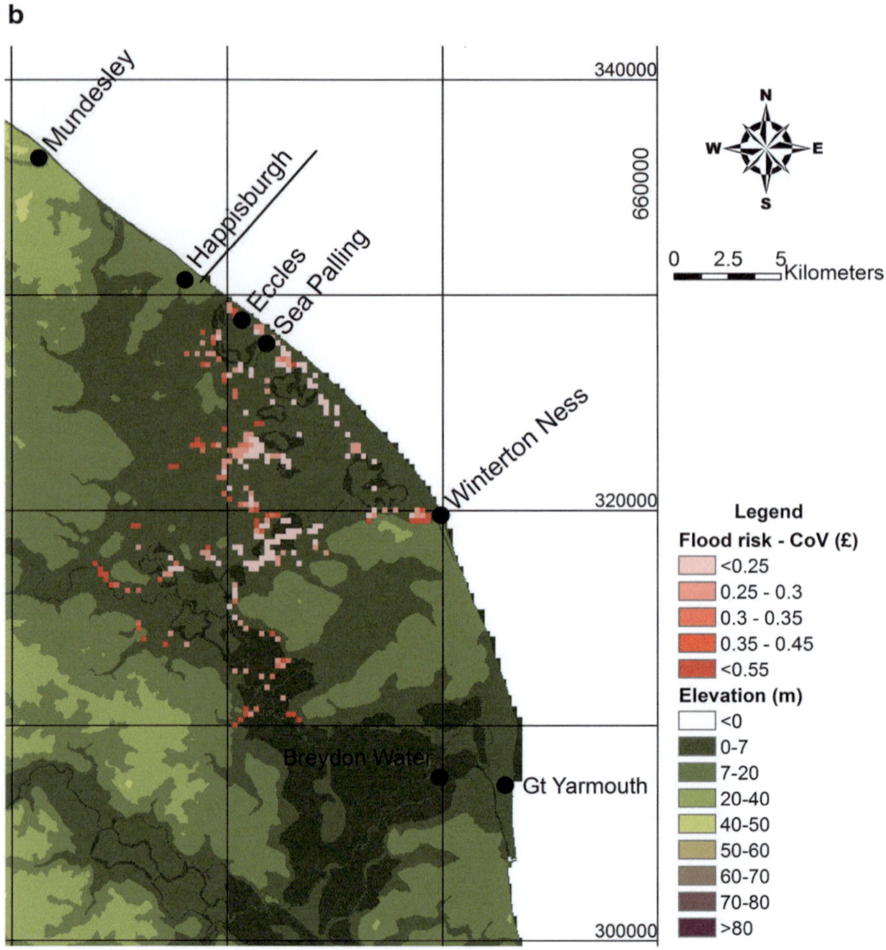

Fig. 9.11 (continued)

is applied to the system. On the other hand, removing the cliff protection up-drift of the site supplies considerable additional sediment and contributes to a reduction in flood risk. In this case the overall volume of beach nourishment required to ensure beach levels remain above the critical threshold is reduced as shown in Fig. 9.14.

All of the risk estimates presented above have been for the BL (baseline) socio-economic scenario. Different socio-economic scenarios (Chap. 4) scale the magnitude of flood damages up or down, resulting in a corresponding increase or decrease in flood risk. A comparison of the effect of different socio-economic scenarios for Scenario 111 is shown in Fig. 9.15. However, the difference between these scenarios is actually rather modest, because of the slowly changing nature of vulnerability in the flood zone.

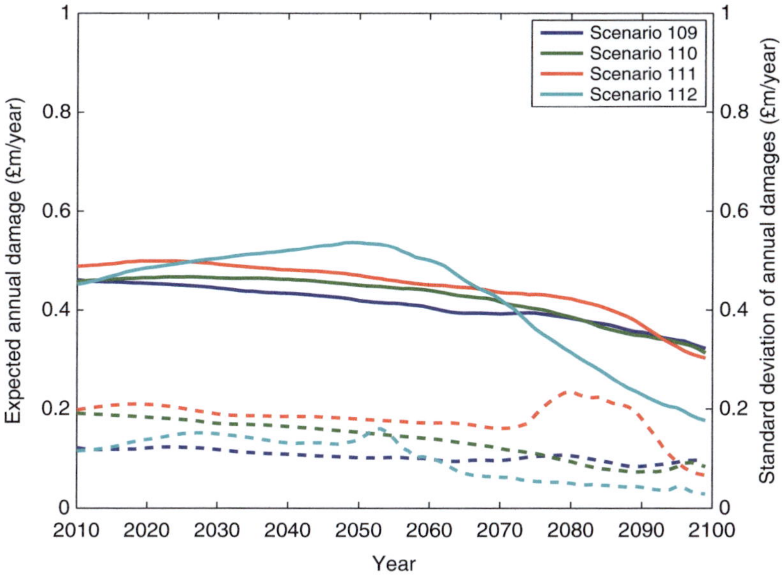

Fig. 9.12 Time series of flood risk for a range of sea-level rise scenarios, under management strategy M4. *continuous line* expected annual damage; *dashed line* standard deviation of EAD

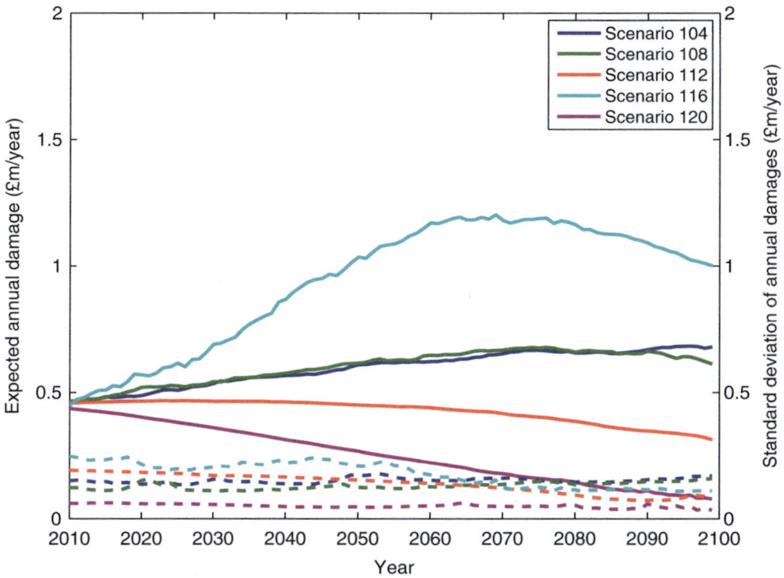

Fig. 9.13 Time series of flood risk under various management strategies for the 'High++' sea-level rise scenario. *continuous line* expected annual damage; *dashed line* standard deviation of EAD

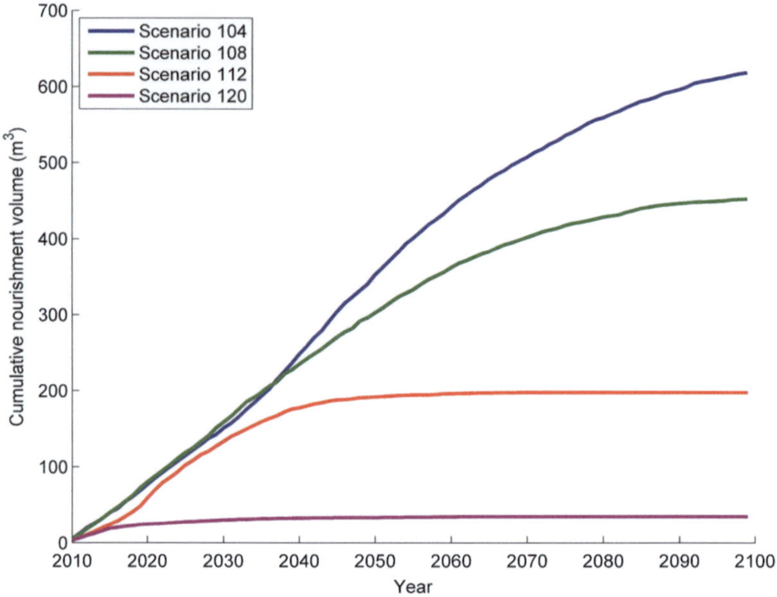

Fig. 9.14 Cumulative beach nourishment volume required over the twenty-first century for different management strategies, for the 'High++' sea-level rise scenario, to ensure beach levels remain above the critical threshold

9.5 Conclusions and Next Steps

A systematic appraisal of the effects of climate and socio-economic change and coastal management policy on coastal flood risk has been described. This has brought together a number of aspects of the analysis described in this volume, including analysis of the uncertainties in sea-level rise and multiple simulations of potential future morphological change. The results of coastal management and climate change scenarios for two natural hazards have, for the first time, been presented quantitatively in terms of changes in expected annual damage. This enables direct economic evaluation of local effects of modifying coastal cliff erosion compared with potential broader-scale effects on flood risk. The analysis indicates that the main drivers for flood risk over the twenty-first century in northern East Anglia are a consequence of a combination of:

- Rising sea levels
- Local and broad-scale natural and anthropogenically driven morphological change that may lower or raise beach levels
- Changes in exposure to flooding and erosion due to socio-economic changes

Increases in wave height are less important owing to attenuation across the nearshore and are only felt on the coast when accompanied by high sea-level rise.

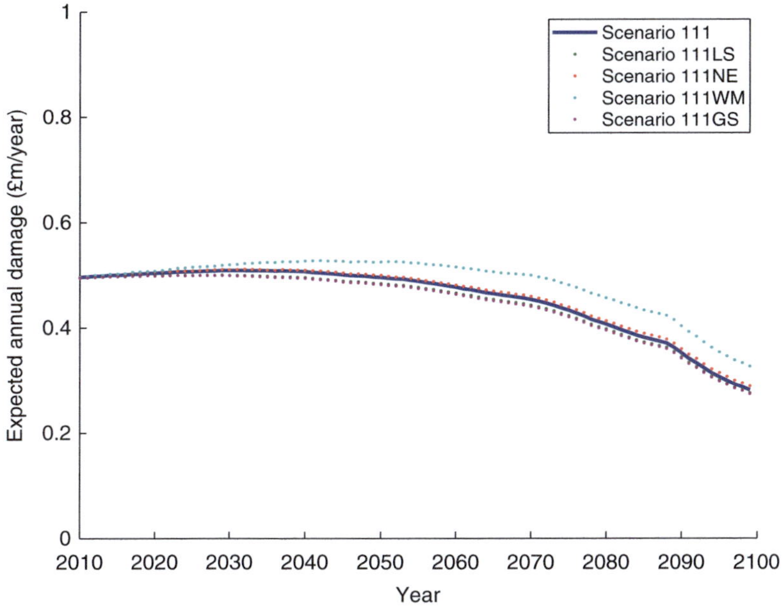

Fig. 9.15 Temporal distribution of flood risk under various socio-economic scenarios for Scenario 111 (management strategy M4 and high sea-level rise)

Possible changes in wave direction have a more noticeable influence on coastal morphology, but this does not translate into significant changes in economic risk. Although the risk profiles will be different for other locations, the key drivers are likely to be the same in similar settings around the world's coasts.

The analysis reveals the fundamental importance of sediment supply in determining the probability of flood defence failure and risk of flooding. In the presence of ample beach levels that are allowed to rise with rising sea levels, the current flood risk can be maintained at a relatively stable level and even decreased in future. However, this relies upon supply of sediment, either from the up-drift eroding coast or from artificial beach nourishment.

In the current simulation study, no attempt has been made to optimise the quantity of beach nourishment with respect to the benefits of flood risk reduction, though the simulation framework established here clearly lends itself to the testing and optimisation of alternative nourishment strategies. These strategies may in any case be regarded as being 'adaptive' as the quantity and location of nourishment can be modified from year to year. Given the concentration of the probability of failure at a few locations along the coast, these should be the focus of monitoring and nourishment efforts and would, in due course, benefit from further strengthening of the flood defences.

The methodology adopted here has been novel in its use of multiple long-term simulations of beach volume and level, which reflect both the uncertainty in sea-level rise projections and the natural fluctuations in beach levels. A nonstationary extreme value analysis has been used to estimate the probability of erosion undermining the flood defences, leading to breaching and coastal inundation. Though not described in detail in this chapter, the joint analysis of extreme values was conducted using a copula function. The approach has enabled computation of small failure probabilities from a relatively small sample (250) of beach level simulations. The methods used here could be applied more widely to support erosion and flood management.

References

Coles, S. (2001). *An introduction to statistical modeling of extreme values*. London: Springer.

Dawson, R. J., & Hall, J. W. (2006). Adaptive importance sampling for risk analysis of complex infrastructure systems. *Proceedings- Royal Society of London A, 462*(2075), 3343–3362.

Dawson, R. J., Dickson, M. E., Nicholls, R. J., Hall, J. W., Walkden, M. J. A., Stansby, P., Mokrech, M., Richards, J., Zhou, J., Milligan, J., Jordan, A., Pearson, S., Rees, J., Bates, P., Koukoulas, S., & Watkinson, A. (2009). Integrated analysis of risks of coastal flooding and cliff erosion under scenarios of long term change. *Climatic Change, 95*, 249–288.

Dickson, M. E., Walkden, M. J., et al. (2007). Modelling the impacts of climatic change on an eroding coast over the 21st century. *Climatic Change, 81*(2), 141–166.

Environment Agency. (2006). Happisburgh to Winterton sea defences stage 3b works. Technical Report. Environment Agency, Bristol. http://www.environment-agency.gov.uk/research/library/consultations/54395.aspx. Cited 20 Dec 2013.

Frew, P., & Guest, S. (1997). Overstrand coast protection scheme. In: Proceedings of the MAFF conference of river and coastal engineers. MAFF, London

Halcrow. (2002). *Happisburgh to Winterton Sea defences: Stage three strategy review. Appendix C*. London: Halcrow.

Hall, J. W., Lee, E. M., & Meadowcroft, I. C. (2000). Risk-based benefit assessment of coastal cliff protection. *Proceedings- Institution of Civil Engineers Water Maritime and Energy, 142*, 127–139.

Nelson, R. B. (2006). *An introduction to copulas*. New York: Springer.

Nicholls, R. J., Tol, R. S. J., & Hall, J. W. (2007). Assessing impacts and responses to global-mean sea-level rise. In M. E. Schlesinger, H. S. Kheshgi, J. Smith, F. C. de la Chesnaye, J. M. Reilly, T. Wilson, & C. Kolstad (Eds.), *Human-induced climate change*. Cambridge: Cambridge University Press.

Thorne, C., Evans, E., & Penning-Rowsell, E. (2007). *Future flooding and coastal erosion risks*. London: Thomas Telford.

Chapter 10
Visualising Potential Coastal Change: Communicating Results Using Visualisation Techniques

Simon Jude, Mustafa Mokrech, Mike Walkden, James Thomas, and Sotiris Koukoulas

Abstract A key challenge facing coastal management organisations is the need to effectively communicate the potential risks posed by coastal change and associated adaptation options, particularly if the potential for misunderstanding and conflict is to be avoided (e.g. in cases which are scientifically complex and involve many interested parties). GIS technology is appropriate and widely used by coastal planners and managers; however, its application in decision-making processes is often hampered by the highly technical output that they tend to produce. This can prove particularly difficult for members of the public to interpret. The dynamic, multidimensional and uncertain nature of the coastal zone presents particular challenges to those wishing to communicate possible future coastal change and management options.

S. Jude (✉)
Centre for Environmental Risks and Futures, Department of Environmental Science and Technology, School of Applied Sciences, Cranfield University, Cranfield, Bedford MK43 0AL, UK
e-mail: s.jude@cranfield.ac.uk

M. Mokrech
Environmental Institute of Houston, School of Science and Computer Engineering, University of Houston Clear Lake, Houston, TX 77058, USA
e-mail: mokrech@uncl.edu

M. Walkden
WSP, Keble House, Southernhay Gardens, Exeter EX1 1NT, UK

WSP Group, Exeter, UK
e-mail: mike.walkden@wspgroup.com

J. Thomas
Experience Prototyping Technician, School of Design, Northumbria University, Newcastle Upon Tyne NE1 8ST, UK
e-mail: James.E.Thomas@northumbria.ac.uk

S. Koukoulas
Department of Geography, University of the Aegean, University Hill, Mytilene 881100, Greece
e-mail: skouk@geo.aegean.gr

R.J. Nicholls et al. (eds.), *Broad Scale Coastal Simulation*, Advances in Global Change Research 49, DOI 10.1007/978-94-007-5258-0_10

This chapter describes how the Tyndall Coastal Simulator project has worked to address this challenge, presenting the visualisation methods that have been developed to assist in the communication of the Tyndall Coastal Simulator's output. These methods range from relatively low-cost landscape visualisation tools underpinned by expert judgement and simple rule-based models, used to communicate managed realignment options, to sophisticated virtual reality systems enabling users to interactively explore future coastal landscapes. The chapter also presents a variety of visualisation approaches explored during the Tyndall Coastal Simulator project that have not only sought to improve the scientific robustness of the visualisations but facilitate the communication of temporal change and uncertainty to end users. Furthermore, the chapter illustrates how the project has also actively developed a number of interactive decision-support tools with embedded visualisation functionality, together with examples of how the research has critically evaluated such tools and visualisations in both applied decision-making and stakeholder engagement processes and experimental studies. Whilst the chapter highlights the potential of such visualisation techniques in coastal management processes, it also discusses the challenges associated with testing such tools in real-world decision-making contexts, particularly the constraints posed by the highly sensitive and emotive issue of managing coastal change. It concludes by arguing that a suite of visualisations may be required by those wishing to communicate coastal change and presents a future research agenda for this work. Crucially, this emphasises the need for further applied and practical testing of visualisation technologies through collaborative research to determine their effectiveness in facilitating improved communication and participatory decision-making as is developed further in Chap. 12.

Keywords Visualisation • Adaptation options • Coastal management • Decision support • GIS

10.1 Introduction

The preceding Chaps. (2, 3, 4, 5, 6, 7, 8 and 9) have considered the generation of large amounts of technical information relevant to coastal management, with Chaps. 6, 7 and 9 most directly influencing human populations and the environment. This chapter considers how such information can be made more accessible to a wide range of stakeholder interests so it can inform the debate surrounding coastal futures. Chapter 11 takes forward aspects of these insights as part of the development of the Tyndall Coastal Simulator interface, and Chap. 12 discusses in more detail the broader social issues concerned with future coastal planning and management.

In light of the risk posed by sea-level rise, coastal change, and the need for the planning and implementation of adaptation responses, a key challenge facing coastal managers is the need for effective communication and engagement with members of the public. Without this, there is a risk of conflict arising, particularly where highly sensitive and emotive coastal management decisions concerning issues, such as the use of 'soft forms' of coastal defence, including managed realignment and the adoption of 'retreating the line' shoreline management plan (SMP) policies are under discussion. However, communicating such issues to members of the public represents a significant challenge, not least because of the highly dynamic, multidimensional and uncertain nature of the coastal environment and coastal processes as explored in Chap. 12. These features present significant challenges to traditional two-dimensional (2D) forms of spatial representation including paper maps and digital mapping (Bartlett and Wright 2000; Raper 2000). Furthermore, such challenges are further compounded by the potential barrier that the complicated map-based information accompanying planning proposals and SMPs can pose to many (Appleton and Lovett 2005).

In recognition of the limitations posed by traditional 2D *'flatland'*, maps and forms of communication have come calls to develop approaches enabling the *'envisioning'* of multidimensional environmental information in three dimensions (3D) (Raper 2000; Tufte 1990). This has led to the development of visualisation technologies and techniques enabling users to explore virtual landscapes that are created from geographical information system (GIS)-based data (Appleton and Lovett 2003). It is recognised that such visualisation techniques may prove valuable in enabling public engagement and participation in decision-making processes (e.g. Al-Kodmany 2001; Orland et al. 2001; Appleton and Lovett 2005), including in participatory coastal management processes (Jude et al., 2003). As such, this chapter outlines the steps that the Tyndall Coastal Simulator project has undertaken to address the challenge of visualising the output that it produces.

10.2 The Tyndall Coastal Simulator: Visualisation Approaches

Facilitating effective communication and dialogue between those engaged in discussions regarding the future management of the coast is of critical importance to those involved in coastal management processes because poor communication can result in conflict regarding coastal management and adaptation policies. To address this communications challenge, a key element of the Tyndall Coastal Simulator project has involved developing and testing methods for visualising the complicated model output that it produces. The aim of this work has been to aid the communication of the outputs from the Tyndall Coastal Simulator in a manner that facilitates dialogue between coastal scientists, decision-makers and members of the public regarding the choices facing the management of the coast in response to coastal change.

A variety of approaches for visualising the output produced by the Tyndall Coastal Simulator are available, from traditional scientific graphics and 2D digital mapping to more advanced 3D virtual reality (VR) and landscape visualisation approaches and methods for presenting temporal change. As noted earlier, the multidimensional and dynamic nature of the coastal zone poses challenges to traditional 2D digital mapping and GIS, and as such, it was felt that the Tyndall Coastal Simulator provided an ideal opportunity to develop and test new techniques for communicating the scientific model output that it produces.

A key consideration in the choice of visualisation approach adopted by the Tyndall Coastal Simulator project was the potential complexity involved with producing visualisations from the wide array of output from the system. In particular, the differing types of model output within the Tyndall Coastal Simulator require different visualisation approaches if they are to be communicated effectively, thus potentially requiring a suite of different visualisation types. The ability to create a variety of visualisations from a range of different data outputs also posed challenges regarding the choice of software to be used. For example, whilst traditional scientific graphics or off-the-shelf visualisation software could have been used to visualise the output from the Tyndall Coastal Simulator, such an approach would not only be expensive but constrained by limited software functionality. Alternatively, whilst researchers such as Stock and Bishop (2005, 2006) have developed custom visualisation software based on games-engines, such an approach was inappropriate for the Tyndall Coastal Simulator project due to cost and complexity. Fortunately, as discussed in Chap. 10, ArcGIS, which was used as the basis for the simulator, not only provides extensive 2D and 3D visualisation functionality, but can be extended by users and importantly is already used by many coastal management organisations.

Whilst ArcGIS underpins the visualisations provided in the Tyndall Coastal Simulator's interface, the wider project has explored a wide range of approaches for visualising the scientific model outputs that the system produces. The evolution and development of these differing approaches and their potential benefits and constraints are now described

10.2.1 Visualising Coastal Change and Adaptation Options

The availability of 3D visualisation functionality in desktop GIS packages, such as ESRI's ArcGIS, together with the increasing availability of low-cost landscape visualisation software packages, such as Visual Nature Studio from 3D Nature (2013), provides new opportunities for communicating future coastal change and adaptation options. The availability of such software provided the foundation for initial research exploring their potential in communicating future coastal change and management options and to facilitate public engagement in coastal decision-making processes (Jude et al. 2006, 2007). This resulted in the successful production of a series of visualisations illustrating the potential impact of sea-level rise and possible managed realignment and defence options for sites at Holme-Next-Sea and Brancaster Marshes, on the north Norfolk coast, using a range of standard GIS datasets that are commonly available to coastal managers (Fig. 10.1). However, one limitation with this original project was the utilisation of simple GIS analyses

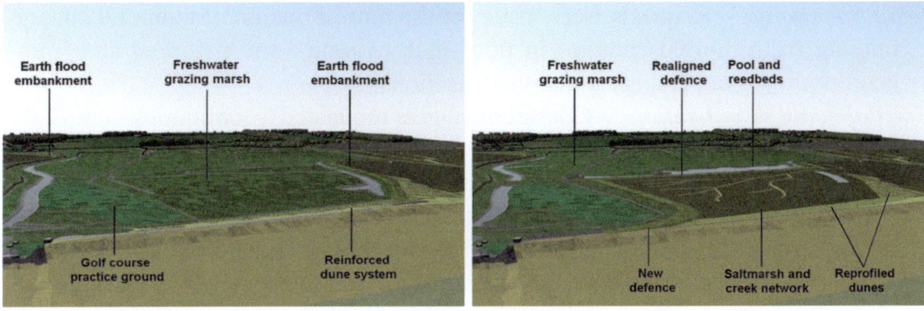

Fig. 10.1 Examples of the static visualisations created to illustrate different potential outcomes of the Brancaster West Marshes managed realignment scheme. These early images were based on management plan documents, simple GIS analyses and expert judgement (Jude et al. 2006)

combined with expert judgement and management plans as the basis of the visualisations. Significantly, this approach limits the scientific robustness of the visualisations, which is a key consideration if such images are to be successfully used in participatory decision-making processes. To address this concern, subsequent work led to the evolution of the visualisation methodology through the development of a simple rule-based methodology for predicting coastal habitat change responses to sea-level rise, which underpins the land cover data used in the visualisations (Jude et al. 2007). Whilst such an approach is simplistic, it has the benefit that it can be applied by coastal management organisations without the need for complicated scientific models.

The development of visualisation methodologies only represents part of this research, with a key element of the work involving the testing and evaluation of the images to determine their potential benefits and limitations. As such, both interviews with coastal management organisations and experimental research accompanying the production of these initial visualisations were undertaken. The results from this work suggest that such approaches have a range of potential uses in coastal management processes and confirmed that they may prove valuable in facilitating communication between coastal management organisations, stakeholder groups and members of the public, particularly as a means of facilitating dialogue between conflicting groups (Jude et al. 2006; Jude 2008). However, whilst the research identified a range of perceived benefits associated with the visualisations, it also highlighted a number of areas where those interviewed felt there was a need for further development of the images. Four key areas requiring further investigation were identified (Jude et al. 2006; Jude 2008):

Uncertainty – Concerns regarding the *'air of certainty'* associated with the visualisation of inherently uncertain coastal processes and management interventions were raised. As a result there was a desire for visualisations depicting long-term coastal change, highlighting the need to develop techniques enabling the communication of the uncertainty associated with the visualisations.

Temporal change – Requests were made for the representation of temporal change, ranging from diurnal changes in tidal state to long-term historical and future geomorphological change in the visualisations.

Coastal-scale visualisations – Coastal managers requested a continuous view of the whole coastline to enable the wider implications of proposed management interventions to be related to larger-scale coastal processes.

Advanced presentation – More sophisticated means of presentation were also called for, including the application of virtual reality theatres to enable users to gain *'a sense of being there and what it will feel like in the future'* (Jude et al. 2006).

As will now be described, the Tyndall Coastal Simulator visualisation research, which has built upon this initial work, has attempted to address the issues raised by the coastal management community.

10.2.2 3D Visualisation of Cliff Recession

The recognition of the need to develop more robust and objective visualisations that overcome the use of expert judgement and simple rule-based approaches coincided with the development of the Soft Cliff And Platform Erosion (SCAPE) model (Walkden and Hall 2002, 2005; Dickson et al. 2005, see also Chap. 7) and the opening of the Social Science for the Environment, Virtual Reality and Experimental Laboratories (SSEVREL) at the University of East Anglia.[1] This not only provided the opportunity to explore possible approaches for visualising the numerical modelling output provided by SCAPE but also the use of the state-of-the-art visualisation facilities provided by SSEVREL that potentially enable users to be immersed within and interactively explore future coastal landscapes. Furthermore, the work offered a test-bed for the wider Tyndall Coastal Simulator project which would require similar linkages between modelling output and visualisation methods to be developed.

SCAPE is a morphodynamic numerical model that determines the reshaping and retreat of shore profiles along the coast and is described in detail by Walkden and Hall (2005) and Dickson et al. (2005) and in Chap. 7 of this book. The numerical output from SCAPE presents challenges for those wishing to apply the model in decision-making processes and to address that this work was undertaken to exploit GIS and visualisation technology to improve its communication. This led to the development of SCAPE GIS (Koukoulas et al. 2005; Pearson et al. 2005), which used ArcView GIS to add visualisation and spatial analysis capabilities to SCAPE and the clifftop prediction models, enabling the models to be integrated with other spatial datasets and thus facilitating the use of SCAPE in impact analysis and decision support. For example, the interface enables users to visualise both present and future scenarios of shoreline evolution using a number of outputs from the SCAPE model. These include the visualisation of clifftop and toe positions and the creation

[1] http://www.uea.ac.uk/environmental-sciences/ssevrel Cited 18 December 2013.

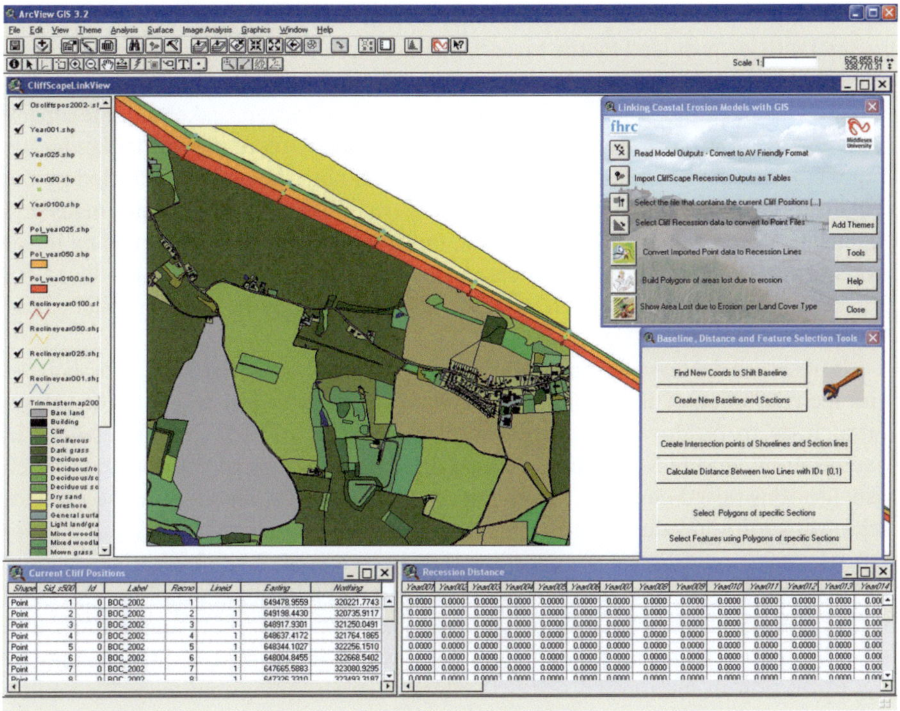

Fig. 10.2 The SCAPE GIS user interface built using ArcView GIS. The clifftop land use is shown together with the uncertainty of clifftop recession

and visualisation of cliff erosion risk zones (Fig. 10.2). Furthermore, the interface also facilitates the calculation of beach volumes, shore platform levels and sediment supply and flux.

Whilst the 2D GIS analyses produced by SCAPE GIS may prove adequate for coastal managers to interpret and understand, this may not be the case for members of the public who may be affected by future coastal change. To address such concerns, work was undertaken to produce a series of interactive VR visualisations of the coast. This offered the opportunity to not only further test and develop methods to visualise the types of numeric output that the Tyndall Coastal Simulator would produce but to evaluate the types of advanced visualisations that had been called for by coastal managers in the earlier Tyndall visualisation research (Jude 2008).

The interactive VR representations of the coast were produced by importing the GIS data into a series of advanced VR visualisation packages. This enabled the production of virtual landscapes that can not only be displayed using desktop or laptop PCs but more advanced portable VR display systems and VR theatres that crucially provide users with a feeling of immersion within the virtual landscape (Fig. 10.3). Such flexibility potentially enables the visualisations to be utilised by coastal management organisations in a wide range of contexts, from scientific and engineering discussions to public consultation and visioning exercises. Furthermore, the interactivity provided by the visualisations potentially overcomes the limitation of viewpoint biases that may be associated with static, animated or structured fly-through visualisations (Jude 2003; Brown et al. 2006). Indeed, stakeholder feedback identified the ability for users to navigate around and explore the virtual landscape (Fig. 10.3) as being a key advantage for end users, with the realism afforded by the visualisations being perceived as being more accessible than traditional maps and documents (Brown et al. 2006). However, this interactivity is accompanied by lower levels of realism than static or animated visualisations, such as those produced for Brancaster West Marshes, which are less computationally intensive to produce (e.g. Appleton and Lovett 2005).

The research did however encounter a number of challenges, particularly associated with the resolution of the 2D cliff recession lines produced by SCAPE GIS. Here, initial tests to produce visualisations from the 2D SCAPE output resulted in visualisations that failed to reflect the detailed morphology of the coastline. Such visualisations raised concerns amongst potential end users, resulting in the interpolation of additional detail in the SCAPE cliff recession lines to produce more realistic representations of the coastline (see Brown et al. 2006 for further details). As Brown et al. (2006) emphasise, such steps to add additional detail not provided by SCAPE result in a tension between the requirement for visualisations that are underpinned by scientific model output – which in the case of SCAPE may be low resolution in nature and result in more abstract, and potentially misleading visualisations (e.g. Tress and Tress 2003) – and non-specialist users, such as coastal managers, who prefer higher levels of realism (e.g. Appleton and Lovett 2003, 2005; Bishop 1994; Lange 2001). Thus, the study highlighted the need for further research exploring the communication of uncertainties, such as the additional

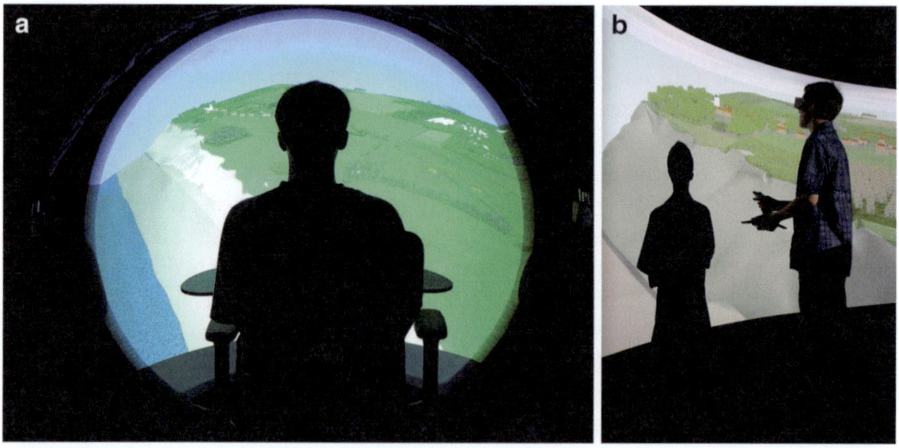

Fig. 10.3 Interactive real-time navigation around the cliff erosion visualisations is not only possible using standard PCs but more advanced and immersive portable VR display systems such as the Elumens VisionStation (**a**) and VR theatres (**b**)

interpolated detail in visualisations, and to investigate issues including the level of acceptable uncertainty (Davis and Keller 1997) in the images and end-user perceptions of uncertainty.

10.2.3 Visualising Temporal Change: iSCAPE

The SCAPE model is a useful tool for predicting the effects of soft cliff, beach and bedrock dynamics under varying natural processes and human interventions. However, as the SCAPEGIS and cliff visualisation project illustrate, difficulties remain when wishing to visualise both the detailed changes in the shore profile that the model produces and temporal change. In particular, as Brown et al. (2006) encountered, the coarse resolution of the SCAPE profiles and the need to modify the underlying elevation data represent particular challenges. Furthermore, whilst it is possible to produce animations illustrating temporal change from specific static viewpoints, the representation of temporal change in real-time visualisation systems, such as those used in the cliff visualisation project, represent significant technical challenges.

Alongside the need to communicate temporal change, challenges also remain in relation to the issue of engrained misconceptions of how the coast reacts under both natural processes and human interventions amongst the general public and some stakeholders. For example, a common expectation is that seawalls represent a permanent solution to erosion, a stance that fails to understand the hidden issues of bedrock lowering and down-drift beach depletion. Whilst these processes and issues can be communicated through literature, there is a need for more persuasive techniques, and as part of the Tyndall Coastal Simulator Project iSCAPE – an interactive version of the SCAPE model, allowing stakeholders to explore the dynamics of the modelled system – was developed.

iSCAPE is an interactive tool developed from a cut-down version of the SCAPE model, consisting of a single cross section of shoreline and two adjacent sections of beach. However, iSCAPE retains the dynamic elements of the SCAPE model that are required for its use as an educational tool, whilst allowing the very rapid simulation needed to quickly explore the system's dynamics. The iSCAPE interface allows users to vary the model inputs, such as sea-level rise, sediment flow and presence of coastal defences, and watch the effects visually on a cross section of coast and graphed over time. Figure 10.4 illustrates the use of the iSCAPE tool to explore the implications of the construction of a sea wall at the base of the cliff.

In practice iSCAPE has primarily been used as a teaching tool to build understanding of coastal morphodynamics. In this role it has allowed both practitioners and students to explore how the shore profile shape, recession rates and beach volumes respond to systemic changes including changes in the balance between available and lost sediment, sea-level rise and anthropogenic impacts such as seawall construction and beach nourishment. A key finding from the practical application of iSCAPE has been that its interactivity stimulates discussion, with the demonstrator

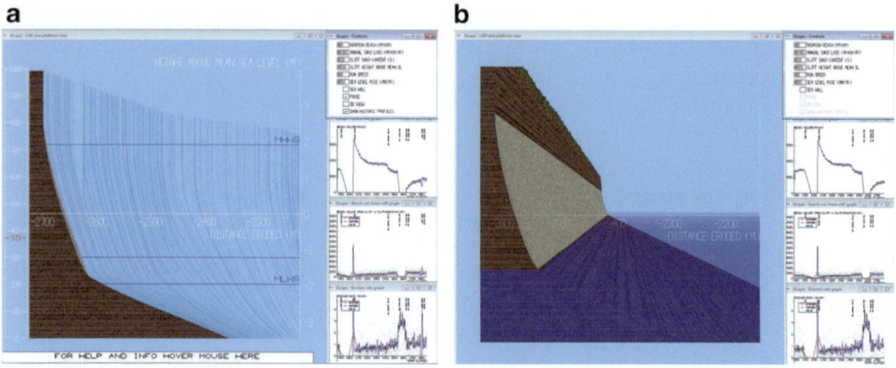

Fig. 10.4 Two examples of the iSCAPE interface, one showing the 2D representation of the cross section (**a**) and the other showing the 3D representation (**b**)

and observer interacting with the model to demonstrate and test perceptions and ideas. At a higher level, iSCAPE has also been used to demonstrate some of the capabilities of the full SCAPE model which lacks a graphical user interface (GUI).

10.2.4 Visualising Uncertainty

A recurrent theme throughout the Tyndall visualisation research has been the issue of uncertainty (e.g. Brown et al. 2006; Jude et al. 2006) and the need for techniques to communicate the inherent uncertainties associated with modelling future coastal change. In particular, the tension between the requirement to interpolate extra detail and the lack of established procedures for communicating the resulting uncertainty, as identified by Brown et al. (2006), when attempting to visualise the outputs from SCAPE, typifies the *'uncertainty communication challenge'*.

In response to the challenges posed by uncertainties, the Tyndall Coastal Simulator project has explored a variety of techniques for communicating uncertainties, ranging from simple representations of the zone of uncertainty associated with the SCAPE model, using colour-coded zones to demarcate such areas in landscape visualisations, to the use of animations illustrating the uncertainties associated with differing habitat change scenarios in the Biodiversity Requires Adaptation in Northwest Europe under a CHanging climate (BRANCH) project.[2] One major challenge associated with this research has been the often limited software functionality available to explore novel means of representing uncertainties, particularly in real-time visualisation packages, requiring the development of custom-built interfaces or software extensions. Such limitations resulted in the creation of custom software enabling the Tyndall Coastal Simulator project to

[2] http://webarchive.nationalarchives.gov.uk/20081125212930/http:/www.branchproject.org/ Cited 18 December 2013.

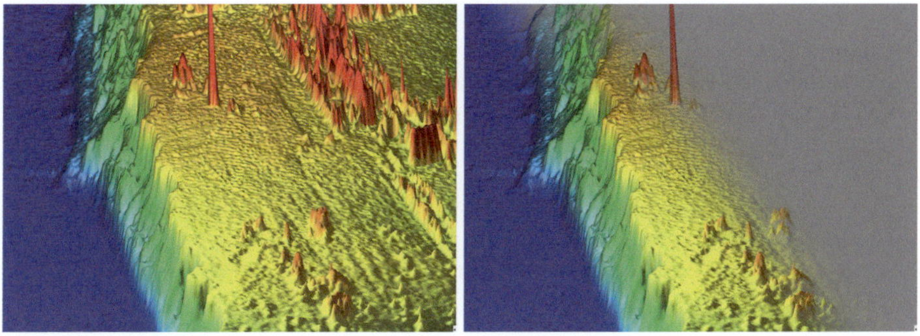

Fig. 10.5 The use of fogging to represent areas of uncertainty in the rate of cliff erosion along the Walton-on-the-Naze coastline

experiment with differing innovative approaches for communicating uncertainties within a VR theatre environment. Figure 10.5 illustrates the software in use, with fogging being used to represent the level of uncertainty associated with predicted cliff erosion on the Walton-on-the-Naze coast.

The experiences from attempting to address the issue of communicating uncertainty in the Tyndall Coastal Simulator project suggest that there remains a need for considerable applied and experimental research effort to address this issue. Indeed, the research to-date has only touched upon the wide range of potential methods for communicating uncertainties. For example, the cliff erosion visualisation study highlights the need for further research evaluating alternative uncertainty communication methods, including the use of abstraction, transparency, fuzzy and animated means of presentation, error representation and sound, for example (Appleton et al. 2004; Brown et al. 2006), and the extent to which such methods affect end-users' understanding and perceptions of the issues and information being communicated.

10.3 Exploring Coastal Futures: Testing the Visualisations

Whilst the development of the visualisation techniques and tools described in the chapter forms a key element of the development of the Tyndall Coastal Simulator, it only represents the first stage of the Tyndall Centre visualisation research. Of potentially greater significance has been the research evaluating the use of the technology in decision-making processes and stakeholder engagement. As described below, this has involved working with a wide range of coastal management organisations to not only assess the utility of the visualisations and to identify any benefits and potential challenges that may influence their use in real-world decision-making contexts but to identify end-user requirements for incorporation in the Tyndall Coastal Simulator interface. Allied to this applied research has been experimental research, testing the application of visualisation techniques in the economic valuation of coastal landscape change and the potential effect of the visualisations on decision-making behaviour (Bateman et al. 2009; Jude 2003). This experimental research (described in detail in Bateman et al. 2009) suggests that the technology could result in improved economic valuation of landscape change overcoming the limitations associated with traditional forms of communication used in economic valuation studies.

Alongside the experimental research have been a number of projects, and these are described below in the following case studies.

10.3.1 Using iSCAPE to Aid Shoreline Management Planning

In parallel to its use as an educational tool, iSCAPE has been tested in real-world coastal management processes. Once such example has been its use in informing the development of a coastal strategy for Clacton and Holland on Sea, in Essex, UK (Royal Haskoning 2011), where the geological, geomorphic and hydrodynamic conditions were sufficiently similar to those embedded within iSCAPE to allow the tool to be used without modification other than some scaling.

It is widely appreciated that whilst coastal defence structures prevent cliff erosion, they may be unsustainable, not only inhibiting coastal processes but becoming increasingly expensive to maintain. To address such concerns, iSCAPE was used by coastal managers during the development of the coastal strategy to explore alternative options for maintaining coast protection structures in the face of uncertain sea-level rise. In particular, iSCAPE was used to simulate different maintenance strategies under alternative scenarios of sea-level rise and to project the consequential cliff recession rates.

The use of iSCAPE in the development of the coastal strategy highlighted the significant value of the visual nature of the tool. Firstly, it facilitated the exploration of the problem, minimising the time required to learn how to operate the tool and interact with it. Secondly, the visualisations provided useful material to communicate the reasoning behind the selected management strategy. More fundamentally, the visualisations were found to build confidence amongst the expert geomorphologists using the iSCAPE tool. Their high confidence in the model was not built on systematic comparison between model behaviour and measured examples, which are often very rare, but instead on a more intuitive response to the nature of the dynamic visualisations. Significantly, the model *'felt'* to be realistic, which proved to be as important as the scientific basis of the model being visualised in enabling the tool being put to practical use.

10.3.2 Visualising Coastal Habitat Change and Adaptation Responses

A key aim of BRANCH was the development of transferable tools that spatial planners can use to assess and communicate the impact of climate change on coastal biodiversity. This included the creation of a visualisation methodology to communicate the results from the wider GIS modelling of coastal habitat change, which was undertaken by the project, to spatial planners and coastal stakeholders. This methodology utilised 3D Nature's Visual Nature Studio (3D Nature 2013) to produce a number of static and animated visualisations representing the potential impact of coastal habitat change for a range of climate change and adaptation scenarios. The visualisations were subsequently used to communicate the study's results to stakeholders, through a series of workshops, and to stimulate debate regarding the adaptation and policy issues arising from the research findings.

The testing of the visualisations in the stakeholder workshops illustrated that they provided a stimulus for debate regarding the challenges that climate change potentially poses to coastal habitats and their future management. Furthermore, many participants felt that the visualisations provided opportunities for organisations and members of the public to begin to consider the future of particular areas or sites. In addition, there was considerable interest in applying the visualisations in spatial planning and coastal management processes, with numerous participants noting that the visualisations may prove valuable in communicating key messages regarding coastal change to members of the public and decision-makers alike. However, as found in other research (Jude et al. 2006; Brown et al. 2006), concerns were raised about the apparent 'certainty' of the visualisations. These were accompanied by calls for more flexible visualisations, the inclusion of additional information including socioeconomic pressures and local landmarks and the presentation of temporal change through the use of animation. Interestingly, given the concerns regarding uncertainties and previous Tyndall research to develop credible and robust visualisations, described earlier in this chapter, some participants called for the creation of dramatic visualisations to engage those who are less familiar with the concerns, issues and policies under discussion.

The findings from the BRANCH project highlight the importance of conducting evaluative research exploring the potential benefits and limitations associated visualisation technologies and their use in coastal management and wider decision-making contexts. For example, whilst a range of visualisation approaches, including static, animated and interactive formats, were evaluated for use in the research, the general consensus amongst the members of the stakeholder-led research team was that static images were preferred over animated and interactive visualisations, which were deemed unnecessary and overly complicated to present to stakeholders. This reflects previous research that has questioned the perceived benefits of animated and interactive forms of visualisation (e.g. Jude 2003; Dockerty et al. 2006). Such findings also reinforce Perkins and Barnhart's (2005) belief that visualisation presentation and information needs require carefully planning early on in a project where stakeholder interaction is involved, to be true, with the visualisations forming part of a wider suite of information that must be provided as part of carefully planned stakeholder events. Such findings are of importance to the wider development of the Tyndall Coastal Simulator because they not only enable such systems to be tailored to end-user needs but can help to target time and resources towards key areas of interest. For example, as a result of demand from stakeholders, the visualisations

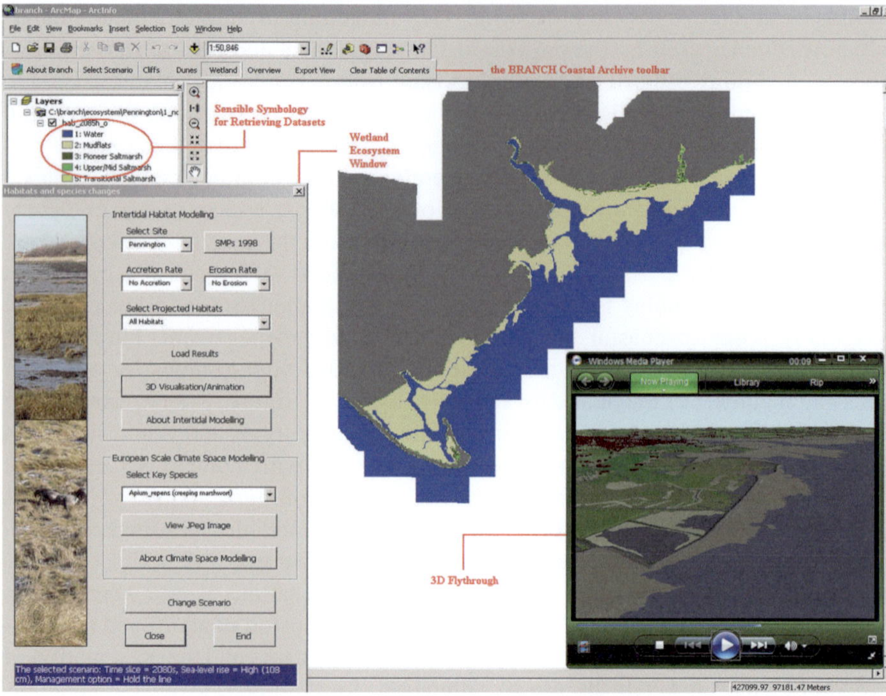

Fig. 10.6 The BRANCH Archive was developed in response to stakeholder needs and was built as a plug-in extension to ArcGIS. The screenshot illustrates the main elements of the BRANCH Archive's interface including the integration of animated fly-though visualisations that were produced as part of the project

were also incorporated in the BRANCH Archive GIS tool to accompany the results from the traditional 2D GIS analyses and to aid the interpretation of the project's finding by end users (Fig. 10.6).

10.3.3 Envisioning and Exploring Coastal Change with Stakeholders

The testing of visualisation techniques in real-world participatory decision-making contexts is key to their successful evaluation and potential application. Thus, a core element of the Tyndall visualisation research has been the emphasis on engagement with coastal decision-makers to test the visualisations in real-world contexts. However, whilst projects such as BRANCH and the Clacton Coastal Study reflect successful engagement with stakeholders, many attempts to apply the visualisations in real-world processes have met significant challenges that have precluded such engagement. Indeed, the most significant challenge encountered during the Tyndall

Coastal Simulator project has been associated with the sensitivities surrounding the coastal management issues that the project and the visualisations have been developed to address. As a result, many attempts to engage with stakeholders, including members of the public, to evaluate and test the visualisations in real-world coastal decision-making processes have been constrained due to the emotive and politicised nature of the issues under discussion (e.g. coastal erosion and managed realignment). A prime example of this was the cliff visualisation study which occurred during the revision of the north Norfolk SMP6 (Kelling Hard to Lowestoft Ness). Here coastal managers were concerned that the level of detail and awareness raising value of the visualisations could potentially undermine the ongoing discussions regarding the sensitive issue of coastal erosion and the development of the revised SMP for the coastline (Brown et al. 2006). Similarly, plans to test the visualisations in the valuation of coastal landscape change on the north Norfolk coast were hampered by both logistical challenges, associated with the requirement to present the visualisations to large numbers of experiment participants, and the concerns of coastal management organisations. Finally, a community coastal visioning exercise, aimed at exploring future coastal management options for an area of at risk from coastal erosion and flooding, was thwarted by stakeholders determined to 'hold the line' of current defences protecting their community (Milligan et al. 2009). The experiences from the Tyndall Coastal Simulator project are not isolated, with other studies encountering similar concerns (e.g. Appleton and Lovett 2005; Dietz et al. 2004), and pose a significant constraint to the practical testing and evaluation of decision-making technologies such as visualisations and the Tyndall Coastal Simulator.

10.4 Conclusions and Next Steps

The Tyndall Coastal Simulator project made significant steps towards addressing the challenges associated with visualising and communicating both the wide array of outputs produced by the simulator and the complex multidimensional nature of coastal change. Crucially, the research has required multidisciplinary approaches, not only in relation to the development of the visualisation tools and methods but during their testing and evaluation in both experimental and real-world contexts.

The research provides considerable evidence that the visualisation methods developed during the Tyndall Coastal Simulator project are of potential value in facilitating improved communication and discussion of coastal change issues. These are often highly sensitive in nature and will increasingly assume greater importance in future if the challenges posed by climate change and associated sea-level rise are to be addressed by coastal management organisations. However, the experience from the Tyndall Coastal Simulator project has also identified a number of challenges facing the practical application of the technology that were unforeseen at the beginning of the research, most notably the concerns voiced by coastal managers regarding the level of detail afforded by the visualisations, that may be more difficult to address than the technological limitations identified.

In terms of the application of visualisation technologies, a consistent theme arising from the Tyndall Coastal Simulator project is that a suite of complementary 2D and 3D visualisation tools and techniques may be required to address both the technological challenges and end-user needs associated with communicating coastal change. Indeed, the use of a range of both 2D and 3D visualisation techniques, including static, animated and real-time techniques, is necessary, depending not only on the type of coastal change being communicated but the purpose and stage of the communication process (Jude et al. 2005; Brown et al. 2006).

In terms of avenues for further research, rapid advances in visualisation and communication technologies open exciting research opportunities. In particular, the Tyndall Coastal Simulator project did not apply the advanced interactive visualisation and display systems that other researchers have developed to enable interactive real-time *'what if?'* type envisioning of environmental change (e.g. Stock and Bishop 2005, 2006; Stock et al. 2007). Furthermore, the availability of low-cost desktop VR systems, interactive touch tables and tablet computing and the development of augmented reality tools have the potential to enable coastal change to be visualised, communicated and interacted with in ways that were inconceivable at the time of the Tyndall Coastal Simulator project. Thus, there is a need for research comparing the practicalities and effectiveness of such technologies, particularly as potential limitations with more advanced forms of presentation, that were called for at the stakeholder events, have been encountered by other studies (Stock et al. 2007).

Despite the best efforts of the research and the rapid development of the types of technologies outlined above, many of the challenges identified by the original visualisation research that produced simplistic representations of coastal change on the north Norfolk coast remain. In particular, the need to communicate uncertainties and to visualise long-term coastal change, including complicated multidimensional geomorphological change, represents significant research challenges (Jude et al. 2006) that should not be overlooked in favour of studies experimenting with the latest technology. Similarly, further applied and practical testing of visualisation technologies, though collaborative projects involving both academics and coastal management practitioners, is required if the effectiveness and implications of the technologies in potentially facilitating improved communication, participation and decision-making are to be determined.

References

3D Nature (2013) 3D Nature home page. http://3dnature.com/. Cited 18 Dec 2013.

Al-Kodmany, K. (2001). Visualization tools and methods for participatory planning and design. *Journal of Urban Technology, 8*, 1–37.

Appleton, K., & Lovett, A. (2003). GIS-based visualisation of rural landscapes: Defining sufficient "sufficient" realism for environmental decision making. *Landscape and Urban Planning, 65*, 117–131.

Appleton, K., & Lovett, A. (2005). GIS-based visualisation of development proposals: Reactions from planning and related professionals. *Computers, Environment and Urban Systems, 29*(3), 321–339.

Appleton, K., Lovett, A., Dockerty, T., & Sunnenberg, G. (2004). Representing uncertainty in visualisations of future landscapes. In: *Proceedings of the XXth international society for photogrammetry and remote sensing (ISPRS) congress*, Istanbul.

Bartlett, D. J., & Wright, D. J. (2000). Epilogue. In D. J. Bartlett & D. J. Wright (Eds.), *Marine and coastal geographical information systems* (pp. 309–315). London/New York: Taylor and Francis.

Bateman, I. J., Jones, A. P., Jude, S. R., & Day, B. H. (2009). Reducing gains/loss asymmetry: A virtual reality choice experiment (VRCE) of land use change. *Journal of Environmental Economics and Management, 58*(1), 106–118.

Bishop, I. (1994). The role of visual realism in communicating and understanding spatial change and processes. In H. M. Hearnshaw & D. J. Unwin (Eds.), *Visualization in geographical information systems* (pp. 60–64). London: Wiley.

Brown, I., Jude, S. R., Koukoulas, S., Nicholls, R. J., Dickson, M., & Walkden, M. (2006). Dynamic simulation and visualisation of coastal erosion. *Computers, Environment and Urban Systems, 30*, 840–860.

Davis, T. J., & Keller, C. P. (1997). Modelling and visualizing multiple spatial uncertainties. *Computers and Geosciences, 29*, 397–408.

Dickson, M. E., Walkden, M. J. A., Hall, J. W., Pearson, S., & Rees, J. (2005). Numerical modeling of potential climate change impacts on rates of soft cliff recession, northeast Norfolk, UK. In: *Proceedings of coastal dynamics*. ASCE, New York.

Dietz, T., Tanguay, J., Tuler, S., & Webler, T. (2004). Making computer models useful: An exploration of expectations by experts and local officials. *Coastal Management, 32*, 307–318.

Dockerty, T., Lovett, A., Appleton, K., Bone, A., & Sünnenberg, G. (2006). Developing scenarios and visualisations to illustrate potential policy and climatic influences on future agricultural landscapes. *Agriculture, Ecosystems and Environment, 114*, 103–120.

Jude, S. R. (2003). *The application of visualization techniques for coastal zone management*. Unpublished PhD thesis, University of East Anglia, Norwich

Jude, S. R. (2008). Investigating the potential role of visualisation techniques in participatory coastal management. *Coastal Management, 36*(4), 1–19.

Jude, S. R., Brown, I., Koukoulas, S., Dickson, M., Walkden, M., Jones, A. P., & Watkinson, A. R. (2005). Exploring future coastlines – Initial steps towards the development of an integrated coastal simulation and visualisation system. In E. Buhmann, P. Paar, I. J. Bishop, & E. Lange (Eds.), *Trends in real-time landscape visualization and participation* (pp. 69–76). Heidelberg: Herbert Wichmann Verlag.

Jude, S. R., Jones, A. P., Bateman, I. J., & Andrews, J. E. (2003). Developing techniques to visualise future coastal landscapes. In E. Buhmann & S. M. Ervin (Eds.), *Trends in landscape modeling* (pp. 228–238, 274pp.). Heidelberg: Herbert Wichmann Verlag.

Jude, S. R., Jones, A. P., Andrews, J. E., & Bateman, I. J. (2006). Visualisation for participatory coastal zone management: A case study of the Norfolk Coast, England. *Journal of Coastal Research, 22*, 1527–1538.

Jude, S. R., Jones, A. P., Watkinson, A. R., Brown, I., & Gill, J. A. (2007). The development of a visualization methodology for integrated coastal management. *Coastal Management, 35*(5), 525–544.

Koukoulas, S., Nicholls, R. J., Dickson, M. E., Walkden, M. J., Hall, J. W., Pearson, S. G., Mokrech, M., & Richards, J. (2005). A GIS tool for analysis and interpretation of coastal erosion model outputs (SCAPEGIS). In: *Proceedings of coastal dynamics*. ASCE, New York

Lange, E. (2001). The limits of realism: Perceptions of virtual landscapes. *Landscape and Urban Planning, 54*, L163–182.

Milligan, J., O'Riordan, T., Nicholson-Cole, S. A., & Watkinson, A. R. (2009). Nature conservation for future sustainable shorelines: Lessons from seeking to involve the public. *Land Use Policy, 26*, 203–213.

Orland, B., Budthimedhee, K., & Uusitalo, J. (2001). Considering virtual worlds as representations of landscape realities and as tools for landscape planning. *Landscape and Urban Planning, 54*, 139–148.

Pearson, S. G., Dickson, M. E., Walkden, M. J., Hall, J. W., Nicholls, R. J., Mokrech, M., Koukoulas, S., & Spencer, T. (2005). *Towards an integrated coastal sediment dynamics and shoreline response simulator* (Tyndall centre technical report number 38). Norwich: Tyndall Centre for Climate Change Research.

Perkins, N. H., & Barnhart, S. (2005). Visualization and participatory decision making. In I. Bishop & E. Lange (Eds.), Visualization in landscape and environmental planning – Technology and applications (pp. 241–250). London, Oxford: Taylor and Francis.

Raper, J. (2000). *Multidimensional geographic information science*. London: Taylor and Francis.

Royal Haskoning. (2011). *Clacton and Holland Clacton and Holland on sea flood and coastal erosion risk management strategy*, on behalf of the Environment Agency and Tendring district Council. Peterborough: Royal Haskoning.

Stock, C., & Bishop, I. D. (2005). Helping rural communities envision their future. In I. D. Bishop & E. Lange (Eds.), *Visualization in landscape and environmental planning – Technology and applications* (pp. 145–151). London/New York: Taylor and Francis.

Stock, C., & Bishop, I. D. (2006). Linking GIS with real-time visualisation for exploration of landscape changes in rural community workshops. *Virtual Reality, 9*, 260–270.

Stock, C., Bishop, I. D., & Green, R. (2007). Exploring landscape changes using an envisioning system in rural community workshops. *Landscape and Urban Planning, 79*, 229–239.

Tress, B., & Tress, G. (2003). Scenario visualisation for participatory landscape planning- a study from Denmark. *Landscape and Urban Planning, 64*, 161–178.

Tufte, E. R. (1990). *Envisioning information*. Cheshire: Graphics Press.

Walkden, M. J., & Hall, J. W. (2002). A model of soft cliff and platform erosion. In: *Proceedings of the 28th international conference on coastal engineering* (Vol. 3, pp. 3333–3345). Cardiff.

Walkden, M. J., & Hall, J. W. (2005). A predictive mesoscale model of soft shore evolution. *Coastal Engineering, 52*, 535–563.

Chapter 11
GIS Platforms for Managing, Accessing and Integrating Model Results: The Tyndall Coastal Simulator Experience

Mustafa Mokrech, Robert J. Nicholls, Sophie A. Day (née Nicholson-Cole), Richard J. Dawson, Simon Jude, and Sotiris Koukoulas

Abstract The Tyndall Coastal Simulator comprised a wide range of modelling activities across disciplines and institutions. This activity generated diverse results, described in the preceding chapters. Organising this body of knowledge into a coherent and accessible form for use by stakeholders and for additional ad hoc analyses emerged as a substantial task in its own right.

This chapter reflects upon the process of designing, developing and demonstrating a user-orientated 'interface' – a geographic information system (GIS) coupled with coastal models – allowing users to explore results of the research and interactions between them (Mokrech et al. 2011). The process of integration was led by the project team, with input from a group of external stakeholders who were given

M. Mokrech (✉)
Environmental Institute of Houston, School of Science and Computer Engineering, University of Houston Clear Lake, Houston, TX 77058, USA
e-mail: mokrech@uncl.edu

R.J. Nicholls • S.A. Day (née Nicholson-Cole)
Tyndall Centre for Climate Change Research, Faculty of Engineering and the Environment, University of Southampton, Southampton SO17 1BJ, UK
e-mail: r.j.nicholls@soton.ac.uk; day_soph@yahoo.co.uk

R.J. Dawson
School of Civil Engineering and Geosciences, Tyndall Centre for Climate Change Research, Newcastle University, Newcastle upon Tyne NE1 7RU, UK
e-mail: richard.dawson@newcastle.ac.uk

S. Jude
Centre for Environmental Risks and Futures, Department of Environmental Science and Technology, School of Applied Sciences, Cranfield University, Cranfield, Bedford MK43 0AL, UK
e-mail: s.jude@cranfield.ac.uk

S. Koukoulas
Department of Geography, University of the Aegean, University Hill, Mytilene 881100, Greece
e-mail: skouk@geo.aegean.gr

R.J. Nicholls et al. (eds.), *Broad Scale Coastal Simulation*, Advances in Global Change Research 49, DOI 10.1007/978-94-007-5258-0_11

opportunities to learn about and test the interface. This interaction helped to give practical context in the identification of broad questions at the beginning of the project and the detailed questions that emerged during the research process – particularly around integration and user potential.

Two interfaces, built upon the same technology, are presented. Both involved the organisation of model results, metadata and additional documentation into a library that can be accessed and queried and developing effective visualisations that meet user needs and requirements. The Norfolk interface development highlights the important role an interface can play in providing a focus for integration and discussion within the research team about how to bring together complex systems analyses. Furthermore, it provided a core platform for stakeholder engagement including using the visualisations described in Chap. 10 and the stakeholder engagement in Chap. 12. The interface developed for the Hampshire case study outputs further demonstrates the potential of such tools by incorporating probabilistic and 'outcome-based' models into a GIS platform.

The development process also highlighted broader possibilities for developing future integrated frameworks that are multi-scale and capable of linking various models in order to simulate complex coastal processes in ways that answer management questions and stakeholder needs. Finally, limitations of the current interfaces, and the future opportunities for development possible with emerging technologies, such as web-based approaches, are considered.

Keywords Geographic information system • User interface • Integration • Stakeholder engagement • Data management

11.1 Introduction

The results and output from the Tyndall Coastal Simulator, presented in earlier chapters, provide significant volumes of spatial and nonspatial data that are crucial for the coastal management community to understand and explore long-term coastal change. Archiving, organising and having access to this scientific information are important in order to explore and understand better the consequences of management decisions that aim to reduce risks and/or reduce possible degradation of social, economic and environmental assets and associated ecosystem services within coastal zones. It also allows stakeholders to engage with the project as a whole, rather than just with individual components. Lastly, it ensures the legacy of the project, enabling results to remain accessible at the end of a research project.

Throughout the project, the term 'interface' was used to describe a bespoke geographic information system (GIS) front end to the Tyndall Coastal Simulator (or any other assessment). This interface not only encompasses spatial queries and visualisations but is loosely coupled with the coastal models using a library of results, which allows access, analysis functionality and interpretation of these results. Collectively, these allow users to explore the Tyndall Coastal Simulator research results and the interactions between them. Hence, the main goals of developing a user interface were fourfold:

- Store relevant results and models across the project in an accessible form
- Allow additional ad hoc queries
- Act as a shared platform for collaborative decision making in the coastal zone
- Engage with relevant stakeholders from different organisations and backgrounds

The importance, and technical challenges, associated with meeting these goals became increasingly apparent over the course of the project. What started, seemingly as a tool for presenting our results, developed into a substantial activity in its own right. Furthermore, developing the interface emerged as an important integrative activity within the research consortium because it encouraged an iterative process of debate and refinement regarding the component models, how they were coupled and which outputs were of most management insight. The interface was the primary mechanism for engaging with stakeholders, acting as a mechanism for presenting research results thereby stimulating feedback on how the research was mapping to their needs and stimulating co-production of knowledge about coastal change. These regular interactions influenced the design and scope of the interface and to some degree also helped to shape the research within the component modules, particularly those that addressed human impacts such as erosion and flooding (Nicholls et al. 2008; Mokrech et al. 2007).

The interface development was carried out in two regions allowing some generic insights. These are subcell 3b in Norfolk, UK (now SMP6), and coastal subcell 5 in Hampshire (now SMP13). The main focus of this chapter, however, is on the interface developed for SMP6: the case study area for the Tyndall Coastal Simulator. The SMP13 sites concern coastal wetland management (see Chap. 6) and illustrate the challenge of archiving data and strategic planning.

Section 11.2 briefly describes the underlying principles for the choices that were initially made in developing the Tyndall Coastal Simulator interface. The actual process of integration happened within the project team, but was guided by a range of external stakeholders who provided feedback around the broad questions identified at the beginning of the project and detailed questions that emerged during the research process – particularly around integration and user potential. The following sections present the development process and the design of interface and associated tools. Stakeholder critique of the interface is also presented. Finally, opportunities for future research and recommendations for future programmes seeking to develop an interface for integrated modelling activities are discussed.

11.2 Basic Principles

The challenge of effectively organising data and model information is widespread and generic, but examples in the literature for coastal systems are limited. Simple approaches such as layers, pre-generated maps (or map caches) placed on an easy-to-use web-based platform,[1] are seen to be effective in communicating flood risk and vulnerability issues to the general public. A similar approach has been adapted by Yoskowitz et al. (2012) in a recent study aimed at a wider range of audiences looking into the assessment of changing ecosystem services provided by marsh habitats using the Sea Level Affecting Marshes Model (SLAMM). They have developed an ecosystem services valuation database and made it available online.[2] In the UK, the *CoastRanger MS* tool (Pontee and Morris 2011) was developed to simulate public debate and improve stakeholder understanding of issues that are relevant to the purpose of achieving sustainable shoreline management. It has been designed to be an educational tool where a synthetic stretch of coastline within a virtual gaming-type environment was used.

When cross-sectoral analysis and integrated assessment models (IAMs) are involved, the above challenge becomes more significant due to the level of modelling sophistication and potential experienced audience involved. A literature review by Dickinson (2007) identified two groups of IAMs that focus either on evaluation or on optimisation of policies. Evaluation-focused IAMs tend to simulate societal and environmental impacts of selected policies under differing scenarios; they follow the pattern of opening with assumptions of socio-economic futures and climate scenarios which are then evaluated and analysed under a predefined policy such as 'hold the line' or 'managed realignment'. On the other hand, optimisation-focused IAMs aim to identify optimal policies of climate change control options that may lead to specific outcomes such as 'maintaining baseline flood risk' or 'maintaining the baseline stock of coastal wetlands'. Another dimension that is essential in this review is the potential role of using libraries of runs (from complex/slow/distributed models) versus the potential inclusion of (more simple/faster) dynamic models within the IAMs. This may lead to substantial resources being dedicated for building a platform that allows dynamic modelling and effective communication. The Regional Impact Simulator (RegIS2) model applied to UK regions (Holman et al. 2008), the Dynamic and Interactive Vulnerability Assessment (DIVA) integrated model of coastal systems developed by the Dynamic and Interactive Assessment of National, Regional and Global Vulnerability of Coastal Zones to Climate Change and Sea-Level Rise (DINAS-COAST) project (Hinkel and Klein 2009), and more recently the CLIMSAVE integrated assessment platform (Harrison et al. 2013) applied to the entire European Union[3] are examples that are multi-scale and are designed for a wide range of audiences.

Drawing on this type of experience, the development of the Tyndall Coastal Simulator interface and other interfaces described here has shown that a linked set of models of the coastal system can be designed and applied and the results made available to a range of stakeholders via a graphic user interface. This provides a more integrated picture of the impacts of both climate change and potential management decisions. Developing an integrated assessment platform for knowledge

[1] For example a sea-level rise tool for the Gulf of Mexico at http://gom.usgs.gov/slr/slr.html

[2] www.GecoServ.org

[3] www.CLIMSAVE.eu

transfer to a range of stakeholders is a challenge. Core to our approach was to develop a framework which describes the model system, processes and impacts and hence key indicators – as explained in Chap. 1. These processes include erosion and flood risk, ecosystem change and land-use change in the face of climate and socio-economic change and different shoreline management choices (see Mokrech et al. 2011; Dawson et al. 2009; Leake et al. 2007).

Designing an interface for an integrated assessment model requires the development of a generic framework (a simplified version of the overall Tyndall Coastal Simulator framework presented in Chap. 1, Fig. 1.11) that is rather like a map of the underlying contents and modelling linkages. It recognises the main physical processes involved and identifies the drivers and underlying pressures, which in this case will lead to physical and environmental changes that may influence flood and erosion risks. Essentially, the interface must enable data to be stored and processed, with a graphical user interface component to make that data and the results of processing accessible to a user (Fig. 11.1). The framework is used to design key features of the interface, including its desktop appearance. It is also used to define how and which of the models' elements are to be linked and expressed through the exploration of the IAM via the interface.

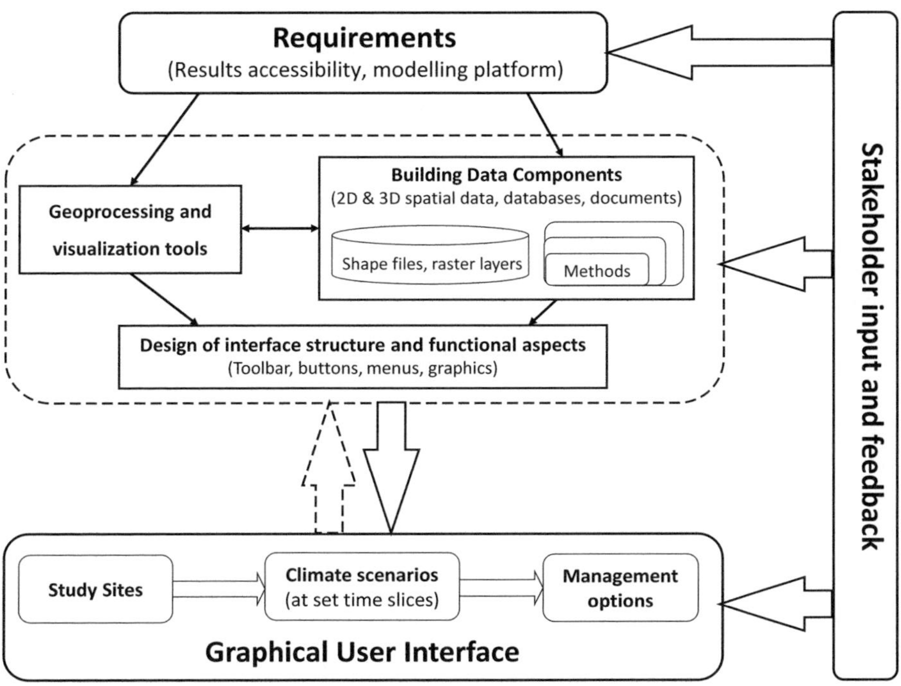

Fig. 11.1 A conceptual model for developing the Tyndall Coastal Simulator interface. The objective of building an interface is defined, and then a spatial database is developed and prepared using GIS functionalities including 2D/3D visualisations. The interface design is produced by iteration and feedback from stakeholders. The final graphical user interface allows user interactions and accessibility to the developed spatial database

Principally, the Tyndall Coastal Simulator interface comprised pre-modelled data organised in a library of deterministic scenarios and visualisations that meet user needs. The main results presented through the interface in Norfolk were erosion and flood risk (Chaps. 7–9), and in Hampshire the focus was on habitat change (Chap. 6).

A key requirement of the interface was that it should allow users to explore and query the modelling results so that it could be useful for the decision-making process. This requires the modelling results to be communicated to users in different ways that make them understandable. Communicating the results of such explorations is a major challenge. To this end, the Tyndall Coastal Simulator was not only designed to incorporate traditional two-dimensional (2D) maps and tabulated results; it also builds on previous Tyndall Centre research (e.g. Jude et al. 2006, 2007; Brown et al. 2006) and includes three-dimensional (3D) visualisations (which were found to be powerful communication tools in stakeholder events, as explained in Chap. 10).

11.2.1 Interface Platform

The interface developed for the Tyndall Coastal Simulator was built in a GIS environment to allow a range of spatial and nonspatial analyses of the different model results, plus 2D and 3D visualisations of the results (as discussed in more detail in Chap. 10). This built on initial investigations that showed that a GIS environment was the most appropriate platform available at the time for developing a graphical user interface (GUI) (Koukoulas et al. 2005).

The Environmental Systems Research Institute (ESRI) ArcGIS Desktop environment was selected not just because it was widely available, but because it was importantly used by our stakeholder organisations and provided powerful spatial analysis tools, visualisation capabilities and an open system for development. The Tyndall Coastal Simulator is compatible with all versions of ArcGIS 9.1–10.2. However, as information technology is progressing rapidly, future ArcGIS versions may not support Visual Basic for Applications (VBA). The current emphasis is on using Python as an alternative programming language for ArcGIS development and customisation. Therefore, it is likely that the Tyndall Coastal Simulator interface will not remain compatible with later versions of ArcGIS. This issue is general to any platform development, and in Sect. 11.4 we consider alternative approaches, such as web services, that have become available since the original interface platform was developed.

11.2.2 *Stakeholder Involvement in the Development of the Tyndall Coastal Simulator*

Stakeholder engagement was a critical element during the development of the Tyndall Coastal Simulator, from its earliest inception until the finalisation of the user interface towards the end of the project. The interface design was an iterative process, and stakeholder input and feedback was very constructive. Section 11.3.3 goes into more detail about some of the actual feedback gained.

This process enabled stakeholders with an interest in coastal planning and management to advise on the policy options that were considered in the Tyndall Coastal Simulator and the types of outputs that would be meaningful and useful. Moreover, it enabled the researchers to elicit stakeholder knowledge and familiarity with the coastal system under study. The investment of time and effort in actively engaging with many coastal stakeholders proved to be of mutual value.

A diverse range of stakeholder interests representing different institutional scales and types of organisation were involved and are summarised in Table 11.1. They

Table 11.1 A summary of the stakeholder organisations that engaged in the process of developing the Tyndall Coastal Simulator

	Institutional scales of stakeholder representation		
Stakeholders involved	National (i.e. the UK)	Regional (i.e. Eastern England)	Local (i.e. Norfolk)
	Department for Environment, Food & Rural Affairs (DEFRA)	Anglian Water	CCAG (Coastal Concern Action Group)
	Environment Agency	Environment Agency	Members of Parliament
	Natural England	Natural England	North Norfolk District Council
	Halcrow	National Trust	National Trust
	HR Wallingford	Royal Society for the Protection of Birds (RSPB)	RSPB
	Met Office	CoastNet	Great Yarmouth Borough Council
	UKCIP (UK Climate Impacts Partnership)	East of England Climate Change Partnership	King's Lynn and West Norfolk District Council
		East of England Regional Assembly	Norfolk Coast Partnership
		English Heritage	Norfolk County Council
		GO East	SCEG (Scratby Coastal Erosion Group)
		Marinet	Suffolk Coastal District Council
		The University of East Anglia	The Broads Authority
			Waveney District Council

included a government department, government agencies with strategic environmental responsibilities, consultancies, non-governmental agencies, research organisations, infrastructure providers, charities, not-for-profit coastal organisations, regional and local government offices, local action groups and so on. The stakeholders who contributed to the project represented local, regional and national scales and different types of institutions including those concerned with both the technical and policy aspects of coastal change planning and management. Many others were more remotely involved, being kept up to date with briefing notes and other project outputs.

There were several stages at which stakeholders from the coastal management community were involved and in different ways. For example:

- Organised stakeholder engagement through planned meetings to discuss the Tyndall Coastal Simulator research and organise data provision (e.g. with the Environment Agency, North Norfolk District Council and the Meteorological Office)
- Specific stakeholder workshops in 2006, 2007 and 2008
- Individual discussions and meetings between people working on the project and practitioners or 'real-world' stakeholders at conferences or other events, i.e. identification of a gap and a need for a robust scientific integrated assessment, accessible for practitioners and policy makers involved in coastal change planning and management

Project workshops (which included presentations of the work involved, practical demonstrations, practical exploration time and group discussion for feedback and individual feedback) were particularly important in the evolution of the Tyndall Coastal Simulator research. It was important that a diverse set of stakeholders who may potentially use this kind of tool (many from a non-technical background) had a chance to explore, test and critique it. Having the opportunity to interrogate and visualise the modelling results via the interface was helpful in offering insights into how the coast might change into the future. Stakeholder feedback was important in order to produce an interface that could ultimately be applicable and beneficial to coastal management practitioners. By responding to stakeholder feedback, where possible incorporating or justifying why we were unable to do so, stakeholders maintained their involvement throughout the project.

The stakeholder events also became fora for constructive, neutral debate between many different stakeholder interests about how to deal with current and future coastal change in an era of acknowledged uncertainty. The events enabled discussion of common objectives and the opportunity to put various interests, agendas and pressures to the table relatively openly. The research did not act in a lobbying capacity; rather we sought to contribute to the debate and to a broader learning process about how to more clearly articulate many of the complex issues, interactions and trade-offs associated with future adaptation that can be contentious and emotive. Engagement with the project gave some stakeholders a better understanding of coastal dynamics in the case region, the potential long-term implications of various types of management strategies and the long-term uncertainties associated with coastal change. Within the local, regional and to some extent national context, the project contributed to a greater urgency for investment in adaptation on dynamic (eroding) coasts. A particular example was the allocation of national funding for the UK Pathfinder programme, an opportunity for local authorities to win funding to explore ways of helping coastal communities to plan and adapt to coastal change.[4]

11.3 The Tyndall Coastal Simulator Interface for North Norfolk

The Tyndall Coastal Simulator is described fully in the preceding chapters of this book, but the most relevant issues regarding the development of the interface itself are summarised here. As outlined in Chap. 1 (e.g. in Fig. 1.2), the Tyndall Coastal Simulator recognises three scales:

1. The global (GCM) scale
2. The regional scale (in this case the North Sea Region)
3. The Tyndall Coastal Simulator domain (defined by discrete physiographic unit(s) – in this case SMP6)

Within the nesting, the larger scale provides the boundary conditions for the smaller scale. The models feed into each other and describe a range of relevant processes: sea level, tides, surges, waves, sediment transport, coastal morphology and relevant socio-economic processes such as urbanisation.

The preliminary, demonstrator, interface is described by Koukoulas et al. (2005) and provided a visualisation and analytical capability for just the cliff erosion results produced by the process-based SCAPE (Soft Cliff and Platform Erosion) model, with a focus on evaluating cliff erosion in terms of building and land-use loss. This demonstrated the generic utility of such interfaces and was a first step towards building a multi-sectoral interface that also includes flooding and land-use change and allows more detailed analysis of costs and benefits of cliff retreat. Starting with downscaled global climate models, regional climate change including sea-level rise, storm surges and waves is considered. These outputs are then linked to shoreline erosion and profile evolution using the SCAPE model and coupled with a flood model to conduct an integrated coastal flood and erosion risk assessment under the full range of scenarios.

11.3.1 *The Interface Design*

The Tyndall Coastal Simulator interface was developed as a toolbar in ArcGIS using VBA scripting language. The toolbar includes a number of windows and controls where users can interact and select options organised in buttons: pull-down menus

[4] North Norfolk District Council successfully secured £3million in 2009 to trial its projects. More information is available here: http://www.northnorfolk.org/pathfinder/.

and/or typing in text boxes. The window-based design was selected because of its familiarity to end users. The windows on the toolbar correspond to the elements defined in the generic framework shown in Fig. 11.1, which include drivers, coastal management choices and impacts. The flow of information and user interaction is designed to match the defined links in this framework. The main interface design objective was to allow easy and efficient access to the library of results. This required:

- Compatibility with a standard GIS licence, namely, ArcGIS 9.x, and later releases as far as possible.
- Simple, familiar design.
- Self-explanatory windows and controls enhanced with tooltips.
- Logical sequence, including an intuitive interrogation route.
- Comprehensive inclusion of results, including 2D and 3D visualisations of results where available.
- Dedicated custom tools for easy navigation and data handling.
- Comprehensive background information as PDF documents. These describe the climate and socio-economic scenarios, the coastal management options and the underlying models.
- Audio help that is played automatically for each window to provide a brief description of content and the main steps to be followed by the user.

The main elements of the Tyndall Coastal Simulator toolbar (Fig. 11.2) include:

(a) Scenario and coastal management policy selection: climate and socio-economic scenarios and coastal management options.
(b) Erosion risk: a library of the SCAPE erosion runs (Chaps. 7 and 9).
(c) Flood risk: a library of the flood model runs (Chaps. 8 and 9).
(d) Regional coastal morphology (Chap. 5).
(e) Habitat change: Other tools are also developed to facilitate the process of retrieving data and information from the interface, for example, spatial intersection of flood model outputs with land use to establish risk (as described in Chap. 9). The discussion and examples presented here are relevant to the erosion and flood elements.

The Tyndall Coastal Simulator (North Norfolk) includes a large number of combined scenarios of sea-level rise, changing wave climate and management of the cliffed coast (see Table 1 in Appendix A) under the baseline socio-economic conditions. Additionally, there are four different socio-economic scenarios at three time

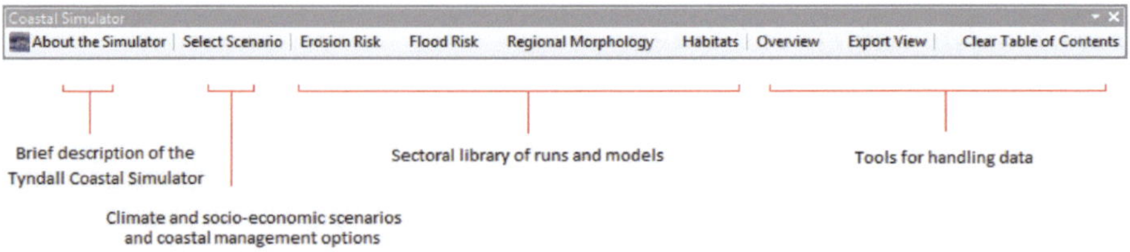

Fig. 11.2 The main elements of the toolbar of the Tyndall Coastal Simulator interface (North Norfolk), illustrating its structure

slices. Thus, predictions of cliff erosion and geomorphic change and inundation frequency, and ultimately erosion and flood risk analysis, are made for the current century at the baseline year 2000 and at three time slices: 2020s, 2050s and 2080s.

The accessibility to the library of results produced by the erosion and flood models provides a range of mechanisms for more effective analysis by interrogating the controls on the toolbar (Fig. 11.2), including the following:

1. Combinations of climate, socio-economic and management option can be selected at a single time slice – in this option 2D and 3D GIS layers and summary statistics of erosion and risk assessments can be viewed.
2. Combinations of climate, socio-economic and management options can be accessed at time slices, including the baseline year, in the form of statistical summaries on a graph.
3. Evolution through time of a range of management options, under a single climate and socio-economic scenario, can be compared graphically.
4. Evolution through time of a range of climate scenarios under a single management option can be compared graphically.

For example, Fig. 11.3 shows the erosion window in action where a number of properties at risk due to cliff erosion are identified for a single climate and manage-

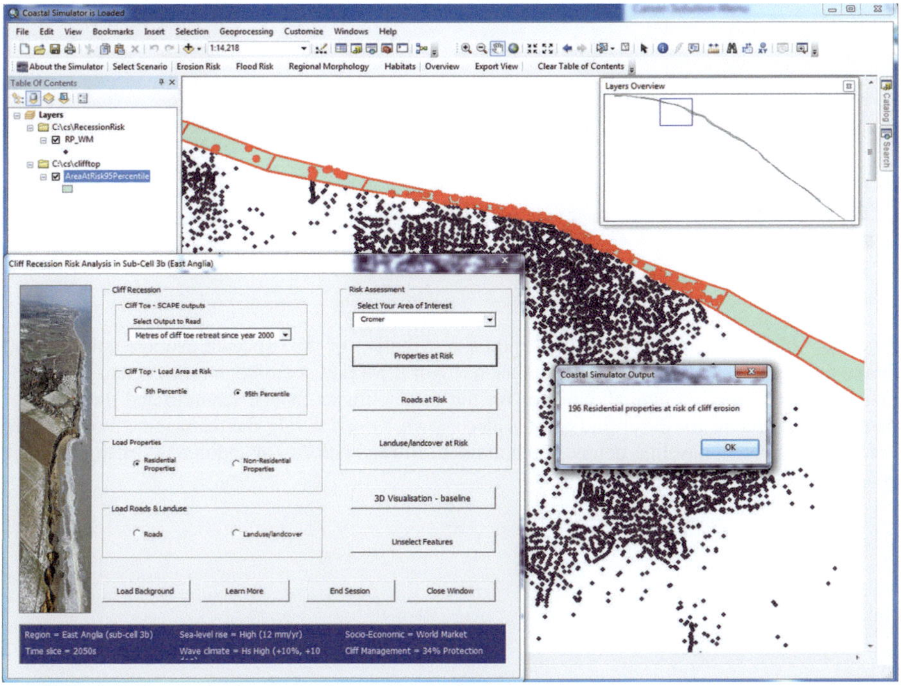

Fig. 11.3 Erosion risk assessment window, showing where areas of risk along the coastline of Cromer have been constructed by overlaying erosion polygons over layers of residential and non-residential properties. Automated spatial intersection selects properties at risk, as described in Box 11.1. The impacts on other land-use classes and infrastructures (e.g. roads) can also be evaluated

ment scenario at a single time slice. Figure 11.4 shows another example of a flood risk map where the spatial distribution of flood damage under a selected scenario is displayed, while a summary of flood damages under five investigated management options at time slices is displayed on a graph for more effective and comparative analysis. In addition, the integration of multiple sectors into the Tyndall Coastal Simulator interface allows cross-sectoral analyses to be conducted that may inform the shoreline management process such as the trade-off between cliff erosion risk and flood risk. An example of this type of analysis is illustrated in Box 11.1.

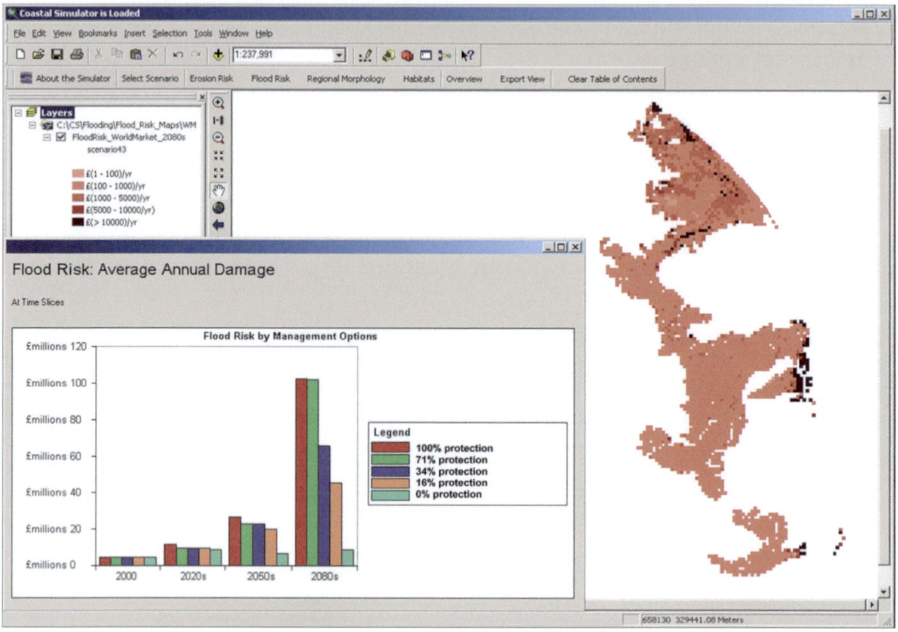

Fig. 11.4 Flood risk map showing the spatial distribution of the expected average annual flood damage for Scenario 43 in the 2080s under the World Market socio-economic scenario. A summary graph of expected annual damages for the same climate change scenario and the different management options for the cliffed coast is also shown. Descriptively, the management options are defend everywhere, baseline defences maintained, limited defence abandonment, major defence abandonment and no defences

Box 11.1 Example of Interface Operation: Assessing Erosion Risk

The Tyndall Coastal Simulator interface has some functionality to calculate outputs requested by stakeholders. For example, a user might be interested in the number of properties, or infrastructure assets, at risk under a given land use and climate scenario. Users may also call up more detailed information in the form of PDF documentation. To do this, a user would select a management option, by way of example, M2 where 71 % of cliffed coastline is protected. The user subsequently selects a climate change scenario, for example, a high sea-level rise scenario of a rate of 12 mm/year combined with low change in offshore wave height (Scenario 15; see Appendix 1). To calculate a range of erosion-related outputs, the interface then calls the appropriate data layer from the outputs of the cliff erosion model (Chap. 7) and integrates this with the appropriate data layer from the land-use model outputs (Chap. 4). ArcGIS has standard functions for establishing spatial relationships, in this case an intersection of the erosion hazard and number of properties from the land-use layer, to calculate the number of properties that fall within the hazard zone (Table 11.2).

Table 11.2 Summary of the impacts of cliff erosion on residential properties and roads by the 2080s for Scenarios 15 and 42 (but assuming no socio-economic growth)

	Erosion (5th percentile)			Erosion (95th percentile)		
Management	No of RP	No of NRP 64	Roads lost (km)	No of RP	No of NRP 98	Roads lost (km)
M2	133	64	35	236	98	40
M7	929	301	55	1,026	333	58

RP residential properties, *NRP* non-residential property, M2 71 % cliff protection, M7 0 % cliff protection

A user may also query the associated flood risk with this scenario at this time, in this case £82 and £9 m for M2 and M7, respectively. Of course, this could be achieved by preprocessing a series of outputs and allowing a user to query them. However, by embedding some of these, computationally fast, calculations within the interface, it provides additional flexibility – for example, if different land-use scenarios were required, only the land-use layer needs to be changed.

The Tyndall Coastal Simulator interface not only incorporates traditional 2D GIS maps, tabulated results and graphs, but it also contains three different types of visualisations, which have been found to be powerful communication tools: (i) standard time series (e.g. lines representing cliff recession over time), (ii) 3D visualisations and (iii) representation of some uncertainties (see also Chap. 10).

In the Tyndall Coastal Simulator interface, users are able to access the visualisations through a standard 2D map interface, by interactively selecting individual

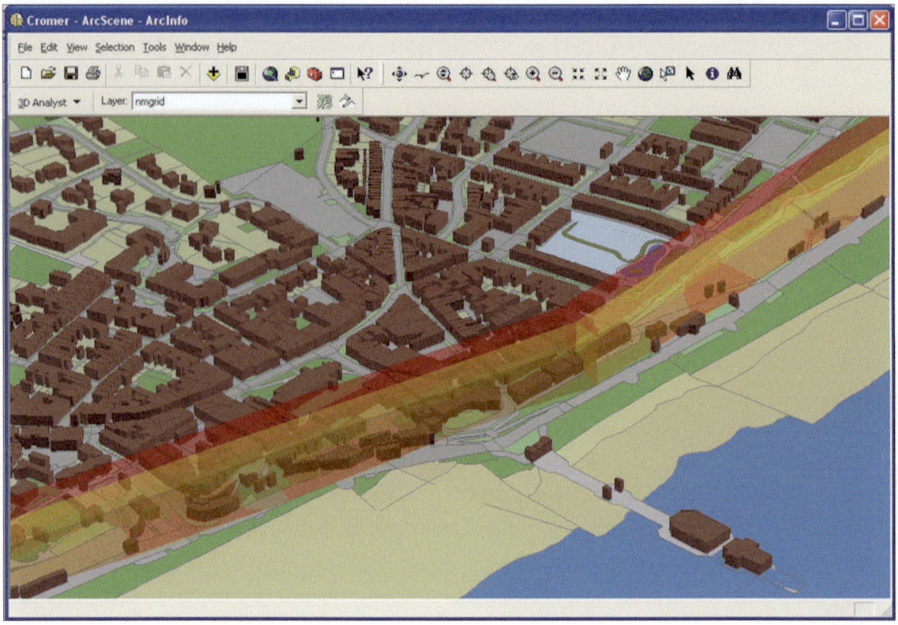

Fig. 11.5 3D visualisation shows properties and erosion risk bands (95th and 5th percentiles) along the coastline of the town of Cromer under the low climate change scenario and no defences option (M7) (see Table 1, Appendix A)

settlements or management choices as described above. As Spiegelhalter et al. (2011) note, interactive visualisations of data in multiple types of representation that can be adjusted to user needs and capabilities are becoming easier to produce and present. In the context of communicating uncertainties about the future changes, they can be a useful tool. Instead of using specialist visualisation tools (e.g. Brown et al. 2006), the Tyndall Coastal Simulator utilises the 3D visualisation capabilities provided by ESRI ArcScene/3D Analyst. This not only enables the 3D representation of coastal features (e.g. terrain, buildings and risk zones) but also provides users with interactive navigation and query tools via a familiar GIS interface. The interface platform also allows uncertainty information, where available, to be presented. Figure 11.5 and the results in Box 11.1 present some of the uncertainties associated with the process of cliff failure and long-term erosion.

11.3.2 Other Interface Examples

Separate GIS-based interfaces based upon the same technology were developed in the form of toolbars to facilitate user access to a library of simulations, including flooding and habitat impacts, as well as the 'outcome-driven' modelling (Chap. 6) to support coastal habitat decision making (BRANCH Partnership 2007; Gardiner et al. 2007).

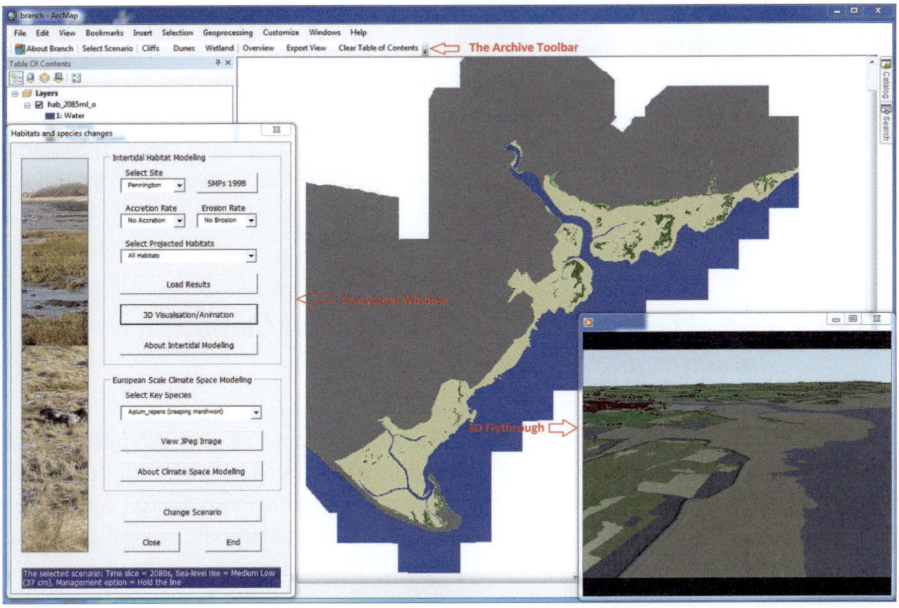

Fig. 11.6 An example screenshot of a habitat data archive interface showing some of its main elements: archive toolbar, habitat map, ecosystem window and 3D animation video. The illustrated study site is the Lymington Marshes on the West Solent, Hampshire, UK. The relevant scenario appears in the *blue box* at the *bottom* of the figure

As described in Chaps. 6 and 10, the potential coastal habitat and species changes due to a wide range of climate change and coastal management scenarios were explored through the twenty-first century at six sites in Hampshire and Dorset on the central English Channel coast, UK. Changes to soft cliffs, sand dunes and mud-flat/salt marsh/coastal grazing marsh/saline lagoon habitats and selected indicative species were also investigated. Overall, approximately 400 scenario combinations of climate, coastal management and morphological change were produced (Gardiner et al. 2007). Figure 11.6 shows a screenshot of the interface with the habitat data archive toolbar and other elements such as the wetland window, 2D prediction of habitats and associated 3D fly-through. This interface includes:

- *Geospatial databases* to represent raster and vector graphics that can be displayed in 2D and 3D. Additional 3D visualisations include animated videos and static images for selected scenarios.
- *Graphical user interface* (GUI) where user-archive interactions take place and the final outputs are presented as a usable product.
- *Documents* (PDF files) describing the BRANCH project, the scenarios and various methods employed.
- *Linkages* between the archive's windows and the associated controls for easy data retrieval using dedicated code.

The 'outcome-driven' interface (Chap. 6) allows users access to the results of the hundreds of simulations stored in a geospatial database. The user can identify a number of possible outcomes, and their likelihood of occurrence, for a selected climate scenario and coastal management option. Alternatively, the user can select a climate scenario and identify the suitable management options to meet given habitat management targets or objectives (e.g. maintaining salt marshes at a particular site) to whether that target is achievable.

11.3.3 A Stakeholder Critique of the Tyndall Coastal Simulator and Interface for Norfolk

Stakeholder engagement was integral to the development of the Tyndall Coastal Simulator, particularly in supporting the design and improvement of the interface. Engaging with coastal management practitioners also revealed potential for further development of the Tyndall Coastal Simulator as a tool to support coastal planning and management. Stakeholders provided useful criticism and highlighted potential for wider application. In terms of the interface, for example, their feedback supported the streamlining of the visual interface, making it more automated with better help functions to support the user. Feedback gained from meetings, workshops and postworkshop questionnaires was actively incorporated into the evolution of the interface. Table 11.3 summarises the positive features of the Tyndall Coastal Simulator interface perceived by stakeholders who had opportunities to experiment with it.

There was a broad range of stakeholder expertise and interest (e.g. technical, management, political), and we found that the interface could not satisfy all needs or levels of understanding; some found the interface intuitive and understandable, whereas others found it rather confusing and complex. The balance of user input and automation was appropriate for some, whereas others wanted more or less control. Suggestions for improvement included aesthetic preferences (e.g. the inclusion of aerial photography as an alternative to base maps and 3D 'fly-through' representations to illustrate future outcomes), process points (e.g. including a flowchart alongside the help function to act as a step-by-step guide or prompt sheet to show their progress through the process of making choices of different scenarios and options) and technical queries relevant to the underlying concepts and structure of the interface. All such suggestions were considered, many shaped the evolution of the interface, while those that were impractical to implement within the constraints of available resources have since become topics for ongoing and future research activity.

On the whole, users were impressed by the capacity of the Tyndall Coastal Simulator to offer scientifically robust outputs in seconds. A particular strength of the interface is its ability to translate national sea-level rise scenarios to what these

Table 11.3 A summary of some of the positive features of the Tyndall Coastal Simulator interface as defined by trial users

Positive features	Examples
Range of scenarios and projections which made it easy to make different choices	Ability to explore different scenarios at different points in time
	'*Ease of making choices re scenarios and outputs*'
	'*Real-time operation and more frequent/wider variable intervals would be good, if possible in future*'
Visual accessibility of the science through the interface	The visualisation capacity could be really helpful for communicating issues around coastal change and helping people to understand future uncertainty
	'*The strength is the visualisation. Could be used in raising awareness/understanding of issues and future scenarios by professionals for non-professionals, e.g. elected members, general public*'
Usability of the interface	A successful face for a complex model. The GIS environment is generally familiar and easy to use. Good accessibility to different components provided by the interface and the simplicity of the process of selecting modelling components. Users were impressed by the capacity of the Tyndall Coastal Simulator to offer scientifically robust, powerful outputs in seconds
	'*Ability to explore different scenarios at different points in time*'
	'*Seemed intuitive, familiar look as per ARCGIS*'
Links national sea-level rise scenarios to coastal change	Using the interface helps to fill the gap between national sea-level rise scenarios (e.g. Appendix A) and what these might mean in terms of local coastal change, including the capacity of local management decisions to influence the outcomes
	'*It would neatly fill the current gap between what UKCIP02 sea-level rise estimates and what they actually mean in terms of coastal hazards/vulnerability*'
	'*Although the implications of long-term scenarios may not feed easily into current asset management planning, they do at least push erosion risk higher up the agenda*'
Understanding the range of drivers	The configuration of the Tyndall Coastal Simulator via the user interface helped many stakeholders to understand better the underlying drivers of coastal change (i.e. waves, sediment movement and shore morphology)
	'*Liked GIS and range of decisions/subjects available; gave realisation of complexity*'
Underlying science	While inevitably complex, the quality of the underlying scientific content was appreciated and given credit. Much effort was taken to help them to understand the various models and components alongside relevant demonstrations
	'*The most up-to-date and accurate evidence base that can be maintained*'

could actually mean in terms of local coastal change. Stakeholders were impressed with the visualisation capacity (particularly the 3D visualisations – see Chap. 10), especially its potential for communicating and helping people ('expert' or otherwise) to understand uncertainty.

The interface requires users to make selections and decisions (e.g. of sea-level rise scenario and management option) in order to understand and explore potential outcomes. Preparatory effort was made to explain the various underlying models and components beneath the interface alongside demonstrations of its use. The functional design of the user interface helped stakeholders to better understand the underlying drivers of coastal change (i.e. water levels, waves, sediment movement and changing shore morphology). By quantifying spatial change in the context of different management options, the Tyndall Coastal Simulator interface is able to demonstrate different possible outcomes over different time slices so that the user can consider these and gain insights into the long-term consequences of different management options in relation to climate change scenarios. It was considered to have the potential to offer a sound basis for exploring the qualitative consequences of various options that might be considered in any planning process, thereby robustly supporting decision making.

Involving coastal management practitioners in the development of the Tyndall Coastal Simulator and its interface was valuable in gaining insights into how it might be used operationally. This aspect would have to be carefully considered before further development was carried out: Who could this tool be useful for? What are its objectives? How should it be accessed? Stakeholders could see the potential value of the Tyndall Coastal Simulator in supporting decision making and coastal planning (e.g. what happens if we decided to do this?). Having an overview of the losses and benefits arising from long-term options is important and fundamental to how we approach coastal planning in the future. For example, the Simulator demonstrates that erosion leads to benefits elsewhere in the coastal system. Accessed via the interface, the Tyndall Coastal Simulator could help to visualise many of these trade-offs and may also help to more accurately assess the optimum timescales of interventions when coupled with information on other factors such as engineering and asset (e.g. roads) relocation direct and indirect costs, insurance implications for losses, environmental gains/losses and tourism gains/losses. Working with the Tyndall Coastal Simulator highlights the inevitability of change and the need to prepare for it in an adaptive manner (the subject of Chap. 12). Some stakeholders noted that we need to talk about coastal evolution more in the context of proactively adapting to change, rather than reacting to events. Some potential opportunities and applications raised by stakeholder participants are summarised in Table 11.4.

Inevitably, no tool can cover the complete range of interests and needs associated with long-term planning in the coastal zone. Table 11.5 presents a summary of some of the more specific challenges that were drawn out of the stakeholder workshops in terms of using the Tyndall Coastal Simulator in practice. These were stimulated by interactions with the interface, but many are quite general and consequently further explored in Chaps. 13 and 14.

Table 11.4 Potential applications for the Tyndall Coastal Simulator identified with the stakeholder group

Asset management and emergency planning	Strategic management and planning of public and private (vital) coastal assets; infrastructure providers (e.g. the water sector) and the insurance industry would benefit from employing a tool like this. The Tyndall Coastal Simulator could be used to help identify areas of strategic coastal change risks and benefits in order to safeguard/relocate existing assets and locate new ones. Also potential for emergency planning
	'*Visualisation of likely impact of coastal processes on NT [National Trust] land and adjacent properties. Helping to identify assets at risk and when they are to be affected and where they may be located*'
	'*Asset management planning – decision making in location of new assets (e.g. water treatment works, water resources, distribution of water and wastewater services)*'
Integrated Coastal Zone Management (ICZM)	Useful in supporting integrated coastal zone management processes across Europe and further afield
	'*Integrated coastal zone management – allows options to be explored*'
Exploring policy options	To aid in the communication of realistic implications of climate change, and to help develop adaptive climate change policies, not necessarily as part of the Shoreline Management Planning process. Informing planning for resilience and measures for adaptation
	'*Identifying areas of strategic coastal risk; identifying, safeguarding assets; communicating realistic implications of climate change helps to develop adaptive climate change policies*'
	'*Long-term options, media use, public communications, looking at timescales of when habitat needs moving*'
Supporting Shoreline Management Planning	Potential for a tool of this kind to be developed and used to support the Shoreline Management Planning process more broadly – in assessing potential management options and associated future change for all sections of coast
	'*Certainly could be a decision-support tool for SMPs. The visualisation aspect would seem to serve a different community. Mainly see it as being useful as a scenario testing tool (e.g. what if this happens or I do this?). It can only ever support not make decisions*'
	'*Not convinced that the simulator would have changed the outcomes of the existing SMP3B. Its value may not be in deciding the policy but in visualising the options for future coast*'
Supporting land-use planning	Potential for some of the outputs to be incorporated into Local Development Frameworks as 'evidence base' in planning future housing allocations and infrastructure. Could also be of use regionally in determining long-term spatial planning scenarios.
	'*Planning future tourism infrastructure – optimising social/economic benefit*'
	'*Input into Local Development Frameworks as 'evidence base', i.e. in future planning for housing allocations, etc.*'
Education and communication	Potential to contribute to aiding public understanding of coastal change and long-term uncertainty and support the development of adaptation options. For example, at present there are still broad expectations of stability rather than change. Visualisations could be used as a tool to help people envisage the implications of various policy options and the ways in which management options and coastal processes interact over time
	'*Demonstrating risks and benefits of change processes; improving understanding of ranges of uncertainty*'
	'*Possibly in a wider educational role? E.g. use in schools to encourage longer-term awareness/understanding. If it could be web based, could be linked from many organisations. Educating in uncertainty*'

Table 11.5 The Tyndall Coastal Simulator interface in practice – key challenges defined by trial users in stakeholder workshops

Practical challenges	
Multiple needs	There is no single end-user audience; different stakeholders have different requirements. No single tool can possibly demonstrate everything we would like to know about coastal futures
	'Links to National Coastal Erosion Risk Mapping. Needs linking to industry practice and Defra/ Environment Agency (EA) functions'
	'Flyovers/3D models helpful to demonstrate to the public; input screens look like they will require technical knowledge – not existent at local authority level'
Handling uncertainty and representing change	Uncertainty is problematic and is difficult to represent. Some stakeholders felt that they would like to have seen more visualisation to help explain and illustrate uncertainty. Communicating change to the public (e.g. scenarios of possible future cliff erosion) is a sensitive subject
	'Visualisation of degrees of uncertainty/probability'
	'It seems to me that the simulator is very useful for giving scenarios of coastal hazards (e.g. erosion and flooding – and what will the future coastline look like?). The issue of translating these to risks would appear to be much more difficult, involving many factors (social, economic, cultural, etc.) and ultimately perhaps value judgements. Hard to create national (or even regional) model that can capture all of this'
Technical understanding	Beware of misuse and inaccurate interpretation of outputs arising from a lack of understanding of what the models and outputs show and mean. Improve help functions
	'Stick to the management options in the SMP guidance. Keep it simple. People don't understand the 'technical' options'
	'It is still too complex for a user to use it – a lot of manual choices and selections required'
	'Appeared to be little in the way of 'step-by-step' guidance, e.g. what does this button do? What does 'world markets' mean?'
Intangible and qualitative elements	The Tyndall Coastal Simulator cannot show many socio-economic effects of management policies, for example, health impacts. It is important to understand that the outputs can only ever support decision making and management planning rather than guide it (Chap. 12)
	'I would like to see stronger links to the demographic consequences of coastal change, e.g. through agent-based modelling'
	'Vital assets and their interrelation with wider environment, i.e. one pumping station lost may affect many more properties than just those shown at risk from erosion or flooding'
Flexibility within the Tyndall Coastal Simulator domain	There was a perceived need for more detailed assessments of the degrees or risk – e.g. extreme events and hazard planning. Users also wanted to have a chance to understand the impacts of taking temporary measures (e.g. to slow erosion in some places for a limited time). How adaptable can the model be to different management approaches?
	'Yes, include flexibility to be able to apply variation to suit different client's needs'

(continued)

Table 11.5 (continued)

Practical challenges	
Transferability to other sites	How easily could the Tyndall Coastal Simulator be applied in other places, for extreme events and for estuaries? When might this be possible?
	'Need to look at this in England, Wales and Northern Ireland. Useful for a national organisation to be able to compare regions/properties – again, important for planning'
Funding	Tools like this are really costly to develop and maintain. There are issues of ongoing funding as the basic models and scenarios need to be updated
	'All well and good planning to manage the coast, but it is dependent upon levels of funding. Will future funding go to coast or will priority go to urban flooding and floodplains?'
	'An awful lot of work has gone into developing the Coastal Simulator. It is important that the tools are accessible and applicable to the coastal/shoreline management process. Due to funder constraints, it may be worthwhile trying to develop an industry facing parallel R&D project (through EA/Defra) to ensure this research is captured and utilised within the public domain'
Scenario familiarity	Scenarios are common but not always widely understood amongst the coastal management community. Additionally, the 'standard' scenario frameworks and boundary conditions which are set by government and (generally) widely used are periodically changed
	'Need consistency (content and naming) with other scenarios (e.g. UKCIP02, UKCIP08, EA scenarios)'
Interface platform	Some feedback suggested other platforms, such as web-based hosting for a tool like the Tyndall Coastal Simulator
	'Web-based interactivity for public interface – "characteristic" trends of certain scenario outcomes; adapt presentations for different types of audience'
	'Would welcome the ability of the CS to be accessed by the general public to help understand how certain scenarios relate to them. Possible front-end web facility?'

11.4 Future Considerations for Interface Development

There were many technical challenges involved in developing the Tyndall Coastal Simulator interface. When considering its potential longevity, additional issues arise. A particular problem is that of obsolescence – a common issue owing to the pace of scientific progress and information technology in the twenty-first century. Maintaining and upgrading an interface in order to maintain its usefulness depends on ongoing funding, to maintain and upgrade related software and technology. As an example, the FutureCoast interface on long-term coastal change in England and Wales (DEFRA 2002; Burgess et al. 2002) still contains valuable content, but has become increasingly difficult to use as the wider software environment evolves. Funds are critical to allow the migration of an interface from one system to another if necessary (e.g. as institutional preferences dictate). This aspect is an important lesson for future applications of this type; repeated upgrades should be expected and planned when developing tools of this kind.

Access issues are also important. Data licence issues aside, web-based platforms would offer real opportunities for a wide potential market including institutions like the Environment Agency, coastal local authorities, non-governmental organisations and the private sector (e.g. water companies and insurance companies). Having such tools embedded in coastal management institutions with the commitment and resources to sustain these tools can be a practical option for handling the issue of fund availability. Many stakeholders felt that some sort of online/web-based format would offer real opportunities. As web-based technology continues to improve, the feasibility of this option increases. In particular, the move to national shoreline management planning in the 1990s and national coastal monitoring of England completed in 2011 coordinated by the Channel Coastal Observatory (CCO) (http://www.channelcoast.org/) (Bradbury et al. 2002; Nicholls et al. 2013) provides a structure where the Tyndall Coastal Simulator interface and its future developments may find a natural home and long-term role. As an example, the national and regional (for SMP7 and SMP22) results of the integrating COASTal Sediment Systems (iCOASST) project concerning coastal geomorphic frameworks and future predictions (Nicholls et al. 2012) will be made available via the CCO website, integrating these results with the more established monitoring data.

More generally, web delivery is far preferable to a desktop interface, but funding for maintenance of web functions will still be required. A server-based GIS provides a potential solution for developing the Tyndall Coastal Simulator interface where model results in the form of maps and graphs can be shared with research partners and stakeholders through the Internet. This solution includes a hosting server where resources (i.e. modelling materials and results) can be placed and client applications can be used to interact with these resources using GIS functionalities embedded in them.

Currently, cloud computing offers an alternative to other methods for integrating models and delivering data. It could be seen as an ideal platform for modelling and visualising coastal changes as it combines GIS software on cloud infrastructure to provide accessibility on the web. Future developments of a GIS-based coastal simulator may also involve a migration to open-source platforms both in terms of modelling and GIS tools. Currently there is a plethora of open-source GIS software (e.g. GRASS GIS, PostGIS, QGIS, SAGA GIS) with full capabilities which would help to minimise maintenance costs. Open-source solutions for web mapping are currently available and blend easily with desktop solutions (e.g. OpenLayers, GeoServer) and could enable users to interact with the Tyndall Coastal Simulator from any web-enabled device. Currently, the integrated assessment is based upon loose coupling, but with protocols such as OpenMI (Gregersen et al. 2007), which is designed to model-model interaction, becoming increasingly powerful and versatile direct coupling should be feasible. The development of a stand-alone version using open-source GIS libraries (e.g. GDAL and OGR) could also prove to be a viable solution. Other alternative options and mechanisms will doubtless present themselves in the future.

From the perspective of improving user experience, the Tyndall Coastal Simulator could be further developed to have greater functionality for decision support. This would require transfer of expert understanding into a knowledge database and the development of an intelligent query engine for the non-expert user. The latter allows for training of the users in otherwise unfamiliar scientific topics, the incorporation of complex 'what-if' scenarios and production of knowledge.

Scenarios themselves are common but not always widely understood amongst the coastal management community. Additionally, 'standard' scenario frameworks established by government departments or international bodies such as the Intergovernmental Panel on Climate Change (IPCC), and consequently widely used, are periodically changed. Consistency of scenario use *between* initiatives and projects is sometimes a problem. It is critical within an integrated assessment to make sure that the baseline scenarios are up to date and consistent. Ideally an interface would enable data to be quickly and easily updated or new layers added. Designing an interface so that different aspects of the analysis can be updated as and when it is necessary is important – the decoupled or modular nature of the Tyndall Coastal Simulator is one of its strengths in this respect. Different components and data can be recalculated without having to update the entire system.

If another, similar interface were to be developed, stakeholder engagement would once again be considered an important part of the process and would be built into the design of the project. It is a critical part of the learning process and one way of testing the underlying conceptual model. At least one cycle of interaction and feedback is necessary – but more are often desirable if not essential. In the context of the Tyndall Coastal Simulator interface, involvement in its design provided stakeholders with a sense of ownership and familiarity and consequently an appreciation of the utility of the final product.

11.5 Conclusions

The experience of building these interfaces has shed insight on a number of aspects of integrated assessment of coasts and perhaps more generally. Importantly, it has shown a method which can be applied to develop a rich set of predictions and insights concerning the present and future coastal system. While coastal systems are inherently complex, they are tractable in terms of analysis, and important aspects were quantified, such as the trade-off between erosion and flood risk. As importantly, we have considered communication and stakeholder engagement issues.

The development of the Tyndall Coastal Simulator has shown that a set of models of the coastal system and their results can be made accessible through an integrative user interface. Ultimately the interface enables users to explore how a section of coastline may look and behave in the future according to alternative climate and socio-economic scenarios and under different management strategies. The user-friendly interface presents information on a range of relevant variables in a variety of forms which can be easily accessed, understood and discussed/debated by the non-specialist user. Not only was this important in supporting stakeholder engagement in the Tyndall Centre Coastal Research Programme but also contributed to the process of integrated coastal management and planning more generally. While such integration requires significant additional effort when compared to more traditional analyses, this approach delivers more comprehensive results of direct relevance to coastal planners that would otherwise not be available. Hence, as with coastal monitoring (Nicholls et al. 2013), the benefits can outweigh the costs.

Without the interface, the Tyndall Coastal Simulator would have been a more technical research exercise with less insight into its potential for application, and the stakeholder aspects would have been less effective. Thus, the discussion around the interface and stakeholder engagement informed the integration process. Based on this experience, it is now recognised that all broad-based projects of this type require a core integrating activity if they are to achieve the goal of being more than the sum of their parts. Hence, specific resources directed towards integration not only generate research value but also maximise its impact outside the research team and immediate field. It also facilitates co-production of knowledge with relevant groups of stakeholders. These are all issues that research funders and users have come to stress much more over the last decade. This aspect is further discussed in Chaps. 13 and 14.

A critical input to the process of design, development and demonstration was the involvement of a wide range of stakeholders throughout the project. Their feedback at various stages provided constructive input to the interface development, much of which is generally applicable to the development of interfaces for IA modelling systems.

Although not recognised at the outset of the research, the interface development had the useful side effect of acting as a powerful collaboration mechanism by providing a final output around which the team could focus. This encouraged an iterative process of debate and refinement about the component models, how they were coupled and which outputs were of most management insight.

A key issue is the choice of software platform. ArcGIS was selected as the best choice during the life of the project as it was being widely used by researchers and practitioners alike; so for most of our stakeholders, the interface was familiar, accessible and aligned with their institutional software. However, technologies in this area have advanced rapidly, with cloud computing, powerful open-source GIS software and web services for mapping that provide alternative options for the future.

References

Bradbury, A. P., McFarland, S., Beck, R., & Horne, J. (2002). Development of a strategic coastal monitoring programme for southeast England. In *International coastal engineering conference*. Cardiff: ASCE.

BRANCH Partnership. (2007). *Planning for biodiversity in a changing climate – Branch project.* Final report. Natural England, UK. http://webarchive.nationalarchives.gov.uk/20081125212930/http:/www.branchproject.org/. Cited 18 Dec 2013.

Brown, I., Jude, S. R., Koukoulas, S., Nicholls, R. J., Dickson, M., & Walkden, M. (2006). Dynamic simulation and visualisation of coastal erosion. *Computers, Environment and Urban Systems, 30*, 840–860.

Burgess, K. A., Orford, J., Dyer, K. R., Townend, I. H., & Balson, P. (2002). Futurecoast - The integration of knowledge to assess future coastal evolution at a national scale. In *Proceedings of the 28th international conference on coastal engineering* (Vol. 3, pp. 3221–3233). New York: ASCE.

Dawson, R. J., Dickson, M. E., Nicholls, R. J., Hall, J. W., Walkden, M. J. A., Stansby, P., Mokrech, M., Richards, J., Zhoh, J., Milligan, J., Jordan, A., Pearson, S., Rees, J., Bates, P., Koukoulas, S., & Watkinson, A. (2009). Integrated analysis of risks of coastal flooding and cliff erosion under scenarios of long term change. *Climatic Change, 95*, 249–288.

DEFRA. (2002). *The Futurecoast project* (Project Code FD2002). London: Department of Environment, Food and Rural Affairs.

Dickinson, T. (2007). *The compendium of adaptation models for climate change* (1st ed.). Toronto: Environment Canada.

Gardiner, S., Hanson, S., Nicholls, R. J., Zhong, Z., Jude, S., Jones, A., Richards, J., Williams, A., Spencer, T., Cope, S., Gorczynska, M., Bradbury, A., McInnes, R., & Ingeby, A. (2007). The Habitats Directive, coastal habitats and climate change – Case studies from the south coast of the UK. In *International conference on coastal management, ICE proceedings* (pp. 193–202). London: Cardiff, Thomas Telford Ltd. ISBN 978-07277-3593-5.

Gregersen, J. B., Gijsbers, P. J. A., & Westen, S. J. P. (2007). OpenMI: Open modelling interface. *Journal of Hydroinformatics, 9*(3), 175–191.

Harrison, P. A., Holman, I. P., Cojocaru, G., Kok, K., Kontogianni, A., Metzger, M., & Gramberger, M. (2013). Combining qualitative and quantitative understanding for exploring cross-sectoral climate change impacts, adaptation and vulnerability in Europe. *Regional Environmental Change, 13*(4), 761–780.

Hinkel, J., & Klein, R. J. T. (2009). The DINAS-COAST project: Developing a tool for the dynamic and interactive assessment of coastal vulnerability. *Global Environmental Change, 19*(3), 384–395.

Holman, I. P., Rounsevell, M. D. A., Cojacaru, G., Shackley, S., McLachlan, C., Audsley, E., Berry, P. M., Fontaine, C., Harrison, P. A., Henriques, C., Mokrech, M., Nicholls, R. J., Pearn, K. R., & Richards, J. A. (2008). The concepts and development of a participatory regional integrated assessment tool. *Climatic Change, 90*(1–2), 5–30.

Jude, S. R., Jones, A. P., Andrews, J. E., & Bateman, I. J. (2006). Visualisation for participatory coastal zone management: A case study of the Norfolk Coast, England. *Journal of Coastal Research, 22*, 1527–1538.

Jude, S. R., Jones, A. P., Watkinson, A. R., Brown, I., & Gill, J. A. (2007). The development of a visualization methodology for integrated coastal management. *Coastal Management, 35*(5), 525–544.

Koukoulas, S., Nicholls, R. J., Dickson, M., Walkden, M., Hall, J. W., Pearson, S., Mokrech, M., & Richards, J. (2005). GIS tool for analysis and interpretation of coastal erosion model outputs (SCAPEGIS). In *Proceedings of coastal dynamics 2005*, Barcelona.

Leake, J., Wolf, J., Lowe, J., Stansby, P., Jacoub, G., Nicholls, R., Mokrech, M., Nicholson-Cole, S., Walkden, M., Watkinson, A., & Hanson, S. (2007). Integrated modeling for coastal impacts. In *Proceedings of the 10th international conference on Estuarine and coastal modelling, November 2007, Newport, Rhode Island.* New York: ASCE.

Mokrech, M., Nicholls, R. J., Richards, J., Watkinson, A. J., Jude, S., Milligan, J., Hall, J., Walkden, M., Stansby, P., Wright, J., Rounsevell, M., Fontaine, C., Acosta-Michlik, L., Lowe, J., & Wolf, J. (2007). The development of an integrated coastal simulator for supporting long term coastal management. In *Proceedings of the international conference on coastal management, Cardiff* (pp. 203–217). London: Thomas Telford.

Mokrech, M., Hanson, S., Nicholls, R. J., Wolf, J., Walkden, M. J. A., Fontaine, C., Nicholson-Cole, S. A., Jude, S. R., Leake, J., Stansby, P. K., Watkinson, A. R., Rounsevell, M. D. A., Lowe, J. A., & Hall, J. W. (2011). The Tyndall coastal simulator. *Journal of Coastal Conservation, 15*, 325–335.

Nicholls, R. J., Hanson, S., Mokrech, M., Stansby, P., Chini, N., Walkden, M., Dawson, R., Roche, N., Hall, J., Nicholson-Cole, S., Watkinson, A., Jude, S., Lowe, J., Wolf, J., Leake, J., Rounsevell, M., Fontaine, C., & Acosta-Michlik, L. (2008). The Tyndall coastal simulator and interface. In *International conference on coastal engineering*, Hamburg.

Nicholls, R. J., Bradbury, A., Burningham, H., Dix, J. K., Ellis, M., French, J., Hall, J. W., Karunarathna, H. U., Lawn, J., Pan, S., Reeve, D. E., Rogers, B., Souza, A., Stansby, P. K., Sutherland, J., Tarrant, O., Walkden, M. J. A., & Whitehouse, R. (2012). iCOASST - Integrating coastal sediment systems. *ICCE2012 international conference on coastal engineering*, Santander, 1–6 July 2012.

Nicholls, R. J., Townend, I. H., Bradbury, A. P., Ramsbottom, D., & Day, S. A. (2013). Planning for long-term coastal change: Experiences from England and Wales. *Ocean Engineering, 71*, 3–16.

Pontee, N. I., & Morris, K. (2011). CoastRanger MS: A tool for improving public engagement. *Journal of Coastal Research, 27*(1), 18–25.

Spiegelhalter, D., Pearson, M., & Short, I. (2011). Visualizing uncertainty about the future. *Science, 333*, 1393–1400.

Yoskowitz, D., Carollo, C., Beseres-Pollack, J., Welder, K., Santos, C., & Francis, J. (2012). *Assessment of changing ecosystem services provided by marsh habitat in the Galveston Bay Region.* Final report to the Gulf of Mexico Foundation and the Habitat Conservation and Restoration Team of the Gulf of Mexico Alliance.

Chapter 12
Many Stakeholders, Multiple Perspectives: Long-Term Planning for a Future Coast

Sophie A. Day (née Nicholson-Cole), Tim O'Riordan, Jessica Bryson (née Milligan), Peter Frew, and Robert Young

Abstract Current planning for the future of coastal zones in England is occurring at a time of great change and uncertainty. Alongside the expectation of increased storminess and impending sea-level rise associated with climate change, coastal decision-making is subject to a whole host of institutional shifts and the legacy of past coastal management decisions. Changing official policy, administrative arrangements and jurisdictions, the need to create conditions for community involvement and local and national level budgetary constraints are all issues in the melting pot.

This chapter summarises recent and current decision-making practice as it applies to Norfolk, UK, and suggests how coastal management for a changing coastline may generally become more adaptive, socially fair and effectively implemented. In Norfolk, coastal change is a complex and emotive issue, the management

S.A. Day (née Nicholson-Cole) (✉)
Tyndall Centre for Climate Change Research, Faculty of Engineering and the Environment, University of Southampton, Southampton SO17 1BJ, UK
e-mail: day_soph@yahoo.co.uk

T. O'Riordan
School of Environmental Sciences, University of East Anglia, Norwich, Norfolk NR4 7TJ, UK
e-mail: t.oriordan@uea.ac.uk

J. Bryson (née Milligan)
Geography and the Lived Environment Research Institute, School of Geosciences, University of Edinburgh, Drummond Street, Edinburgh EH8 9XP, UK
e-mail: jessica.bryson@ed.ac.uk

P. Frew
Coastal Strategy for North Norfolk District Council, Coastal Management Consultant, 17 Main Street, Little Thetford, Ely CB6 3HA, UK
e-mail: pf@peterfrew.co.uk

R. Young
Coast and Community Partnerships, North Norfolk District Council, Holt Road, Cromer, Norfolk NR27 9EN, UK
e-mail: Robert.Young@north-norfolk.gov.uk

R.J. Nicholls et al. (eds.), *Broad Scale Coastal Simulation*, Advances in Global Change Research 49, DOI 10.1007/978-94-007-5258-0_12

of which has evolved significantly over the last two decades. This chapter specifically addresses the ways in which national, regional and local stakeholder interests have interacted during this tumultuous time in North Norfolk to negotiate pathways for adapting to coastal change. The case example presented in this chapter, and more broadly in this book, illustrates that there are social limits and barriers, which hinder the conditions most likely to enable progressively more adaptive coastal governance.

Many lessons have been learned from this experience and these are outlined in the context of how coastal governance in England and further afield may seek to become more adaptive. Integrated approaches to the management of future coasts are needed, encompassing robust science, genuine stakeholder commitment and inclusion, and funding arrangements (as adaptation always comes at a cost). In this regard, the deployment of the Tyndall Coastal Simulator approach can help to support this need.

Within the Tyndall Coastal Simulator, this work provided an important context for all the model development. With the visualisations (Chap. 10) and interface (Chap. 11), it also provided a conduit from the models to a true two-way stakeholder interaction and feedback. This experience showed that regular stakeholder engagement activities should always be a core part of integrated assessment exercises.

Keywords Adaptive coastal governance • Coastal change • Stakeholder engagement

12.1 Introduction

This book outlines an approach to the integrated assessment of coastal change – the Tyndall Coastal Simulator – that was based principally on a case study in North Norfolk, UK. Part of its aim was to bring a number of decision elements together into a scientifically robust tool that could be accessible for a wide range of stakeholders involved in managing and planning for the future of the coastline of Eastern England. Chapters 2–11 have given the reader a clear impression of the complexity of the coastal system in the case study area; this chapter gives an insight into the challenge of adapting to a changing coast in a political and social context.

North Norfolk has become iconic in the UK as to how people are responding to a changing coastline, deciding how it should be managed and testing options for adapting to coastal change. Planning for coastal futures is complex and characterised by great dynamism and uncertainty. In this chapter, we draw upon the case of North Norfolk to reveal some real challenges experienced and lessons learned in preparing for future coastal change. It is impossible to generate a 'one-size-fits-all' approach or tool to guide coastal management and adaptation to long-term coastal change. This is because the characteristics of each case location and set of stakeholder interests are different. Equally, taking a location-specific or discipline-based approach is problematic because of the complicated nature of our coastal zones where physical systems, economies, cultures and societies are interlinked and unique. In our presentation of this particular case, we aim to draw out a set of useful lessons about adaptive coastal governance that have emerged from real experience and that could be relevant in other coastal contexts.

In the UK, the approach historically taken to managing flood and coastal erosion is not tenable. For five decades, hard engineered defence was seen as an appropriate way to manage a dynamic coastline. As is evident in the preceding chapters, we now have a much greater understanding of the physical system, and consequently a broader portfolio of choices as to how we may adapt to future change. The idea that hard defence systems can stop coastal change is a fallacy; the financial commitment alone involved in protecting the entire coast would be enormous. Throughout history, people have lived by the sea and accepted the reality of coastal change; however, in recent decades, our efforts to control natural coastal processes (both flooding and erosion) have morphed into a widely held expectation of protection. There is a growing understanding that coastal change needs to be accepted, but critically that it must also be managed to enable adaptation to this change to take place. How our future coastline will evolve is dependent on coastal governance arrangements learning to take a more adaptive and long-term approach.

This chapter begins by offering a picture of the sociopolitical puzzle in relation to the changing physical nature of Eastern England's coastline. It addresses the concept of adaptive coastal governance and explains what has been done in practice in North Norfolk to tackle present and future coastal change. We also consider the potential for a tool like the Tyndall Coastal Simulator to aid future coastal change management. There are lessons that can be learned from what has happened in the case study area and we feel that these can be applicable in other contexts, both in the UK and abroad. The chapter ends with a short reflection of these lessons in terms of future adaptive governance for a more sustainable coastline. These issues are also developed in Chaps. 13 and 14.

12.2 The Challenges of a Changeable Coast

The coastline of North Norfolk has a long history of retreat over thousands of years – it is an inherently dynamic coastline. Only within relatively recent history (i.e. since Victorian times and particularly since the building of large-scale defence works following the 1953 North Sea storm surge) has it been considered stable. The construction of hard engineered erosion and flood defence works over many decades brought instant and tangible 'solutions' to coastal change. Since the 1950s and until relatively recently, official policy has been to defend the coast by 'holding the line' with coastal defences.

Anthropogenic alterations to natural patterns of sediment supply, transport, deposition and defence have, however, led to an array of new coastal instabilities (e.g. French 2004; Leafe et al. 1998). For example, the existence of structures that disrupt sediment movement around the coast may starve downdrift beaches of material and much needed beach protection (Dawson et al. 2009[1]). Additionally, coastal defence works are engineered structures, which do not have an infinite lifespan. Many are reaching the end of their design life and will require substantial investment to maintain their effectiveness and standard of protection.

The existence of coastal protection works has to some extent also created a perception that resisting the action of the sea will achieve a static coastal state within which we can act with certainty about the future. Some coastal residents appear not to appreciate that they are living on a dynamic coastline, a coast that has been retreating for thousands of years and will continue to do so. Because of more than five decades of defence and protection, many people understand it as being stable and appearing to be permanent (Frew 2012). There is a culture of affection for, and familiarity with a benign coast, one that is defended and where the defences themselves are seen as being part of the character of the place (e.g. O'Riordan and Nicholson-Cole 2010; Tunstall and Penning-Rowsell 1998). Alongside a deeply held popular view of coastal stability, there remains an expectation of entitlement to the maintenance of a fixed and defended shoreline (e.g. Milligan et al. 2009). A Canadian study by Freisinger and Bernatchez (2010), found that despite an awareness of coastal hazards and the causes of erosion, the adaptive methods preferred by residents are not always appropriate for the coastal system; respondents in their study favoured massive and hard defence structures even whilst recognising that such methods have a major financial cost and a significant impact on coastal dynamics that can reduce the coast's resilience.

In reality, flood and coastal erosion protection is neither a legal right nor an entitlement, and the way coastal change is being planned for and managed is shifting. The concept of adaptation to coastal change has for the last decade become more common parlance, as an alternative to resisting change through expensive structural defence measures (Nicholls and Lowe 2004). Recent policy recommends that in some places, investment in defence will not continue long term (East Anglian Coastal Group 2012). This is primarily for two reasons – a move towards a more strategic whole-system thinking (partly driven by a better understanding of the impacts of climate change) and the sheer cost associated with an engineered coast.

One of the biggest challenges for those involved in the future planning and management of changing coasts is how to build awareness of coastal change and negotiate the opportunities for adaptive responses to them in a managed and measured way. Chapter 7 in particular demonstrates the uncertainty associated with rates of erosion; indeed, there are multiple possible futures, which are partly dependent on

[1] Also appended within Chap. 1 of this book

management decisions taken elsewhere in the coastal system. In order to explore what the options may be for adapting to a changing coast, we must understand the existing social and economic conditions of any complex coastline, as well as the environmental and geomorphological context. Using the North Norfolk case study, this chapter explores the notion of adaptive coastal governance and examines the potential for broad-scale coastal simulation to play a role in its delivery.

12.2.1 Adaptive Governance and Limits to Adaptation

The planning and implementation of adaptation to coastal change are essentially governance issues, involving deliberate and planned action or inaction. Governance involves processes through which we engage with our environment and the rest of society. It involves activities which make a purposeful effort to guide, steer, control or manage sectors or facets of societies (Adger et al. 2009). These processes occur in relation to different values and at scales ranging from the collective to the individual. Nicholson-Cole and O'Riordan (2009) propose eight key conditions for adaptive coastal governance as illustrated in Fig. 12.1. These conditions are independently challenging and complicated, but without an integrated approach, which recognises that they are all important factors, it is impossible realistically to generate an effective adaptation pathway or plan.

Schmidt et al. (2012) propose a similar model developed as a result of examining the challenges around managing coastal change in Portugal. They identify three key arrangements that are required to be in place for any evolution of adaptive coastal governance to take place:

Fig. 12.1 Eight conditions for adaptive coastal governance (Nicholson-Cole and O'Riordan 2009)

- *A credible science* for predicting future coastal changes backed by broad scientific consensus but also understood and supported by a wide array of interests
- *Evolving institutional capability* with policy clarity and political will, with comprehensive management arrangements and decision structures
- *Clear goals* for identifying courses of action at various management levels backed by appropriately designed financing, powers and accountable management procedures (inclusive participatory procedures being part of this)

Huitema et al. (2009) and Olsen et al. (2009) also address the issue of adaptive coastal governance. Both highlight the need for social learning to be achieved through progressive trust building and the development of resilient coping structures based on a common understanding of the science of coastal evolution, brokered by effective engagement of local and national interests. All of these examples conclude that coastal management problems cannot effectively be addressed by taking a segmented or disciplinary view. In dealing with long-term problems that link the environment, society and economic interests, an integrated approach is critical.

The process of adapting to change is not costless in terms of time, effort and money; it is an active process and one which requires great forethought and ongoing investment. During the study period in North Norfolk, it became evident that many of the factors that ideally underpin adaptive coastal governance were not in place. As was observed in practice, the absence of these underpinning factors and concepts can constitute limits or barriers to planning adaptively and hinder efforts towards implementing more adaptive coastal management visions and practices. Adger et al. (2009) discuss this concept in terms of these being social limits (as opposed to biological thresholds as limiting factors) to adaptation, for example, the ways in which we perceive risk, policy inertia and so on. The experience in North Norfolk demonstrated that these limits are indeed likely to be inhibiting our capacity to adapt to coastal change at local to national scales.

12.2.2 Reactions to Coastal Change in North Norfolk, UK

In the policy arena, the coastal status quo (tending to defend and protect) changed in 2005 with a reorientation of strategic thinking, heralded in a policy paper produced by the Department for Environment, Food and Rural Affairs. This was provocatively (in some people's eyes) entitled, 'Making Space for Water' (Defra 2005). A better understanding and recognition of the impacts of climate change and an aspiration to return to a more naturally functioning coastal system, towards a state of dynamic equilibrium in the sediment budget were the key drivers for this report.

This change was implemented at the local level predominantly via the shoreline management planning process,[2] administered by local government authorities and the Environment Agency (a government agency in England and Wales charged with having a strategic overview of coastal matters and delivering the environmental priorities of central government). A shoreline management plan (SMP) provides a large-scale assessment of coastal evolution and presents a policy framework, which determines how a section of shoreline will be managed, i.e. which sections of the coast are to be protected and if not, what the policy options are. It also includes an economic assessment of the costs and benefits of the approach recommended for each coastal frontage. It projects the short-, medium- and long-term effects of the policy, indicating which areas are likely to be at risk of coastal erosion and coastal flooding.

As explained in Chap. 1, Norfolk is covered by three shoreline management plans (SMPs): SMP 4 (Gibraltar Point to Old Hunstanton), SMP 5 (Hunstanton to Kelling Hard) and SMP 6 (Kelling Hard to Lowestoft Ness). SMPs 5 and 6 relate to two physical units of the North Norfolk Coast. SMP6 (Kelling Hard to Lowestoft Ness) (East Anglian Coastal Group 2012) is predominantly the subject of this book and contains the case study area. The original SMP for this area (SMP 3b, published in the 1990s) contained policies purporting to maintain the status quo, a state of ongoing protection from coastal erosion and flooding (or 'holding the line') for those places that had previously been defended. However, this plan has undergone much iteration since the review process began in 2004. The draft 'second-generation' SMP 6 was published in 2005 and proposed that only in some places should the previous 'hold-the-line' status be maintained. The changes reflected the emergence of a more strategic and long-term approach to looking at the future of the coast. In some cases, the policy recommendations changed in order to allow shoreline retreat and a more naturally functioning coastline through policies of 'managed realignment' or 'no active intervention'.

The SMP policies were illustrated as coloured zones on maps showing sections of the coast, to represent the projected extent of erosion and coastal retreat over different time frames (initially 25, 50 and 100 years) associated with the proposed policy for each unit of coastline. An aerial view interpretation of these zones is shown in Fig. 12.2. This gives a poignant illustration of the implications of this policy on land and property in the village. The publication of the maps in the second-generation SMP 6 was combined with rapid change in coastal conditions. Defences had begun to degrade and were not being repaired. In 2007, for example, there was a storm event during which the cliffs noticeably retreated at Happisburgh. All this had a strong negative impact on community confidence and people's certainty about the economic integrity of their property values and businesses. Box 12.1 gives a descriptive account of how the changes have and are likely to manifest themselves in the future in the village of Happisburgh.

---

[2] Shoreline management plans (The Environment Agency): http://www.environment-agency.gov.uk/research/planning/104939.aspx Cited 19 November 2013.

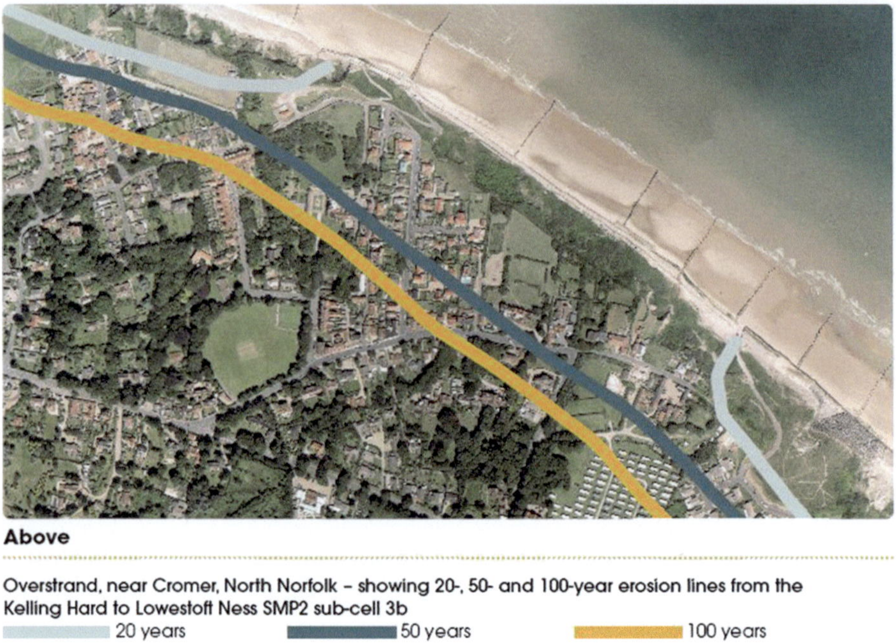

Fig. 12.2 A portion of the projected erosion rates transposed onto an aerial photograph, showing what could be at stake under the policy presented in SMP6 (Young 2009, p. 420)

Box 12.1 The Happisburgh Case

From 1999 onwards, the combination of a series of decisions and policy changes, coupled with storm events and changing coastal processes, led to the failure of the existing sea defences (with no prospect for replacement) and the loss of beach at Happisburgh.

The direct impacts were the loss of approximately 16 chalets (approximately half of which were in permanent residential use) together with the closure of the café and guest house and the loss of the beach access ramp (providing access for the lifeboat and beach users). Aside from the actual property loss, environmental degradation has resulted from the remains of properties and infrastructure on the cliff top and the beach, the remnants of coastal defences scarring the coastal landscape and the redundant (uninhabitable) dwellings close to the cliff edge.

The predictions in the emerging shoreline management plan show the loss of a further 11 properties up to 2025 as well as the caravan park, lifeboat house, coastguard station, public car park and site of the public toilets (recently demolished). Further, the houses, the village pub and the Norman church are predicted to be lost between 2025 and 2105, at which point the main coast road (providing access to and beyond the village) is also likely to be lost.

(continued)

Box 12.1 (continued)

Figure 12.3a, c show aerial views of Happisburgh in 1998 and 1996. Cliff-top properties are visible but disappearing. Figure 12.3b, d show Happisburgh in 2013 and 2012 before and after demolition works to remove properties at risk, close to the end of the beach road. Disappearance of other cliff-top activity including the lifeboat station and loss of land on the campsite can also be observed. The comparisons clearly show the degradation of sea defences and consequent expansion of the bay. Figure 12.4 shows houses on the beach road, which once extended all the way along the cliff top where the bay now exists.

Source: North Norfolk District Council[3]

Fig. 12.3 (**a**) Aerial view of Happisburgh from the southeast (1998) (Source: Mike Page http://www.mike-page.co.uk/). (**b**) Aerial view of Happisburgh from the southeast (2013) (Source: Mike Page http://www.mike-page.co.uk/). (**c**) Aerial view of Happisburgh from the northwest (1996) (Source: Mike Page http://www.mike-page.co.uk/). (**d**) Aerial view of Happisburgh from the northwest (2012) (Source: Mike Page http://www.mike-page.co.uk/)

[3] http://www.northnorfolk.org/pathfinder/happisburgh_intro.asp (last accessed 13th November, 2013).

(continued)

Box 12.1 (continued)

Fig. 12.3 (continued)

(continued)

Box 12.1 (continued)

Fig. 12.4 Houses along beach road which once ran along the cliff top at Happisburgh, where the bay exists now (Image courtesy of Norfolk County Council Library and Information Service)

The proposed changes in the 2005 draft SMP came as a shock to many coastal communities, organisations and individuals. Many felt isolated and let down by the decision-making process. In coastal towns around Eastern England and in other parts of the country, community groups self-organised in order to try and get their concerns heard. They felt desperately unsettled, uncertain and worried about what the plans meant (see, for example, Fig. 12.5). Following the release of the draft SMP, a turbulent period of stakeholder interaction ensued. This involved government officials from national to local levels, representatives of government bodies such as the Environment Agency and Natural England (previously English Nature) and local community groups and individuals, to name a few. Upon publication, the policies proposed in the draft SMP imbued a widespread sense of local unease and uncertainty about the future of people's homes and communities.

As documents, SMPs are unable to determine a social response to the consequences of their policies (Frew 2012). From a community perspective, the SMP in this case made policy decisions based on scientific predictions but contained no mechanisms for managing the changes and impacts that would result. There were no policies, funding or any official documentation at national, regional or local scales as to how the consequences of the proposed policy changes could be handled.

Fig. 12.5 An aerial view of protesters at Happisburgh, UK (Source: Mike Page http://www.mike-page.co.uk/)

Predictions or projections without implementation and management plans (for mitigation of consequences, adaptation, investment or whatever) risk only every being interpreted by those experiencing the adverse consequences as 'scaremongering'. When the impacts of long-term coastal change are communicated, typically the effect is felt not at the end of that period when a property is eventually lost, but as soon as the prediction is made. The result, in Eastern England, was a fundamental loss of trust in the institutions responsible for producing SMPs, which is extremely difficult to rebuild. A simple depiction of this situation is illustrated in Fig. 12.6.

Coastal systems will undergo long-term changes, depending on various factors and inputs that need to be understood if we are to plan adaptively for this change. Adaptive management is necessarily long term and its benefits can be far less apparent in a local context (especially those with property imminently at risk) than they are strategically or nationally. This is, however, anathema to the process of representative democracy, which by its very nature has short and local horizons. Long-term environmental change is inherently difficult to comprehend (Lorenzoni et al. 2007; Tonn et al. 2006) and accordingly to plan for. People tend to relate things to their own (relatively short) lifetimes, and the decisions of householders and businesses relate to the even shorter timescales of mortgages and investment plans.

There are also many sensitive social and economic aspects that cause coastal communities a tendency to be somewhat marginal and vulnerable. The House of Commons Community and Local Government Committee (2007) identified some of these, including physical isolation, high deprivation levels and lack of affordable housing, inward migration of older people, high levels of transience (adults of working

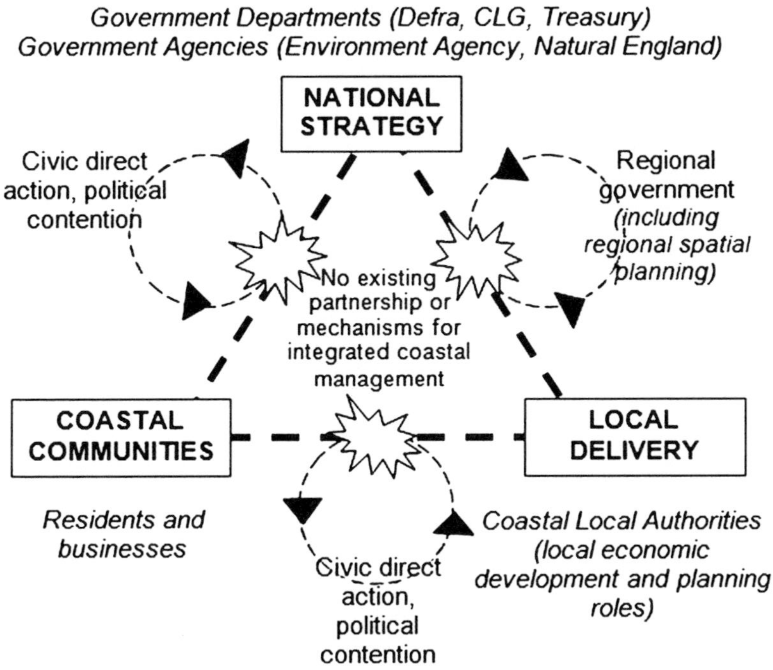

Fig. 12.6 An illustration of tensions between key institutions that were involved in coastal governance in England in 2005 (there is no longer a regional tier of governance)

age and overseas migrant workers moving on a frequent basis), outward migration of younger people, poor quality housing and fragile, isolated coastal economies. These factors are particularly important in the North Norfolk context and were compounded by the uncertainty and anxiety of many people and their communities in response to the draft SMP being published.

Changing patterns of physical risk are demonstrably part of a more complex set of contributors to the vulnerability of coastal communities. The greater the vulnerability of coastal communities, the less capacity they have to respond to and cope with change. Without strategic planning for likely future change, coastlines and their communities will be left in an even more vulnerable state – socially, economically and physically in terms of actual erosion and flooding.

12.2.3 Local Progress Towards Adaptation

In an attempt to discover more about coastal adaptation options, in 2009, a programme called the Coastal Change Pathfinders was announced by Defra (Department for Environment, Food and Rural Affairs). Local authorities were invited to apply to the GBP £11 million fund; consequently 15 authorities were given funding to explore new ways of adapting to coastal change (Defra 2011a). The geographical spread of the Pathfinders, together with the wide range of projects being explored, were intended to provide insights into different approaches to supporting community adaptation to coastal change. Working in partnership with their communities, the Pathfinders road tested innovative approaches to planning for, and managing coastal change. The Pathfinder initiative has sought to improve understanding of how coastal communities can adapt to coastal change and what the costs and benefits of different approaches. The aim was also to provide practical lessons and examples that can be shared with other practitioners, particularly on community adaptation planning and engagement and delivery of adaptive solutions. Tompkins et al. (2010) looked at examples of adaptation in practice to climate change in the UK and found that to date adaptation has principally occurred in the form of research into impacts and potential adaptation policy and activities. This suggests that the Pathfinder projects are good examples of explicit adaptation activities being put into practice; it is anticipated that the lessons learned from these will be really critical in the coastal planning and management arena. The Norfolk example is summarised in Box 12.2. Frew (2012) gives a detailed account of the original vision and the actual project including the finance and management aspects.

Box 12.2 The North Norfolk Pathfinder Project: An Overview

North Norfolk District Council – awarded £3 million, some of which was spent on projects in and around the village of Happisburgh.

Until recently, Happisburgh had been a community plunged into despair because of the failure of its sea defences, severe weather causing rampant cliff erosion and predictions of long-term change threatening village facilities, businesses and people's homes. The name Happisburgh became synonymous with erosion and 'blight'.

Community engagement

Engagement with local businesses affected by coastal change impacts to provide short- and long-term tailored business support and advice on issues and adaptation options available. In Happisburgh, a community project to understand, record and manage the impacts of coastal change on their heritage has led to the community producing a book about its history.

Adaptation planning

Facilitating the rolling back of businesses, infrastructure and community facilities aimed at restoring confidence and reversing the spiral of decline caused by blight and disinvestment in 'at-risk' coastal settlements. Thanks to local people campaigning for 'social justice' and helping to fund 'temporary' rock defences to slow the rate of erosion, the council was able to work with the local community to develop an alternative solution. A substantial proportion of the funding has been used to deal with the consequences of erosion in Happisburgh and ensure that processes are in place to manage the impacts of coastal change in the future. This has restored confidence in the future of this seaside village, which is already becoming known again for its fantastic environmental assets.

(continued)

Box 12.2 (continued)

Delivering adaptive solutions

Plans have been put into practice to support the development of a sustainable future for the community of Happisburgh and address the dereliction caused by coastal change. Thanks to the hard work of the community and the huge effort from North Norfolk District Council:

- The most vulnerable houses have been purchased (from the willing owners) and those households have been able to move on with their lives.
- The buildings have been demolished and the blighted cliff-top environment is being restored with grass and wildflowers.
- A planning application for replacement houses on a suitable inland site is hoped to bring new investment to the village.
- The beach has been restored by the removal of debris left by failed defences and collapsed buildings.
- A wonderful new car park has been created, with creatively designed new public conveniences and a new ramp directly onto the beach.
- Businesses have been provided with support to maintain investment and find appropriate ways out of their predicament.
- Elsewhere projects include replacing a coastal footpath.

More information is available at: http://www.northnorfolk.org/pathfinder/ Cited 19 November 2013

Following the publication of the draft SMP in 2005 (which for the first time proposed policies to allow coastal change), a lot of work was done by North Norfolk District Council (the local government office) to bring stakeholders together to work out how to proceed. Over time, the various stakeholder interests came together to more constructively discuss and pave a way through the fog of distrust and uncertainty that had settled. The work done in the case study area to tackle such a difficult situation has made the case of North Norfolk a flagship example which is nationally recognised and has locally resulted in a more united approach to adaptation to coastal change. This partly enabled them to put together such a coherent and successful Pathfinder bid. In developing a bid for Pathfinder funding, the council sought to generate projects that would have immediate as well as lasting benefits. The much publicised purchase and demolition of the houses that were most immediately at risk on the Happisburgh cliff front was in many ways the least important of those measures. More important were some other projects such as support for businesses, the creation of a new visitor's car park not at risk, an alternative footpath away from a risk zone and the relocation of a village hall. All these options have a lasting benefit to the surrounding communities and in many cases can be directly managed by them.

Unfortunately, since 2010, much institutional change nationally, regionally and locally has led to an even more fragmented and disjointed situation for coastal management. For example, at national government level, commitment to adaptation to climate change is noticeably weaker since a change of political administration in 2010. The nongovernmental bodies (Environment Agency and Natural England) have undergone severe budget cuts and staff numbers have decreased dramatically with roles changing and staff moving around. A change in the political administration of North Norfolk meant that a lot of knowledge and experience amongst councillors has been lost. In sum, the collegiate relationships between government, public bodies and coastal communities are now much more dilute and responsibility for strategic coastal management remains unclear. These are important hindrances to adaptive coastal governance.

12.2.4 The Tyndall Coastal Simulator: Stakeholder Engagement

Chapter 11 outlined the value brought to the scientific and conceptual development of the Tyndall Coastal Simulator by engaging with a wide range of coastal stakeholders. The relationships that were established during the project, proved to be mutually beneficial by way of knowledge exchange and in fact led to much co-production of new knowledge and understandings about coastal issues in the case study area.

As well as examining the practical challenges surrounding the construction of an interface for an integrated coastal assessment, Chap. 11 presented some broader issues around how a tool such as the Tyndall Coastal Simulator – or certainly that kind of approach – might help to provide a much more robust base for adaptive coastal governance. Integrated assessments – with user interfaces – could be of great value in the process of planning adaptation to coastal change, particularly in really challenging contexts. Providing a strong scientific basis that is informed by socioeconomic and other features would be a great asset in approaching many coastal problems in an adaptive way with long-term foresight.

One item of stakeholder feedback was that it is positive to communicate in a language of 'change' rather than of 'risk', which has more subjective connotations; many stakeholders felt that we need to talk about coastal evolution more in the context of adapting to change, not reacting to risk. Contributing to the development of the Tyndall Coastal Simulator underlined for some the inevitability of change and the need to prepare for it sustainably. More detail is presented in Chap. 11 about the feedback gained from stakeholders during the development of the Tyndall Coastal Simulator.

12.3 Adaptation in Practice: The North Norfolk Experience

The physical and technical complexity of a dynamic coastline such as that of North Norfolk has been demonstrated well in the preceding chapters of this book. Considering future change, how to manage this and how to prepare for change in a 'sustainable' manner clearly presents us with a huge challenge and a number of questions that are hard to answer. Who should act in the face of scientific evidence? How and when should they do this? Whose values count and how can these be prioritised? What lessons can be learned from test cases? What best practice initiatives could be taken forward and applied elsewhere?

Adger et al. (2009) claim that the governance of adaptation is likely to be complex and somewhat messy. Adger et al. also note that the legacy of past modes of operating combined with persisting paradigms makes it difficult to enact effective adaptations to an issue as complex and multifaceted as climate change. The case example presented in this chapter supports this notion. Indeed, nationally, there is no clear institutional set of arrangements for coastal governance or clear funding to support local initiatives. It can be argued that the lack of political and economic 'staying power' is a common feature of the weakness of adaptive coastal governance in England. Adaptive coastal governance needs strong and sympathetic public involvement (rarely provided) as well as progressive political leadership. Stojanovic and Ballinger (2009) have questioned the effectiveness of institutional arrangements and policies for coastal governance. We find by looking at this case study that coastal planning and policy processes are disconnected and unclear at national to local scales. With political shifts and changes in economic circumstances come changes in policy and funding priorities for adaptive coastal planning at all levels. This means that since the early 2000s, coastal governance arrangements have been in a state of near-constant flux at all administrative levels. Falaleeva et al. (2011) identify fragmented governance structures and responsibilities of key stakeholders, lack of coordinated support for ICZM or similar mechanisms at the national level and a relatively weak awareness of the specifics of adaptation at the local level as being barriers to integrated coastal zone management – a concept that we come to next.

12.3.1 Limiting Factors

The Tyndall Centre research conducted largely in North Norfolk identified eight key challenges to effective adaptation planning for a changing coastline (Nicholson-Cole and O'Riordan 2009). These were developed through working with national, regional and local stakeholders involved in the coastal management arena and are presented in Fig. 12.7. Whilst they were identified as a result of case study research in Norfolk, England, we believe that some or all of them are probably acting in other contexts to make adaptive governance difficult to put into practice.

In each context, there are different sensitivities, drivers of change, threats to livelihoods and vulnerabilities. The challenges presented in Fig. 12.6 are not only specific to the case example but can also be interpreted more broadly to the challenge of managing any dynamic coastline. They constitute barriers and limits to adaptation which transcend administrative and management scales, as are presented in Table 12.1. Barriers to adaptation in a coastal context are also explored by Falaleeva et al. (2011) who similarly report that the following factors constitute particular limitations: the fragmentation of governance structures and responsibilities

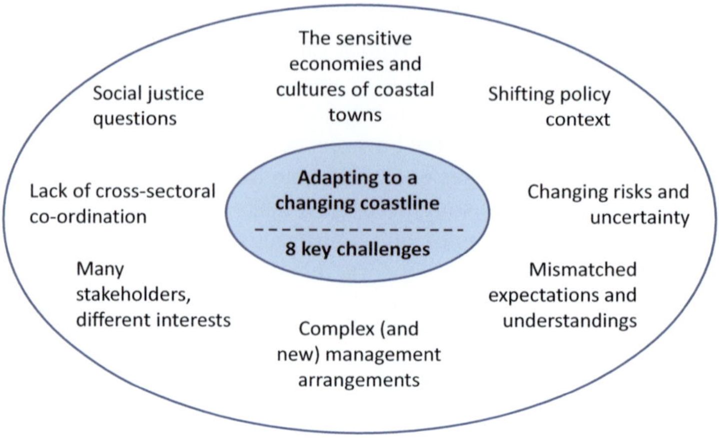

Fig. 12.7 Eight key challenges facing adaptation to a changing coast (Source: Nicholson-Cole and O'Riordan 2009, p. 370)

Table 12.1 Barriers and limits to adaptation

Unclear goals of adaptation	*Individual and social factors*
Adaptation to what exactly, by whom, when, and how?	Legacy of past protection leading to perceptions of entitlement
Comprehensive vs. segmented approaches to managing change and tension between scales	Expectation of compensation if property is lost to erosion
Lack of mechanisms to enable dialogue and take into account diverse values resulting in an impasse on decision-making	Difficulty implementing long-term strategic plans with individual buy-in in the absence of enabling adaptation mechanisms
Uncertain coasts	*Undervalued coastal cultures and icons*
Developing visions of different future coastal conditions is difficult	Familiar coastal landscapes symbolise security, identity and community belonging
Science is probabilistic and depends on a range of uncertain factors that need to be understood	Fear of loss and change
Coastal processes will respond to the management options pursued	Justice and ethics: who has the right to decide if a place is 'not worth saving'?
	What constitutes resilience and adaptation socially, economically and environmentally in any given place?

Adapted from Nicholson-Cole and O'Riordan (2009), p. 378

of key stakeholders, a lack of coordinated support for integrated coastal zone management implementation at the national level and a relatively weak awareness of the specifics of adaptation at the local level. Uncertainty is often cited as a hindrance to decision-making about adaptation, thereby constituting a limiting factor when it comes to adaptive coastal governance. Adger et al. (2009), however, argue that adaptation need not be limited by uncertainties associated with foresight of future climate change and that methods of assessing robust adaptations can provide opportunities for overcoming perceived limits.

Broadly, the challenges and limits illustrated in Fig. 12.7 and Table 12.1 give a sense of why coastal governance during the study period was to all intents and purposes dysfunctional in terms of adaptation to coastal change. The different scales of decision-making and the incongruence between aspirations, aims, priorities and interests are a real problem for taking an adaptive approach to coastal management. There is also a deeper underlying aspect, a culture ill-prepared for coastal adjustment, which is sufficiently entrenched as to form a serious barrier to adaptation.

12.3.2 Lessons Learned

The North Norfolk experience is a very good example of present and looming future coastal change, which will undoubtedly have a huge impact on the communities living on this coast. It demonstrates how people struggle to cope with being told that change will happen and feeling that a policy has been decided without full account being taken of their well-being. Flood and erosion risk are subject to different interpretations, particularly over the question of what constitutes adaptation. There is a pervasive fear of overall economic decline, leading to a sense of social impotence and local political weakness. Funding for coastal protection is being withdrawn and funding for adaptation is not reliable. Having no control over the 'comfort zone' for future conditions is debilitating, even if people understand that change will happen. All of these things stop progress from being made, because of a sense of distrust that develops in the institutions responsible for managing the process of change.

Over the years, a lot has been learned from the North Norfolk experience. The Pathfinder programme is the culmination of years of effort to bring local people together to generate a better understanding of what the future may hold and how it may be managed adaptively. At this stage, little has been done by way of a review as the programme is still in progress. The Department for Environment, Food and Rural Affairs commissioned an evaluation study of the five largest Coastal Change Pathfinder projects (Defra 2011b). However, the findings take a very short-term and immediate stance and there is very little reflection or information available as to how the national government plans to absorb or build upon any of the 'lessons learned'. There are also no guarantees of future funding for coastal adaptation. It will take much greater commitment (through legislation and guidance) and real resources to enable adaptation to coastal change to become a real possibility.

Some very positive lessons can still be drawn here from what has been achieved so far. One appears to be that initiatives which result in a positive legacy for a community in terms of reducing economic and social vulnerability (without requiring a lot of ongoing funding) have a lot of value locally. For example:

- Making changes to the local planning policy to enable relocation of homes lost to the sea

- Buying out homeowners at immediate risk of losing property to offer confidence and reduce blight, whilst longer-term adaptation initiatives are being developed to avoid such crises in future
- Setting up coastal forums to enable ongoing discourse and trust consolidation
- Providing genuine public access points to beaches that can move as erosion happens
- Improving the cliff-top environment for wildlife and aesthetic enjoyment

In the context of the present situation, Box 12.3 gives a brief summary of the key lessons that can be learned from the North Norfolk experience to date.

Box 12.3 Adaptation in Practice – Lessons Learned in North Norfolk

The situation

- Inevitable change and uncertainty over timescales and impacts amongst expectations of stability
- Few resources for coastal defences
- Fragmented institutional commitment to adaptation
- No established process or resources for adaptation
- Fear of economic and social decline, disinvestment and blight

Lessons learned

There are no quick fixes:

- Coastal policy cannot be changed overnight and adaptation takes time and must 'learn to learn'.
- Responses to the impacts of coastal change need sufficient time in order for the local community to understand and 'own' them. Not all communities have the capacity to engage in these complex issues; this takes time, resources and creativity.

Develop, invest in and implement local adaptation initiatives which:

- Help to provide certainty
- Remove immediate fear and anxiety about people's well-being, buying time for adaptive responses to be fully developed and delivered
- Are no regrets and which have co- or multiple benefits for the local community
- Have a long legacy and do not require significant ongoing funding, offering a sense of long-term continuity as things change
- Are part of a wider portfolio for bettering the coast: upgrading the local economy and expectations
- Provide mechanisms for continuing social dialogue

Adaptation may involve some significant transformations as well as incremental changes, some of which will be painful to groups or societies who are reluctant or not able to embrace change (Adger et al. 2009). We have seen this in practice in the Norfolk case, which has become well known partly because of the furore that developed around the publication of the draft SMP6 in 2005. The publication of this draft plan had strong social consequences, because it presented a policy which was designed to allow change but did not include any suggestions for how the consequences of that change (i.e. erosion) would be dealt with. Additionally it was felt by the local communities that their views had neither been invited nor included in the preparation of the plan. In the absence of plans to actively manage and support change (let alone in a way that is perceived to be fair and equitable), widespread reluctance to accept coastal change was hardly surprising. All this breeds distrust and a sense of being let down. These are not conducive for continuing and effective participation, an essential ingredient for successful adaptive coastal governance.

The North Norfolk experience has demonstrated that coastal change can be more effectively planned for by providing leadership and commitment to managing change, showing respect to the range of interests involved and encouraging collaboration, and giving opportunities for gaining information and for participating in the decision-making process. The process was slow, but in time a broader understanding of the difficulties and demands of the situation and a more unified vision of how to move forward have been achieved. Armitage et al. (2009) would likely term this an example of adaptive co-management where explicit attention is paid to the learning and collaboration functions necessary to improve our understanding of and ability to respond to complex social-ecological systems. Gopnik et al. (2012) write in the context of marine spatial planning in the USA. They found that broader stakeholder engagement by policy makers and planners at early stages in a meaningful and transparent manner was important. They highlighted that bringing parties together early in a process to learn and listen from one another was very valuable. In the North Norfolk case, it was particularly important for local communities to have an opportunity to make themselves heard and for issues relating to sense of place, cultural identity and so on to have a place at the table (see, e.g. Adger et al. 2011). A more partnership-based approach is also advocated by Stojanovic and Ballinger (2009), who have looked at many local and regional examples of non-statutory coastal management around the UK.

After much negotiation, SMP6 was finally adopted by all partner authorities and the Environment Agency in 2012. The North Norfolk District Council accepted SMP6 following changes that will enable further studies to be completed, along with assurances that current policies will not change until sound coastal adaptation measures are available. The local council is also embarking upon developing a more integrated approach to coastal management beyond the existing initiatives. The approach will develop appropriate ways of managing the coast with greater reference to communities' needs, coastal processes, policy context and the availability of funding. The approach will offer improved coastal information, development of funding opportunities and implementation of a broader spectrum of initiatives and schemes for managing the coast across the frontage. It will recognise the importance of protecting the coastline, where possible, and will seek to realise the resources to achieve this. It will also develop tools and methods to assist communities affected by coastal change and will seek to sustain and manage coastal infrastructure. Delivery will be made possible through capital and revenue coast protection works, partnership funding initiatives, implementation of the SMP action plans, development of adaptive management measures and long-term infrastructure planning, all within the context of wider coastal issues. Throughout all the work streams, effective stakeholder engagement is considered essential; the Coastal Issues Forum operating within the council is seen to be instrumental in achieving this as well as the establishment of Web-based 'Coastal Hub' and other specific engagement activities.

A more openly available coastal simulator tool has great potential to support adaptation to coastal change at local, regional and national scales. One of the key messages that emerge from this book is that we do not know for sure how or when things will change, but we can understand in an impartial manner that ongoing protection is unsustainable. A hold-the-line strategy is not an effective adaptation measure in terms of how the coastal system functions, regardless of the unavailability of funding. The Tyndall Coastal Simulator is a tool that can help to explore patterns of long-term change and how this is influenced by various management options. Being able to explore future options could support a much more informed process of dialogue amongst decision-makers and other organisations or individuals who have a stake in the future of the coast. It is possible that using a tool like the Tyndall Coastal Simulator could offer a neutral means of reaching shared understandings of coastal change. It could also act as a mechanism for communicating the long-term consequences of various management options and an accompanying dialogue around developing mechanisms for adaptation to future coastal change. In this way, there could be opportunities for an IA of this kind to help build some basic consensus around how the coast could be managed.

Genuine stakeholder involvement is really important if the different parties involved in coastal governance structures can genuinely start managing and planning for the future of the coast. Cooperation does not happen without resources and support; it needs to be enabled. An important lesson is that these resources need to be allocated if adaptation can happen. Whether you continue to defend or you choose to allow change and manage it adaptively, the options are expensive. We know that investment in adaptive management now will have long-term value. Finding the funding is a really difficult issue, particularly in short-term political systems and in times of recession. Not investing in adaptation now is likely to lead to more critical problems later on at much greater expense.

12.4 Conclusions and Next Steps

There is a growing understanding of the need to manage the coast strategically, and in a more integrated manner, in order to reduce future vulnerability to the impacts of ongoing coastal and climate change. The question of how to adapt sustainably to coastal change in the UK has been widely discussed but remains a thorny and somewhat fragmented issue. Adaptation to coastal change in practice still appears to be at a tentative early stage. There remains genuine concern over the future of coastal communities unless socially and economically orientated adaptation plans and investment are developed alongside policies intended to allow coastal change.

Whilst we have a mass of experience in the UK given our relatively large and varied coastline, it is not obvious that lessons have been or are being learned in terms of how to plan for future change in a long-term and integrated way. Some initiatives towards a more adaptive pathway have been undertaken; however, no coherent system for learning from these examples exists nationally. It is becoming increasingly important to be aware of the innovative approaches that are being taken at various scales and to learn from examples (e.g. Tompkins et al. 2010); the UK's national Pathfinder programme and many of the local-scale projects that arose from this and are ongoing offer important learning opportunities.

Certainly within the case study area, coastal governance arrangements are in a transitory place; one which no longer sits comfortably with a defend-and-protect approach yet is ill-prepared to embrace coastal change. Coastal adaptation can constitute massive national projects (such as the construction of barriers in places including St. Petersburg, Venice and London). Equally, it is important for coastal governance structures to recognise the importance of smaller scales and regional and local initiatives, which focus on measures that both reduce vulnerability and implement planned adaptation to coastal change. There appears to be at least for now, a lack of tried and tested tools, processes and support mechanisms available in the UK or elsewhere for supporting adaptation to coastal change on the ground and broad learning from test case examples. In an era of institutional flux and financial uncertainty, there is still much concern and confusion as to what the future holds for this particular coastline and others around the UK and further afield.

How to design and deliver adaptation to coastal change in an effective and fair way remains to some extent a great unknown. The research summarised here particularly highlights the critical role of mechanisms to support a fair and just process of planning for coastal change. Without meaningful mechanisms to enable coastal planners to develop an agreed capacity for adaptation and appropriate funding mechanisms to help communities actively adjust to a changing coastline and seize resulting opportunities, there may never be a tolerated civil recognition of coastal change and hence no effective delivery of central policy. For coastal adaptation initiatives to be appropriate and effective, broad and prolonged stakeholder engagement is necessary in the development of coastal management plans and tools as this book demonstrates. Coastal change is not purely a science issue, or problem. It affects real people's lives, whole communities, economies and regions, ecosystems and so on. Inevitably a broad-scale approach has to be taken but with the capacity to understand what this means at local levels in order to assist in giving these scales the capacity to adapt.

Integrated approaches to the management of future coasts are critical. The Tyndall Coastal Simulator is one type of tool which can help to support sound local, regional and even national coastal adaptation initiatives. The simulator is the first tool of its kind in England to give a broad section of stakeholder interests (both 'expert' and non-technical) exposure to understanding and exploring credible scenarios of coastal change.

References

Adger, W. N., Dessai, S., Goulden, M., Hulme, M., Lorenzoni, I., Nelson, D. R., Naess, L. O., Wolf, J., & Wreford, A. (2009). Are there social limits to adaptation to climate change? *Climatic Change, 93*, 335–354. doi:10.1007/s10584-008-9520-z.

Adger, W. N., Barnett, J., Chapin, F. S., & Ellemor, H. (2011). This must be the place: Underrepresentation of identity and meaning in climate change decision-making. *Global Environmental Politics, 11*, 1–25.

Armitage, D. R., Plummer, R., Berkes, F., Arthur, R. I., Charles, A. T., Davidson-Hunt, I. J., Diduck, A. P., Doubleday, N. C., Johnson, D. S., Marschke, M., McConney, P., Pinkerton, E. W., & Wollenberg, E. K. (2009). Adaptive co-management for social-ecological complexity. *Frontiers in Ecology and the Environment, 7*, 95–102.

Dawson, R. J., Dickson, M., Nicholls, R. J., Hall, J., Walkden, M. J. A., Stansby, P. K., Mokrech, M., Richards, J., Zhou, J., Milligan, J., Jordan, A., Pearson, S., Rees, J., Bates, P. D., Koukoulas, S., & Watkinson, A. (2009). Integrated analysis of risks of coastal flooding and cliff erosion under scenarios of long term change. *Climatic Change, 95*, 249–288.

Defra. (2005). *Making space for water: Taking forward a new government strategy for flood and coastal erosion risk management in England*. First Government response to the autumn 2004. Making space for water consultation exercise. London: Department for Environment, Food and Rural Affairs.

Defra. (2011a). *ARCHIVE: Coastal change pathfinders*. http://archive.defra.gov.uk/environment/flooding/manage/pathfinder/. Cited 19 Nov 2013.

Defra. (2011b). *Coastal pathfinder evaluation: An assessment of the five largest pathfinder projects: A final report by Regeneris Consulting*. London: Department for Environment, Food and Rural Affairs. http://www.defra.gov.uk/publications/files/pb13721-coastal-pathfinder-evaluation.pdf. Cited 19 Nov 2013.

East Anglian Coastal Group. (2012). *SMP 6 – Kelling hard to lowestoft ness*. http://www.eacg.org.uk/smp6.asp. Cited 13 Nov 2013.

Falaleeva, M., O'Mahony, C., Gray, S., Desmond, M., Gault, J., & Cummins, V. (2011). Towards climate adaptation and coastal governance in Ireland: Integrated architecture for effective management? *Marine Policy, 35*, 784–793.

Freisinger, S., & Bernatchez, P. (2010). Perceptions of Gulf of St Lawrence coastal communities confronting environmental change: Hazards and adaptation, Québec, Canada. *Ocean & Coastal Management, 53*, 669–678.

French, P. W. (2004). The changing nature of, and approaches to, UK coastal management at the start of the twenty-first century. *Geographical Journal, 170*, 116–125.

Frew, P. (2012). Adapting to coastal change in north Norfolk, UK. *Proceedings of the ICE – Maritime Engineering, 165*(3), 131–138.

Gopnik, M., Fieseler, C., Cantral, L., McClellan, K., Pendleton, L., & Crowder, L. (2012). Coming to the table: Early stakeholder engagement in marine spatial planning. *Marine Policy, 36*, 1139–1149.

House of Commons Communities and Local Government Committee. (2007). *Coastal towns* (Second Report of Session 2006–7). London: The Stationary Office Limited. http://www.publications.parliament.uk/pa/cm200607/cmselect/cmcomloc/351/351.pdf. Cited 13 Nov 2013.

Huitema, D., Mostert, E., Egas, W., Moellenkamp, W., Pahl-Wostl, C., & Yalcin, R. (2009). Adaptive water governance: Assessing the institutional prescriptions of adaptive co-management from a governance perspective and defining a research agenda. *Ecology and Society, 14*(1), 26.

Leafe, R., Pethick, J., & Townend, I. (1998). Realizing the benefits of shoreline management. *Geographical Journal, 164*, 282–290.

Lorenzoni, I., Nicholson-Cole, S. A., & Whitmarsh, L. (2007). Barriers perceived to engaging with climate change among the UK public and their policy implications. *Global Environmental Change, 17*, 445–459.

Milligan, J., O'Riordan, T., Nicholson-Cole, S. A., & Watkinson, A. R. (2009). Nature conservation for sustainable shorelines: Lessons from seeking to involve the public. *Land Use Policy, 26*(2), 203–213.

Nicholls, R. J., & Lowe, J. A. (2004). Benefits of mitigation of climate change for coastal areas. *Global Environmental Change, 14*(3), 229–244.

Nicholson-Cole, S. A., & O'Riordan, T. (2009). Adaptive governance for a changing coastline: Science, policy and publics in search of a sustainable future. In W. N. Adger, I. Lorenzoni, & K. L. O'Brien (Eds.), *Adapting to climate change: Thresholds, values, governance*. Cambridge: Cambridge University Press.

Olsen, S. B., Page, G. G., & Ochoa, E. (2009). *The analysis of governance responses to ecosystem change: A handbook for assembling a baseline* (LOICZ Reports and Studies No. 34, 87pp.). Geesthacht: LOICZ International Project Office.

O'Riordan, T., & Nicholson-Cole, S. (2010). Turning the tide in north Norfolk: A case study of adaptation to coastal change. In I. Scott & R. Worsley (Eds.), *The return of the tide…on the saltmarsh coast of north Norfolk*. Norfolk: JJG Publishing.

Schmidt, L., Prista, P., Saraiva, T., O'Riordan, T., & Gomes, C. (2012). Adapting governance for coastal change in Portugal. *Land Use Policy, 31*, 314–325.

Stojanovic, T. A., & Ballinger, R. C. (2009). Integrated coastal management: A comparative analysis of four UK initiatives. *Applied Geography, 29*, 49–62.

Tompkins, E. L., Adger, W. N., Boyd, E., Nicholson-Cole, S., Weatherhead, K., & Arnell, N. (2010). Observed adaptation to climate change: UK evidence of transition to a well-adapting society. *Global Environmental Change, 20*(4), 627–635.

Tonn, B., Hemrick, A., & Conrad, F. (2006). Cognitive representations of the future: Survey results. *Futures, 38*, 810–829.

Tunstall, S. M., & Penning-Rowsell, E. C. (1998). The English beach: Experiences and values. *Geographical Journal, 164*, 319–332.

Young, R. (2009). Coastal planning – Plus ça change! *Town and Country Planning, 78*(10), 418–422.

Chapter 13
International Opportunities for Broad Scale Coastal Simulation

Robert J. Nicholls, Richard J. Dawson, Sophie A. Day (née Nicholson-Cole), David Walker, Nobuo Mimura, Melissa Nursey-Bray, Leonard Nurse, Munsur Rahman, Kathleen D. White, and Barbara Zanuttigh

Abstract The preceding chapters of this book have looked at the details of Integrated Assessment on the UK coast, especially in Norfolk. In addition to explaining this analysis in detail, the book aims to look for wider and more generic lessons about Integrated Assessment for coasts. In this regard, this chapter turns the focus to other parts of the world and the 'global' coast in general. Through diverse coastal examples from Australia, Bangladesh, Barbados, Italy, Japan and the USA, the opportunities and challenges associated with transferring the Tyndall Coastal Simulator approach to other locations are critically evaluated.

R.J. Nicholls (✉) • S.A. Day (née Nicholson-Cole)
Tyndall Centre for Climate Change Research, Faculty of Engineering and the Environment, University of Southampton, Southampton SO17 1BJ, UK
e-mail: r.j.nicholls@soton.ac.uk; day_soph@yahoo.co.uk

R.J. Dawson
School of Civil Engineering and Geosciences, Tyndall Centre for Climate Change Research, Newcastle University, Newcastle upon Tyne NE1 7RU, UK
e-mail: richard.dawson@newcastle.ac.uk

D. Walker
School of Civil, Environmental and Mining Engineering,
The University of Adelaide, Adelaide 5005, Australia
e-mail: david.walker@adelaide.edu.au

N. Mimura
Institute for Global Change Adaptation Science, Ibaraki University,
Mito, Ibaraki 316-8511, Japan
e-mail: mimura@mx.ibaraki.ac.jp

R.J. Nicholls et al. (eds.), *Broad Scale Coastal Simulation*, Advances in Global Change Research 49, DOI 10.1007/978-94-007-5258-0_13

These diverse case studies indicate a number of similarities with the tensions that are apparent in the North Norfolk case study. This includes multiple drivers such as increasing population pressures, changing land use, relative sea-level rise, management conflicts and significant/diverse stakeholder concerns. They also highlight important coastal issues that are not addressed within the Tyndall Coastal Simulator but could in principle be added – such as tsunamis, hurricanes, changing marine ecosystems, etc., as well as the range of ecological and socio-economic contexts within which the different coastline study areas are embedded. Despite these contrasts, it is clear that in general terms, the nature of multiple interacting coastal pressures and drivers means that there are numerous coastal locations around the world that would benefit from an Integrated Assessment (IA) approach. Such an approach provides a proactive method to assess present and future problems as well as considering more sustainable responses to both long-term pressures and following episodic extreme events. From this foundation, Chap. 14 examines the way forward for Integrated Assessment of coastal areas.

Keywords Integrated Assessment • International case studies • Transferability • Coastal management • Coastal pressures

M. Nursey-Bray
Geography, Environment, Population, School of Social Sciences,
University of Adelaide, North Terrace, Adelaide 5005, South Australia, Australia
e-mail: melissa.nursey-bray@adelaide.edu.au

L. Nurse
Centre for Resource Management and Environmental Studies (CERMES),
Faculty of Science and Technology, University of the West Indies (UWI),
Cave Hill Campus, St. Michael BB11000, Barbados
e-mail: leonard.nurse@cavehill.uwi.edu

M. Rahman
Institute of Water and Flood Management (IWFM) Bangladesh,
University of Engineering and Technology (BUET), Dhaka 1000, Bangladesh
e-mail: mmrahman@iwfm.buet.ac.bd

K.D. White
Global and Climate Change, Institute for Water Resources, US Army Corps of Engineers,
Portland, OR, USA
e-mail: Kathleen.D.White@usace.army.mil

B. Zanuttigh
DICAM, University of Bologna, Viale Risorgimento, 2-40136 Bologna, Italy
e-mail: barbara.zanuttigh@unibo.it

13.1 Introduction

So far this book has examined the details of IA in the context of the UK coast, especially North Norfolk. Multiple drivers are apparent and include increasing population pressure, relative sea-level rise and other climate change as well as proposals for significant change to the coastal management strategies and widespread stakeholder concern. However, this North Norfolk case study is just one contextual example of a coastal IA. It was designed and tested in the context of a specific set of circumstances and their associated challenges. The benefits of the application in Norfolk are clear because of the multiple and interacting drivers, and pressures lead to trade-offs that can only be resolved using an integrated approach.

This chapter aims to set the lessons of the Norfolk case study into a wider and more generic context about IA for any coastal area. The coastal choices around the world are examined drawing on the North Norfolk example as well as diverse examples of coastal case studies from around the world, comprising Australia, Bangladesh, Barbados, Italy, Japan and the USA. Hence, the opportunities and challenges associated with transferring the Tyndall Coastal Simulator approach to other locations are critically evaluated and considered.

Chapter 13 is structured as follows. Section 13.2 examines coastal problems and choices in the context of the UK and global coastal problems and issues. Section 13.3 then examines the six case studies from around the globe. Section 13.4 assesses this information in terms of challenges and opportunities to transferring the approach of the Tyndall Coastal Simulator and associated methods. Section 13.5 concludes. The following Chap. 14 examines the way forward for IA of coastal areas.

13.2 Coastal Choices Around the World

Compared to Norfolk, coastal locations around the world will have different geomorphological, social, economic and environmental characteristics, together with associated tensions. Figure 13.1 gives examples of key coastal issues around the UK. Mapping these issues to specific places will start to identify where multiple and interacting issues such as erosion and flooding occur and hence where an approach such as the one described in this book could be useful to support planning. Priorities might include tourist resorts and erosion, urban environments and estuaries and critical national infrastructure at risk of erosion and flooding.

The management and social context is also key. Shoreline Management Plans (SMPs) in England and Wales are promoting the consideration of more dynamic coastal adaptation options that work with natural tendencies (Nicholls et al. 2013a). It is becoming clear, as highlighted in this book in the context of North Norfolk, that in the future we face difficult options and choices on the coast. For example, how do we adequately plan to retreat from an eroding coastline in a way that maintains a viable economy and quality of life for the local population, rather than allowing

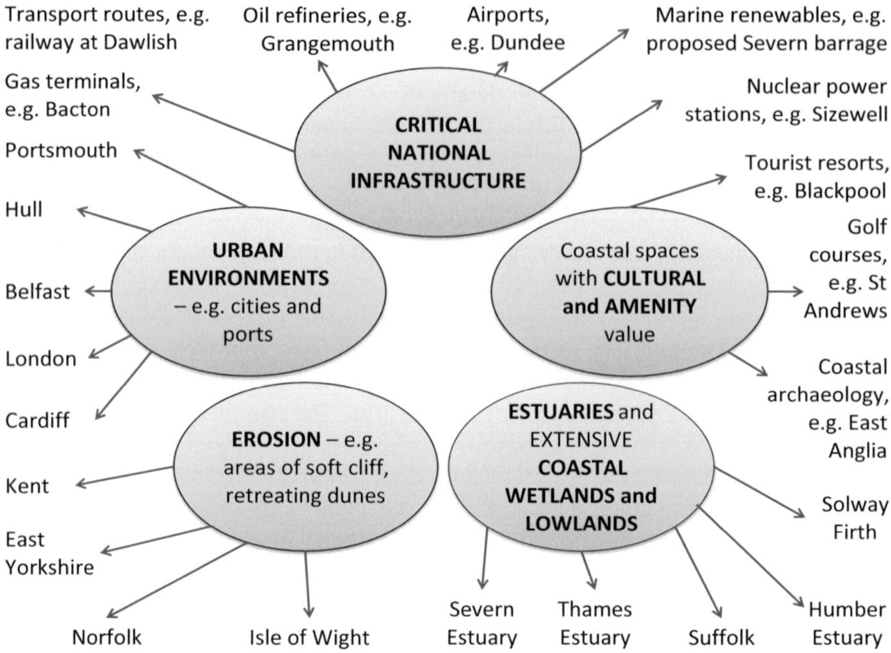

Fig. 13.1 Some examples of key issues on the coast and example locations in the UK

blight and despair to develop? Can we plan to reconfigure urban centres to take account of enhanced flooding or erosion risk in the context of specific social and cultural settings? As well as complex geophysical circumstances, ecosystems and habitats, rural land uses and so on, people's values associated with the coast, local coastal economies and other somewhat intangible aspects are critical when it comes to managing change. But they are often difficult to consider and previously have been ignored. Questions of how future choices can be negotiated in a democracy and what kind of institutional arrangements are required. As in the case of North Norfolk, an IA model can provide insights from a technical perspective, but it becomes much more powerful when linked to stakeholder engagement and participation, as this knowledge can inform and shape governance. This approach is transferable and could assist analysis of other coastal locations in exploring difficult questions and issues.

Coastal societies around the world are facing large and growing pressures relating to population, resources and environmental change (Box 1.3). Whilst the drivers and challenges of coastal management in Norfolk and Britain are significant, more diverse and larger challenges are apparent in other parts of the world. The objectives and constraints of coastal management are also diverse. For example, development to support economic growth is a primary driver, especially in developing countries (Parry et al. 2009; World Bank 2010), whilst in other locations, maintaining human and economic safety is more important (Delta Commission (Deltacommissie 2008),

Thames Estuary 2100 (TE2100) project (see Box 14.1), etc.). Further, environmental protection is now stressed in a way that was not the case 50 years ago with supporting legislation like the EU Habitats Directive: whilst welcome, this makes the job of coastal management more difficult. The increasing concentration of people, assets and economic activity in the coastal zone also exacerbates coastal issues. For example, the impact of both climate change and development on coastal port cities is significant – large populations are already exposed to coastal flooding in port cities, with approximately 40 million people (0.6 % of the global population) exposed to a 1 in 100-year coastal flood event (Hanson et al. 2011). Becker et al. (2012, 2013) suggest that climate change will disproportionately affect ports and port-based economies, and the effects of climate variability and change on global trade raise an interesting set of questions on their direct and indirect impacts. Wildcards are difficult to assess but should not be ignored. For example, Englander (2013) raises the prospect of major declines in coastal property values as people understand the long-term implications of sea-level rise in terms of land degradation and loss. In poorer societies, degradation and loss of coastal ecosystem services linked to livelihoods are of concern. Hence there are profound implications for coastal nations and economies at all scales.

Whilst it is difficult to propose a universally accepted ranking of the threats to the world's coastal and marine areas on the basis of magnitude or severity, it is possible to identify some key vulnerabilities based on their persistence and global prevalence (Brown et al. 2013). These include, inter alia, the risks posed by global climate change; rapid growth of coastal populations accompanied by supporting economic and social infrastructure; habitat degradation, which is often associated with biodiversity damage and loss; and threats from alien invasive species. Together, these constitute a significant global threat to livelihoods and human well-being as well as ecosystem health. Consequently, they are likely to be in sharp focus as adaptation planning and implementation are pursued with the aim of ensuring that the world's coastal areas are developed in a sustainable manner.

Figure 13.2 gives a simple illustration of a few examples of the different types of coastal pressures and threats that are present around the world's coasts. These include different development contexts, geomorphic features such as deltas, atolls and the richness of biodiversity in coral reefs and mangroves (Nicholls et al. 2007a; Wong et al. 2014). The rest of this chapter explores challenges and opportunities that exist for transferring the Integrated Assessment approach described in this book to other coastal contexts.

13.3 International Coastal Challenges Where Integrated Assessment Could Be a Useful Tool

Varying types of coastal challenges around the world are faced by different countries and communities – Norfolk is by no means unique. There is a history of abandonment related to coastal changes around the US coast. For example, Gibbons and

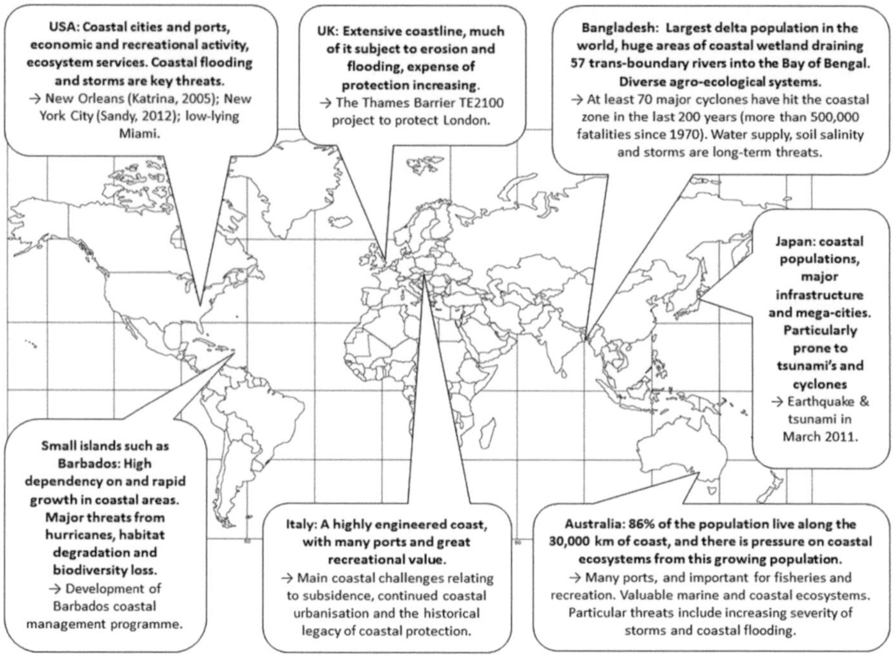

Fig. 13.2 Examples of distinct coastal pressures and threats around the world where IA may provide significant insight (Based on the authors' experience)

Nicholls (2006) found that Chesapeake Bay islands were abandoned due to relative sea-level rise, compounded by steady depopulation causing critical community services to collapse, and loss of faith in the future. This suggests there is potential for other communities around the world to be abandoned long before they become completely uninhabitable. Universal protection of coasts as sea levels rise is not possible or desirable (e.g. Huntington et al. 2012), and this problem is ripe for IA analysis. Other illustrative examples of coastal challenges from around the world, including Japan, Bangladesh, Barbados, Australia, Italy and the USA, are presented in this section. Each provides contextual background and then suggests where and how a coastal IA analysis may 'add value' as a system scale assessment to provide insights to improve future coastal management and planning in that context. Collectively, these examples give a sense of the breadth and variety of global coastal challenges and therefore provide a useful set of studies against which to benchmark our progress in coastal IA. The lessons about IA across these studies are synthesised in Sect. 13.4.

13.3.1 Japan: Tsunami Risks and Large Coastal Cities

Japan is an industrialised country with a strong economy and well-developed infrastructure. Although the overall population of Japan is in slight decline, the coastal megacities and cities continue to grow, with denser occupation of Japan's urban areas leading to increased vulnerability. On the coast, urban hotspots support important industries such as transportation and fishing – with many of the country's most economically significant assets in the bays of Tokyo, Osaka and Nagoya (Nicholls et al. 2008). The geography, topography and climate of Japan mean, however, that it is exposed to frequent typhoons, heavy rain and tsunamis and hence has a long history of serious flooding. Climate change is expected to raise sea levels, intensify storm surges and waves, change precipitation patterns and increase the severity of extreme events. About half the population and 75 % of the country's economic assets are concentrated in flood risk areas, whilst 5.4 million people already live in (protected) areas below mean sea level (OECD 2009).

The three key priorities for coastal management in Japan are to ensure the protection of coasts against natural disasters, protection of the environment and people's rights. These are formalised in the Coastal Act of 1953, amended in 1999. Disaster reduction and prevention relating to earthquakes and tsunamis is an important national goal, particularly following the Great East Japan earthquake and tsunami, which struck on 11 March 2011. This left more than 17,500 people dead or missing. Approximately 13,000 houses were destroyed and 260,000 suffered significant damage. A major accident at Fukushima Nuclear Power Plant was triggered by the tsunami in 2011; recovery from this event will take several decades.

The scale of devastation brought about by the 2011 tsunami event highlighted a need to reconsider the likelihood and potential consequences of future major events and will have a significant legacy on coastal management in Japan for decades to come. Thus far, this has focussed on developing a strategy for managing up to the 150 year event using a multiple 'hard' measures such as coastal structures, sea walls, coastal forests, high roads and so on. Alongside this, a standard to manage bigger and rarer events up to a 500 year (or greater) event (such as the tsunami in 2011) focuses on 'soft' measures such as early warning and evacuation systems, organisation and communication, planning such as moving hospitals, residential areas and other critical infrastructure to high areas and improvement of coastal facilities in terms of tolerance to a disaster. The government at local and national levels is making efforts to increase preparedness for the next event, including estimating impact areas, developing more sophisticated and finer-resolution tsunami simulation models and disseminating disaster plans.

There is scope for an IA to be developed in the Japanese context to help explore the trade-offs between different reconstruction strategies and to support the development of long-term adaptation strategies, for example, in the Tohoku region, or around Tokyo where similar risks exist. In Tohoku after the 2011 tsunami, land subsidence of up to 1.2 m occurred. Recovering from the event needs to take into account the likelihood and potential impacts of future tsunamis and storm surges. Tsunamis, like other coastal risks, can manifest themselves over broad spatial scales so an IA could be developed with a dynamic tsunami simulation model. Of course, a simpler approach that incorporates the spatial extent of the 150 and 500 year tsunami events as a static layer, may be sufficient for certain policy questions. Although tsunami and climate change impacts ensue over different timescales, an IA can consider extreme tsunami events under current and future climatic conditions, alongside other risks. Within an IA, the impacts of land use changes, environmental trade-offs (e.g. associated with higher sea walls) and other factors can also be considered in relation to the risks of a possible tsunami. Whilst rather uncertain, human response to the management strategies might also be considered. Importantly, an IA approach could also be a mechanism for the wide range of stakeholders and coastal authorities to collectively understand and manage these risks and help to maintain awareness of rare real events via simulation environments.

13.3.2 Bangladesh: Populous Deltas and Developing Economies

Bangladesh occupies much of the Ganges-Brahmaputra delta, the largest and most populated delta in the world. It has a rising population that presently numbers 150 million at a density exceeding 1,100 people/km^2. Coastal districts comprise around 32 % of the land area and 28 % of the population (Islam et al. 2006; Wilde 2011). This essentially rural population is supported by agriculture, aquaculture and fisheries, and Bangladesh has the world's largest mangrove forest (the Sundarbans). In the dry season, salinisation can be a major issue. There are land use conflicts between shrimp aquaculture and rice paddy, with aquaculture offering higher profits, but supporting fewer people and increasing salinisation. At least 70 major cyclones, comprising high winds, storm surges and intense rainfall, have hit the coastal zone over the last 200 years, sometimes resulting in catastrophic mortality (most recently in 1970 and 1991) (Nicholls 2006). Since 1991, deaths and damage have significantly reduced through improved forecasting and warning systems, coastal afforestation projects, cyclone shelters and embankments.

It has long been recognised that Bangladesh is vulnerable to climate change, sea-level rise and subsidence: large areas of land could be submerged, causing the displacement of millions of people in the extreme (Milliman et al. 1989; Ericson et al. 2006). In addition to sea-level rise, important drivers include catchment management, where dams and water diversions can reduce the freshwater input and the sediment supply that built the delta. Within the delta, population growth, economic growth and resulting land use change are profound. For example, urbanisation in Bangladesh is producing one of the world's largest cities at Dhaka as well as the large cities of Chittagong and Khulna (Hanson et al. 2011). A transition to a dominantly urban population is likely in the next 50 years. The creation of polders increases agriculture yields in the short term but raises long-term issues about soil fertility and residual flood risk (waterlogging). Hence, climate change needs to be set in this broader context.

Hundreds of diverse development projects (PDO-ICZMP 2004) have been implemented in Bangladesh over the last 20 years. These embrace water management, infrastructure development, forestry, fisheries and agriculture, public health and sanitation, environment, gender issues, etc. Many are run by government agencies and NGOs with external financial support such as The World Bank. However, these projects have generally been implemented in isolation with little strategic overview, for example, of the long-term trends mentioned above, including the impacts of climate change.

The proposed Bangladesh Delta Plan 2100[1] aims to address this challenge and formulate a national plan around scenarios of climate change, with two main foci: water resources management and technological development. The issues of interactions between these sectors and with the wider environment are key concerns which will need to be addressed. Many of the challenges of producing this plan – such as the formulation of an appropriate conceptual model to allow different disciplines and perspectives to work together, development of appropriate scenarios and linking models – are shared with the process that was undertaken to develop the Tyndall Coastal Simulator. An IA framework to support the Bangladesh Delta Plan 2100 would seem essential to fully consider these different drivers and issues. An example of such an approach that is under development is focussed on ecosystem services and poverty alleviation (Nicholls et al. 2013b).

13.3.3 Barbados: Vulnerable Small Islands

Barbados is the most easterly island in the Caribbean island chain, measuring only 430 km^2. It has an estimated population of 284,000 and tourism is a major industry. A large proportion of the island is surrounded by coral reefs, which support fisheries and are the source of most of the sand on the island's beaches. As in many other small island states where there is high dependency on the coastal area and its resources and where tourism is dominant, there is a constant struggle in Barbados to strike a balance between the imperative of economic development and responsible resource exploitation.

Pressures include hurricanes and storm surges, energetic swells, sediment-laden run-off after rainstorms, groundwater seepage through the beach face, effluent discharge, land clearance, construction within the active beach zone and challenges relating to the intense competition for scarce, high-value coastal space. Small tropical islands, which are already vulnerable economically and physically, will be placed under increasing pressure by a variety of threats including the impacts of climate change (e.g. sea-level rise, accelerated coastal erosion, coral bleaching, ocean acidification), coral diseases and alien invasive species such as the recently introduced *Pterois volitans*, also known as the red lionfish (Albins and Hixon 2008; Côté and Maljković 2010).

In Barbados, local authorities have implemented an ongoing integrated coastal management programme since 1983, which is designed to respond to the challenges posed by the multiple stressors on the coast. This is similar to the approach to adapting to climate change in small islands proposed by Tompkins et al. (2005). It is based on the premise that the objectives of integrated coastal management are best achieved by adopting an interdisciplinary, systematic, ongoing programmatic

[1] http://www.bangladeshdeltaplan.org/

approach, rather than through quick fixes. Unsustainable development in the Caribbean has reduced the effectiveness of coastal ecosystems as natural ecological buffers against storms and climate change/sea-level rise (Mycoo and Chadwick 2012). Hence, hard engineering is highly prioritised in Barbados and neighbouring islands because of the vulnerability of economically significant settlements, infrastructure and beaches to erosion. Softer measures such as beach nourishment, revegetation and the application of minimum building setbacks from high water mark are also employed. Opportunities for retreat and relocation are limited and controversial given the small size of the island.

In Barbados, there is already an excellent understanding of the coastal system and institutional capacity for adaptation planning. A robust scientific and technical baseline has been assimilated to improve knowledge about the changing coast (Mimura et al. 2007). These include investigations of possible causes, identification of potential strategies for coastal protection and enhancement and assessment of costs and benefits of these options. Institutional capacity building (including staff allocation and training) with strong and ongoing political support is a major and critical part of the programme. A Coastal Zone Management Plan (CZMP) has been prepared and includes pilot projects involving ‘hard’ and ‘soft’ options and the establishment of a fully functional GIS system as a tool for aiding decision-making, along with training in managing the system. Community involvement in coastal planning is central to the CZMP.

Barbados’ rolling programme of investment and research to increase coastal management capacity is an exemplar for other coastal countries to follow. The initiative presents an excellent foundation and a major opportunity for a range of coastal IA. A coastal IA would offer a formal mechanism for structuring the substantial body of existing knowledge, as well as a way of exploring the interactions between features of the coastal system, and thinking about change over extended timescales. Of relevance, an IA approach could improve understanding of how climate and other changes may interact and affect the island in the long term and how the island may improve its resilience to these changes across a wide range of sectors (cf. Engelen et al. 1995). Importantly, a range of other questions could be posed depending on national priorities and concerns.

13.3.4 Italy: A Highly Engineered Coast

The Emilia Romagna coastal zone, in North East Italy, comprises 130 km of low and sandy coast and is a good example of a coastal management problem. The beaches face the Northern Adriatic Sea, a relatively shallow epicontinental shelf with low tidal amplitude, and the area is subject to flood and erosion risks. The highest tides in Venice for more than 20 years in December 2008 and 2009 made global news (Zanuttigh 2011). This coast includes the major Italian port of Ravenna, and tourism activity is significant, generating a gross annual income of €9.8 billion in 2006 (Preti et al. 2009). In this region, climate change combines with, and amplifies, non-climate stressors on coastal ecosystems (Nicholls et al. 2007a, b) that are already seriously stressed due to enhanced subsidence, intense coastal development and overpopulation, fragmentation and loss of habitat, overfishing, pollution and spread of invasive species.

This coastline faces high anthropogenic pressure but also a lack of strategic coastal governance. The coast is highly engineered – 50 of 130 km is defended by parallel emerged barriers. The construction of coastal defences, often without a proper planning strategy, has in some places had impacts on coastal ecosystems, including increases in water eutrophication, enhancement of nearshore deposition of fine sediment and reduced water circulation in the protected cells (Preti and Zanuttigh 2011). Nourishment is nowadays the preferred technique to tackle both erosion and flooding mainly for reasons that can be ascribed to ICZM principles: maintenance of wide beaches for tourism and recreation, low environmental impact and aesthetic benefits. This is a controversial option, as it requires huge quantities of sand; about 1 million m^3 of offshore dredged sand was placed in 2007 and 2009 along a 9 km long stretch (Preti et al. 2009) of coast. There are concerns over the impact on benthic communities and the short-term nature of this solution. Other kinds of 'innovative' measures such as insurance, evacuation plans, landscape planning and proactive citizen defence (for instance, by means of temporary flood defences) have been used as exceptional measures in a few instances but are not included in a systematic management process.

Coastal management in Italy does not account so far for climate variability (i.e. extremes) and climate change effects in a systematic way. The approach is essentially driven by the consequences of major events, and coastal management initiatives tend to be patchy, reactive and short term. There is no plan to restrict further urbanisation of areas most at risk of flooding or erosion because of high tourism value, low risk awareness and hostility to realignment options. Public support for any adaptation policy is lacking due to low participation and an expectation that the government will defend the coast whatever (De Vries et al. 2011). A key challenge for future coastal management in this context is to develop better long-term planning for the legacy of a highly populated and engineered coast with little remaining natural resilience. The expectation of permanence and control leaves little flexibility for planning alternative options with stakeholders – essentially this is a lock-in situation for a continuation of the status quo. The lack of integration in coastal governance presents additional challenges for sustainable coastal development (De Vries et al. 2011).

As in North Norfolk, there is a strong morphological interdependency; changing the management strategy in an area may lead to changes in the sediment transport patterns in adjacent coastal areas, as in Norfolk. Here, a coastal IA could be used to generate a better understanding of the morphological interdependencies and patterns associated with a range of coastal management strategies, for example, helping to identify options which offer more flexibility or restore natural processes that reduce the future costs of coastal protection. An IA can raise awareness of the implications of different policy decisions – including failure to manage urban development – and provide a communication platform to enable stakeholders to more fully engage with future coastal planning, and each other, in a more holistic manner. A Decision Support System has been developed for this region within the Theseus Project to explore different adaptation options (Zanuttigh et al. 2013).

13.3.5 Australia: Unique Coastal Ecosystems and Expanding Coastal Populations

Australia has a long (about 60,000 km), highly biodiverse coastline. Mangroves cover 18 % of the coastline, which hosts the world's most diverse suite of tropical and temperate seagrasses and the Great Barrier Reef. It also has a growing population of 22 million in 2012, focussed on the coast; 85 % of the Australian population live within 50 km of the coast (DCC 2009; Norman 2009), mainly around the major coastal cities of Perth, Adelaide, Melbourne, Sydney and Brisbane. Australia's population could reach 60 million people by 2100. At the same time, it is experiencing a population migration to non-metropolitan and regional coastal areas (Burnley and Murphy 2004; Beeton et al. 2006; Gurran and Blakely 2007; Gurran et al. 2013). Australia's coast supports a variety of recreational, economic and societal activities. The gross value of fisheries and aquaculture in 2007–2008 was A$2.2 billion (ABARE 2009). Australia is the fifth largest user of shipping with major ports (BITRE 2009; BTRE 2006). Tourism is worth A$35 billion nationally, with important hotspots such as the Gold Coast.

All this is increasing pressure on coastal biodiversity (Nursey-Bray and Shaw 2010), particularly around the five major coastal cities. In addition, there are major concerns about climate change and sea-level rise, including ocean acidification, for these coastal ecosystems, whilst erosion and flooding will increasingly threaten coastal settlements and their economies. Coastal management is complex, involving a three-tier government system (federal, state and local), with different coastal legislation for each state (Harvey et al. 2012; Nursey-Bray and Shaw 2010). There remains a gap between the science about the coast and effective policy frameworks and management. Nationally, a vulnerability assessment of Australia's coastline has been initiated by the government, and eight climate change adaptation networks have been established in the areas of indigenous, infrastructure, social, marine and other dimensions to examine adaptation options (e.g. Pecl et al. 2009). Local governments across Australia have developed climate change adaptation programmes for identifying priorities and implementing appropriate coastal adaptation measures (e.g. Nursey-Bray and Shaw 2010). Whilst these arrangements, to some extent, acknowledge the interactions between offshore systems and the coast, the diversity of jurisdictional frameworks in Australia, which exist separately for both coastal and marine areas, means this relationship could be strengthened. Each state also has its own mandate via state-based legislation and policy frameworks to manage for the coast. For example, in South Australia, the state government works to a policy document called the *Living Coast Strategy*, and coastal acts include the NSW

Coastal Protection Act 1979, in Queensland the Coastal Protection and Management Act 1995 and in South Australia the Coast Protection Act 1972.

IA approaches have the potential to help integrate understanding about coastal interactions. There is a need to consider the effects of long-term socio-economic and climatic changes on coastal cities and ecosystems and their interaction with offshore systems and processes. Australia, with its extensive but intermittently populated coastline, also highlights the need for a multi-scale approach where sophisticated IA might be most suitable for cities and key ecosystems, but more pragmatic approaches to management may suffice along many less populated stretches of the coastline where pressures could be simpler to analyse.

13.3.6 USA: Coastal Cities and Storms

Numerous major US cities (e.g. New York, Los Angeles, New Orleans, Washington DC, Houston, Seattle and Miami) are located within the coastal zone. Forty-nine percent of the output of the US economy is produced in estuarine areas, which encompass less than 13 % of the area of the continental USA (Colgan 2008). US ports handle more than US$800 billion worth of goods annually, with about US$30 billion in output value and supporting about 60,000 jobs (Jin 2008). The US coastal recreation use value is estimated by Pendleton (2008) at between US$20 billion and US$60 billion annually.

Changing sea level, waves, tides, surges and coastal storms will affect both developed and undeveloped coastlines of the USA, resulting in threats to public health and safety and economic impacts (direct and indirect) for developed coastlines which have dense residential, industrial and commercial development and associated transportation, energy, water, wastewater or other critical infrastructure. Strauss et al. (2012) estimate that by 2050, some six million people may be at risk of flooding in low-lying coastal areas. Property values in low-lying coastal areas may decline as people start to understand the implications of sea-level rise on floods and coastal damage (Englander 2013). Areas that have experienced significant relative sea-level rise such as New Orleans and Norfolk, Virginia are in the front line for these issues. For less developed coastlines, the key challenge is to mitigate the impact of changes to enable continued provision of their ecosystem services, many of which are already subject to threats from other sources (e.g. development, pollution, invasive species).

Importantly, defence standards in coastal cities are much lower than in their European counterparts, and North American cities dominate present coastal flood risk (Hallegatte et al. 2013). New York City (NYC) has recognised that it is also threatened by climate change and is attempting to adapt to this and other challenges. The reports of the NYC Panel on Climate Change (2009, 2010, 2013) noted a range of potential hazards, especially extreme events related to increasing temperature, more variable precipitation and changing sea levels. Rosenzweig et al. (2011) summarise the NYC coastal climate change management approach as one that is 'characterized by a multi-jurisdictional stakeholder–scientist process, state-of-the-art scientific projections and mapping, and development of adaptation strategies based on a risk-management approach'. In response to Hurricane Sandy, the US Army Corps of Engineers (USACE) are undertaking a North Atlantic Coast Comprehensive Study (NACCS).[2] This embodies the main principles of the Tyndall Centre IA: anticipation of a changing environment, pursuance of a systems approach that incorporates natural, social and built systems and promotion of increased recognition and awareness of risks and consequences among decision-makers, stakeholders and the public.

Miami and southern Florida have expanded in population during the twentieth century on land no higher than 5 m above high tide. Miami sits between the Everglades and the Atlantic Ocean and is threatened by hurricanes, with significant historic landfalls. The Great Miami Hurricane in 1926 has been estimated to be the most costly hurricane in US history when normalised for socio-economic change (Pielke et al. 2008). Hence, future hurricanes threaten much greater damage due to the huge growth in exposure, the rise in sea level and the potential of more intense hurricanes. Water supply is also threatened with the system of canals at risk of salinisation and 'sunny day' flooding from higher tides (due to sea-level rise) already a regular occurrence. Sediment for beach nourishment is being sourced from further afield as local supplies diminish, with economic implications for existing beach management strategies. The extent to which Miami can continue to develop – in the face of climatic and socio-economic pressures – is a multifaceted question that is ripe for an IA.

NYC and Miami highlight the many issues and sectors that interact within cities. Coastal city IAs could help to plan changes in infrastructure policy and investments, with a better understanding of potential impacts and how more robust response systems can be put in place. However, assessing many of these issues within the urban environment poses a number of challenges – not least the movement of floodwater around and through buildings and streets. Interacting networks of infrastructure, individuals, organisations and economic activity, particularly over the spatial extent and density of a megacity, further increase the complexity of any analysis.

13.4 Challenges and Opportunities for Transferring the Tyndall Centre Approach to Coastal Integrated Assessment

This book has specifically reflected on the experience of the implementation of an IA for the North Norfolk coastal zone, UK. The Tyndall Coastal Simulator incorporates physical processes of waves, wind, tides (Chaps. 2 and 3) and the associated

[2] http://www.nad.usace.army.mil/Missions/CivilWorks/HurricaneSandyCoastalRecovery/NorthAtlanticComprehensiveStudy.aspx

morphological, flooding and environmental impacts (Chaps. 5, 6, 7, 8 and 9) as well as socio-economic pressures including residential and industrial development patterns (Chap. 4) and meaningful and participatory engagement with key stakeholders (Chaps. 10, 11 and 12) in order to provide a valuable contribution to coastal planning and management.

The international case studies presented in Sect. 13.3 provide examples of other situations in which an IA approach could improve understanding of coastal systems and support decision-making. These case studies have introduced a number of different coastal challenges that were either not relevant, or outside the scope, of the UK study. This highlights areas of common ground, as well as the challenges and opportunities for transferring the Tyndall Centre IA approach elsewhere, that are now considered.

13.4.1 Alternative Coastal Impacts

The application of the Tyndall Coastal Simulator in North Norfolk stimulated the development of a number of modelling components that were necessary to deliver appropriate insights. Around the world, coasts are pressured by similar drivers to those affecting North Norfolk: changing land use, rising sea levels and increasing – and often ageing – population. Similarly, the risks of coastal erosion and flooding are widespread. However, from the examples in Sect. 13.3, it is clear that wider transferability of the IA approach requires additional coastal components. As experience develops across a number of sites, a tool box of modules might emerge which would facilitate application to new sites. Although not comprehensive, in light of the international locations considered above, priorities might include:

Extreme events: tsunami and hurricane processes – The USA, Japan, Bangladesh and many other countries are subject to extreme events that can wreak havoc on seemingly unprecedented spatial scales. As indicated above, technically models exist to simulate the impacts of such events and they could be directly or loosely coupled into an IA. In light of the recent extreme events discussed in the USA and Japan, as well as elsewhere (e.g. Typhoon Haiyan in the Philippines, 2013), an obvious potential for an IA is to support redevelopment and long-term planning for an area that has been devastated. Further, we can envisage a future where disasters are stimulated and many responses proactively proceed the real thing.

Complex urban areas and their infrastructure – Urban areas concentrate people and economic activity and assets, making them potential hotspots of climate and coastal risks. Cities are also at the intersection of multiple interests and varied and/or large investments, especially big cities such as New York City (Strauss et al. 2012; Rosenzweig et al. 2010), which can lead to conflict about coastal management priorities. However, urban infrastructure – including roads, buildings, flood defences and drainage systems – can also help to mitigate the impact of extreme events.

Modelling the complex interactions between coastal hazards and dense urban infrastructure in large cities will require a different modelling approach from the Tyndall Coastal Simulator. More generally, re-engineering coastal cities may be an effective long-term mechanism of adapting to global change, but this must be preceded by systemic analysis. The TE2100 project made important advances in taking an IA approach to long-term flood risk management for London (see Box 14.1), but multirisk IAs are required for cities to avoid conflict between sectors (Hall et al. 2009; Walsh et al. 2011).

Environmental systems and biodiversity – Ensuring the viability of coastal resources such as freshwater supply, fisheries and coastal habitats/biodiversity whilst enabling human activity is increasingly recognised. However, experience in many countries shows that we still have a limited understanding of how such systems may respond to future environmental and development pressures. In order to ensure the widest range of impacts and benefits are captured in IA, modelling tools able to capture environmental processes and be integrated within wider coastal analyses are essential and need further development.

13.4.2 Analysis for Decision Support

The North Norfolk case study provides metrics that have been tailored to UK policy needs for risk management (i.e. expected annual damages) (Penning-Rowsell et al. 2013). However, if we are to assess the wider impacts and benefits of long-term change and coastal management policy around the globe, then new approaches are required that capture the wider social and environmental value from sensitive ecosystems in locations such as Australia and the Caribbean. Studies such as the UK's National Ecosystem Assessment (2011) provide important evidence of the value that ecosystem services provide. Whilst this, and similar initiatives, provides a useful advance, there are still requirements for methods and approaches that can help decision-makers reconcile non-economic benefits and impacts with economic ones. Cities, in particular, can distort this assessment, their size and concentration of economic activity justify investment in protection, but whether this is sustainable over the longer term – particularly in cities at the intersection of many coastal pressures such as Miami – remains unclear.

Improved metrics across multiple dimensions are only part of the challenge for coastal decision-making. One particular decision-making challenge, highlighted in the Norfolk case study and many of the international case studies (Sect. 13.3), is dealing with a coast that has been developed and protected for many decades or longer. The legacy of historical decisions, engineered structures and substantial investment can leave little room for manoeuvre in the short term and may ultimately lead to controversial policies – as is now the situation in North Norfolk. The current Tyndall Coastal Simulator can be used to explore how various adaptation options may manifest themselves in the long run. The potential multiple combinations of options and sequences of implementing them require the development of methods that facilitate analysis with decision-makers that identify strategies that are most robust across the uncertainties.

13.4.3 The Role of Spatial Scale

Information on coastal risks and how they might change in the future is required to support the appraisal of policy options, allocation of resources and monitoring performance of investment in coastal management. The amount of resource that is committed to such an assessment, in terms of data acquisition and analysis, should reflect the nature of the decision(s) that the assessment seeks to inform (Hall et al. 2003). Similarly, coastal management decisions can be structured at a number of levels, ranging from national policy decisions, to planning decisions in geomorphic units such as coastal cells or socio-economic units such as cities, down to local design and operational decisions. Hence, a hierarchy of IA methods is required to support the range of coastal management decisions.

This is the case in almost all of the examples above but particularly clear in Australia where much of the population is concentrated in a small number of coastal locations. Although more detailed analyses may be appropriate for urban areas and environmentally sensitive locations, elsewhere heuristic or data-driven approaches may provide sufficient analysis for coastal policy. At the global scale, the DIVA model (Vafeidis et al. 2008; Hinkel et al. 2012, 2013), which analyses the impacts of global sea-level rise in terms of metrics such as the number of people at risk, loss of land and wetlands, etc., provides one possible approach to broader scale risk analysis.

13.4.4 Extreme Events and Long-Term Pressures

Extreme events influence how coasts are managed, often shaping how they evolve over the decades which follow. History is peppered with examples of extreme coastal events (Kron 2013), with the 1953 North Sea storm surge being of particular relevance to our Norfolk case study site and the Bangladesh cyclone of 1991, Hurricane Sandy in 2012 and the Great East Japan earthquake and tsunami in 2011 of relevance to the examples above. Such events often lead to reflection and changes in coastal management policy, with long-term implications.

Although the current focus of the Tyndall Coastal Simulator is on longer-term changes, this review has highlighted the opportunities from considering these issues over shorter timeframes. Extreme events provide windows of opportunity for fundamental change, but typically society is quick to start to rebuild and recover following an extreme event. That this should develop a more resilient future seems obvious, but devising a new plan from scratch is not feasible. Thus a key transferability opportunity for an IA is to provide outputs that enable planners and engineers to proactively prepare post-event policy and rebuilding strategies to lead to a configuration that is ultimately more sustainable.

13.4.5 Stakeholder Processes and Building Adaptive Capacity

Small islands and low-lying deltas are often particularly vulnerable, and geographical and socio-economic factors may limit the range of available management options compared to stretches of coastline like North Norfolk, UK. The examples above highlight the importance of the coastal zone to the whole economy and environment of small islands such as Barbados, whilst in Bangladesh, the delta is changing rapidly and provides a crucial source of food. Managing long-term change in these vulnerable regions, particularly when they have limited resources available to invest in coastal management, is especially challenging.

An IA requires baseline data and investment to initiate its development. Barbados, with its established monitoring systems, is already well placed to implement an IA – although important additional components would need to be developed to model 'whole island' socio-economics and coastal processes. However, as described in the preceding chapters, modelling is only part of the IA process. Supportive governance arrangements are crucial to encompassing the multiple interests and stakeholder groups that have an interest in long-term coastal change. Moreover, these arrangements must join up across spatial and temporal scales – for example, in Australia where there remains a gap between coastal science and effective policy frameworks and management or in Italy where climate change effects are not yet considered in a systematic way.

The Tyndall Coastal Simulator approach has an important role to play in supporting stronger coastal governance, not just via a stronger evidence basis but also by providing a mechanism to support the identification and involvement of relevant stakeholders and provide a platform for collective learning and building consensus about future coastal management choices between them.

13.4.6 Standardisation of Methods

A common theme to emerge from all the international perspectives presented in this chapter is that coastal zones are areas with multiple jurisdictions, multiparty investments, multiple habitats and scales and many competing interests: coastal areas are complex. They are at the interface of multiple, interacting systems and consequently are subject to multiple pressures and threats over different timescales. However, the case studies also highlight that whilst some processes may be amenable to modelling using generic methods and tools, a systems approach to understanding the coast requires consideration of local issues. This inevitably requires consideration of climatic, socio-economic and other factors specific to a region, but limitations of available data, governance and stakeholder arrangements or complex physical and socio-economic processes may require development of bespoke methods and approaches to reflect particular processes or issues. Thus, the potential to transfer the Tyndall Coastal Simulator technology and simulation is more limited than the transferability of the processes by which these were developed and applied. This process is explored in greater detail in Chap. 14, as well as considering the research necessary to develop a common modelling framework that can facilitate transferability.

13.5 Conclusions

This chapter has reflected upon the challenges and opportunities for transferring coastal IAs to other parts of the world. The international case studies reviewed here indicate that there are many relevant opportunities to transfer the Tyndall Coastal Simulator approach, but these often raise challenges for its further development. In some cases, an IA may be of limited utility or require investment that is disproportionate to the problem under scrutiny. Coastal IAs will be most useful in situations where there are multiple pressures, conflict and trade-offs or where proactive use of an IA can enable more sustainable responses to extreme events.

Different locations may require the development of alternative IA components to capture relevant processes: this is an early consideration in such an assessment. Similarly, to provide support to coastal management at a range of scales, there is an opportunity to develop IAs of different spatial and temporal resolution. Alongside modelling developments, there are a number of challenges associated with improving the way IA information about multiple coastal systems is used to inform decisions.

In the light of this review, Chap. 14 concludes this book by describing a generic framework for transferring the Tyndall Centre process and a manifesto for future research.

Acknowledgements The authors are grateful to Andrew Watkinson and Craig Hutton for his review and feedback which greatly improved this chapter.

References

ABARE. (2009). *Australian fisheries statistics 2008*. Canberra: ACT.

Albins, M. A., & Hixon, M. A. (2008). Invasive Indo-Pacific lionfish Pterois volitans reduce recruitment of Atlantic coral reef fishes. *Marine Ecology Progress Series, 367*, 233–238.

Becker, A., Fischer, M., Inoue, S., & Schwegler, B. (2012). Climate change impacts on international seaports: Knowledge, perceptions, and planning efforts among port administrators. *Climatic Change, 110*, 5–29.

Becker, A., Acciaro, M., Asariotis, R., Carera, E., Cretegny, L., Crist, P., Esteban, M., Mather, A., Messner, S., Naruse, S., Ng, A. K. Y., Rahmstorf, S., Savonis, M., Song, D., Stenek, V., & Velegrakis, A. F. (2013). A note on climate change adaptation for seaports: A challenge for global ports, a challenge for global society. *Climatic Change, 120*(4), 683–695. doi:10.1007/s10584-013-0843-z.

Beeton, R. J. S., Buckley, K. I., Jones, G. J., Morgan, D., Reichelt, R. E., & Trewin, D. (2006). *Australia state of the environment 2006*. Independent report to the Australian Government Minister for the Environment and Heritage, Department of the Environment and Heritage, Canberra.

Brown, S., Nicholls, R. J., Woodroffe, C. D., Hanson, S., Hinkel, J., Kebede, A. S., Neumann, B., & Vafeidis, A. T. (2013). Sea-level rise impacts and responses: A global perspective. In C. W. Finkl (Ed.), *Coastal hazards* (pp. 117–149). Dordrecht: Springer. doi:10.1007/978-94-007-5234-4_5.

Bureau of Infrastructure, Transport and Regional Economics (BITRE). (2009). *Australian Sea Freight 2007–08* (Information Paper 64). Canberra: ACT.

Bureau of Transport and Regional Economics (BTRE). (2006). *Container and ship movements through Australian ports 2004–05 to 2024–2025*(Working Paper 65). Canberra: ACT.

Burnley, I. H., & Murphy, P. (2004). *Sea change: Movement from metropolitan to Arcadian Australia*. Sydney: UNSW Press.

Colgan, C. S. (2008). The value of Estuary regions in the U.S. Economy. In L. H. Pendleton (Ed.), *The economic and market value of coasts and Estuaries: What's at stake?* (pp. 37–64). Arlington: Restore America's Estuaries.

Côté, I. M., & Maljković, A. (2010). Predation rates of Indo-Pacific lionfish on Bahamian coral reefs. *Marine Ecology Progress Series, 404*, 219–225.

De Vries, W. S., Zanuttigh, B., Steendam, G. J., Kloosterboer, H., Van der Nat, A., & Graaf, H. (2011). Integrating science and policy for creating tools for safer European coasts in a changing climate. *Irrigation and Drain, 60*(1), 77–83. doi: 10.1002/ird.670. ISSN: 1531–0361.

Deltacommissie. (2008). *Working together with water: A living land builds for its future*. Findings of the Deltacommissie 2008. The Hague: Deltacommissie. http://www.deltacommissie.com/doc/deltareport_full.pdf

Department of Climate Change. (2009). *Climate change risks to Australia's coast – A first pass national assessment*. Canberra: Department of Climate Change, Commonwealth of Australia.

Engelen, G., White, R., Uljee, I., & Drazan, P. (1995). Using cellular automata for integrated modelling of socio-environmental systems. *Environmental Monitoring and Assessment, 34*, 203–214.

Englander, J. (2013). *High tide on main street: Rising sea level and the coming coastal crisis* (2nd ed.). Boca Raton: The Science Bookshelf. ISBN ISBN 978–0615637952.

Ericson, J. P., Vörösmarty, C. J., Dingman, S. L., Ward, L. G., & Meybeck, M. (2006). Effective sea-level rise and deltas: Causes of change and human dimension implications. *Global and Planetary Change, 50*, 63–82.

Gibbons, S. J. A., & Nicholls, R. J. (2006). Island abandonment and sea-level rise: An historical analog from the Chesapeake Bay, USA. *Global Environmental Change, 16*, 40–47.

Gurran, N., & Blakely, E. (2007). Suffer a sea change? Contrasting perspectives towards urban policy and migration in coastal Australia. *Australian Geography, 38*(1), 113–132.

Gurran, N., Normal, B., & Hamin, E. (2013). Climate change adaptation in Coastal Australia: An audit of planning practice. *Ocean & Coastal Management, 86*, 100–109.

Hall, J. W., Dawson, R. J., Sayers, P. B., Rosu, C., Chatterton, J. B., & Deakin, R. (2003). A methodology for national-scale flood risk assessment. *Water and Maritime Engineering, ICE, 156*(3), 235–247.

Hall, J. W., Dawson, R. J., Walsh, C. L., Barker, T., Barr, S. L., Batty, M., Bristow, A. L., Burton, A., Carney, S., Dagoumas, A., Evans, S., Ford, A. C., Glenis, V., Goodess, C. G., Harpham, C., Harwatt, H., Kilsby, C. G., Kohler, J., Jones, P., Manning, L., McCarthy, M., Sanderson, M., Tight, M. R., Timms, P. M., & Zanni, A. (2009). *Engineering cities: How can cities grow whilst reducing emissions and vulnerability?* Tyndall: Centre for Climate Change Research.

Hallegatte, S., Green, C., Nicholls, R. J., & Corfee-Morlot, J. (2013). Future flood losses in major coastal cities. *Nature Climate Change, 3*, 802–806.

Hanson, S., Nicholls, R. J., Patmore, N., Hallegatte, S., Corfee-Morlot, J., Herweijer, C., & Chateau, J. (2011). A global ranking of port cities with high exposure to climate extremes. *Climatic Change, 140*(1), 89–111. doi:10.1007/s10584-010-9977-4.

Harvey, N., Clarke, B., & Nursey-Bray, M. (2012). Australian coastal management and climate change. *Geographical Research, 50*(4), 356–367. doi:10.1111/j.1745-5871.2011.00734.x.

Hinkel, J., Brown, S., Exner, L., Nicholls, R. J., Vafeidis, A. T., & Kebede, A. S. (2012). Sea-level rise impacts on Africa and the effects of mitigation and adaptation: An application of DIVA. *Regional Environmental Change, 12*, 207–224.

Hinkel, J., Nicholls, R. J., Tol, R. S. J., Wang, Z. B., Hamilton, J. B., Boot, G., Vafeidis, A. T., McFadden, L., Ganopolski, A., & Klein, R. J. T. (2013). A global analysis of erosion of sandy beaches and sea-level rise: An application of DIVA. *Global and Planetary Change, 111*, 150–158.

Huntington, H. P., Goodstein, E., & Euskirchen, E. (2012). Towards a tipping point in responding to change: Rising costs, fewer options for arctic and global societies. *Ambio, 41*, 66–74.

Islam, R. M., Ahmad, M., Huq, H., & Osman, M. S. (2006). State of the coast 2006, integrated coastal zone management program, Bangladesh. Water Resources Planning Organization (WARPO), Ministry of Water Resources, Government of the People's Republic of Bangladesh.

Jin, D. (2008). Economic benefits of coastal restoration to the marine transportation sector. In L. H. Pendleton (Ed.), *The economic and market value of coasts and Estuaries: What's at stake?* (pp. 97–115) (p. 103). Arlington: Restore America's Estuaries.

Kron, W. (2013). Coasts: the high-risk areas of the world. *Natural Hazards, 66*, 1363–1382.

Milliman, J. D., Broadus, J. M., & Gable, F. (1989). Environmental and economic implications of rising sea level and subsiding deltas: The Nile and Bengal examples. *Ambio, 18*, 340–345.

Mimura, N., Nurse, L., McLean, R. F., Agard, J., Briguglio, L., Lefale, P., Payet, R., & Sem, G. (2007) Small islands. In M. L. Parry, O. F. Canziani, J. P. Palutikof, P. Van der Linden & C. E. Hanson (Eds.), *Climate change 2007: Impacts, adaptation and vulnerability*. Contribution of working group II to the fourth assessment report of the Intergovernmental Panel on Climate Change. Cambridge, UK: Cambridge University Press

Mycoo, M., & Chadwick, A. (2012). Adaptation to climate change: The coastal zone of Barbados. *Proceedings of the Institution of Civil Engineers – Maritime Engineering, 165*(4), 159–168.

New York City Panel on Climate Change. (2009). *Climate risk information*. New York: The City of New York.

New York City Panel on Climate Change. (2010). *Climate change adaptation in New York City: Building a risk management response* (Vol. 1196, pp. 63–85). New York: Annals of the New York Academy of Sciences, New York Academy of Sciences.

New York City Panel on Climate Change. (2013). *Climate Risk Information 2013: Observations, climate change projections, and maps* (C. Rosenzweig, & W. Solecki, Eds.). NPCC2. Prepared for use by the City of New York Special Initiative on Rebuilding and Resiliency, New York.

Nicholls, R. J. (2006) Storm surges in coastal areas. In M. Arnold, R. S. Chen, U. Deichmann, M. Dilley & A. L. Lerner-Lam (Eds.), *Natural disaster hotspots: Case studies* (Disaster risk management 6) (pp. 79–108). Washington, DC: World Bank Publications.

Nicholls, R. J., Wong, P. P., Burket, V. R., Codignotto, J., Hay, J. E., McLean, R. F., Ragoonaden, S., & Woodroffe, C. D. (2007a). Coastal systems and low-lying areas. In M. L. Parry, O. F. Canziani, J. P. Palutikof, P. J. van der Linden, & C. E. Hanson (Eds.), *Climate change 2007: Impacts* (Adaptation and vulnerability, pp. 315–356). Cambridge, UK: Cambridge University Press.

Nicholls, R. J., Tol, R. S. J., & Hall, J. W. (2007b). Assessing impacts and responses to global-mean sea-level rise. In J. Reilly, J. Edmonds, H. Kheshgi, C. D. Kolstad, J. B. Smith, & M. Schlesinger (Eds.), *Climate impact assessment* (pp. 119–134). Cambridge, UK: Cambridge University Press.

Nicholls, R. J., Hanson, S., Herweijer, C., Patmore, N., Hallegatte, S., Corfee-Morlot, J., Château, J., & Muir-Wood, R. (2008). *Ranking port cities with high exposure and vulnerability to climate extremes: Exposure estimates* (Environment Directorate Working Papers, (ENV/WKP(2007)1)) (62 pp). Paris: Organisation for Economic and Co-operative development (OECD).

Nicholls, R. J., Townend, I. H., Bradbury, A., Ramsbottom, D., & Day, S. (2013a). Planning for long-term coastal change: Experiences from England and Wales. *Ocean Engineering, 71*, 3–16. doi:10.1016/j.oceaneng.2013.01.025.

Nicholls, R. J., Hutton, C. W., Lazar, A. N., Rahman, M. M., Salehin, M., & Ghosh, T. (2013b). Understanding climate change livelihoods in coastal Bangladesh. *Hydrolink, 2013*(2), 40–42.

Norman, B. (2009). Principles for an intergovernmental agreement for coastal planning and climate change in Australia. *Habitat International, 33*, 293–299.

Nursey-Bray, M., & Shaw, J. (2010). Australia, 'Climate change and the sea change phenomenon'. *The International Journal of Environmental, Cultural, Social and Economic Sustainability, 6*, 15.

OECD. (2009). OECD reviews of risk management policies: Japan 2009 – Large-scale floods and earthquakes. Organisation for Economic Co-operation and Development. http://www.keepeek.com/Digital-Asset-Management/oecd/governance/oecd-reviews-of-risk-management-policies-japan-2009_9789264050303-en#page1

Parry, M., Arnell, N., Berry, P., Dodman, D., Fankhauser, S., Hope, C., Kovats, S., Nicholls, R., Satterthwaite, D., Tiffin, R., Wheeler, T., Lowe, J., & Hanson, C. (2009). *Assessing the costs of adaptation to climate change: A review of the UNFCCC and other recent estimates*. http://pubs.iied.org/pdfs/11501IIED.pdf

PDO-ICZMP. (2004). *Inventory of projects and initiatives in the coastal zone*. Program Development Office for Integrated Coastal Zone Management Plan Project, Water Resources Planning Organization, Dhaka.

Pecl, G., Frusher, S., Gardner, C., Haward, M., Hobday, A., Jennings, S., Nursey-Bray, M., Punt Revill, H., & van Putten, I. (2009). *The east coast Tasmanian rock lobster fishery – Vulnerability to climate change impacts and adaptation response options*. Report to the Department of Climate Change, Australia.

Pendleton, L. H. (2008). The economic value of coastal and Estuary recreation. In L. H. Pendleton (Ed.), *The economic and market value of coasts and Estuaries: What's at stake?* (pp. 140–165) (p. 165). Arlington: Restore America's Estuaries

Penning-Rowsell, E. C., Priest, S., Parker, D., Morris, J., Tunstall, S., Viavattene, C., Chatterton, J. B., & Owen, D. (2013). *Flood and coastal erosion risk management: A manual for economic appraisal*. London: Routledge.

Pielke, R. A., Jr., Gratz, J., Landsea, C. W., Collins, D., Saunders, M. A., & Musulin, R. (2008). Normalized Hurricane damage in the United States: 1900–2005. *Natural Hazards Review, 9*, 29–42.

Preti, M., & Zanuttigh, B. (2011). Integrated beach management at Igea marina, Italy: Results of ten-years monitoring. In J. M. Smith & P. Lynett (Eds.), *Proceedings of the International conference on coastal engineering* (Management 33). doi: 10.9753/icce.v32.management.33

Preti, M., De Nigris, N., Morelli, M., Monti, M., Bonsignore, F., & Aguzzi, M. (2009). *Conditions of the Emilia Romagna littoral up to 2007 and ten-year management plan*. Bologna: I quaderni dell'ARPA.

Rosenzweig, C., Solecki, W., Hammer, S. A., & Mehrotra, S. (2010). Cities lead the way in climate-change action. *Nature, 467*, 909–911. doi:10.1038/467909a.

Rosenzweig, C., Solecki, W. D., Blake, R., Bowman, M., Faris, C., Gornitz, V., Horton, R., Jacob, K., LeBlanc, A., Leichenko, R., Linkin, M., Major, D., O'Grady, M., Patrick, L., Sussma, E., Yohe, G., & Zimmerman, R. (2011). Developing coastal adaptation to climate change in the New York City infrastructure-shed: Process, approach, tools, and strategies. *Climatic Change, 106*, 93–127. doi:10.1007/s10584-010-0002-8.

Strauss, B. H., Ziemlinski, R., Weiss, J. L., & Overpeck, J. T. (2012). Tidally adjusted estimates of topographic vulnerability to sea level rise and flooding for the contiguous United States. *Environmental Research Letters, 7*, 014033. doi:10.1088/1748-9326/7/1/014033.

Tompkins, E., Nicholson-Cole, S., Hurlston, L. A., et al. (2005). *Surviving climate change in small islands: A guidebook*. The Tyndall Centre for Climate Change Research. Funded through the Overseas Territories Environment Programme, a joint initiative of the Department for International Development and the Foreign and Commonwealth Office, in partnership with the UK Overseas Territories.

UK National Ecosystem Assessment. (2011). *The UK national ecosystem assessment technical report*. Cambridge: UNEP-WCMC.

Vafeidis, A. T., Nicholls, R. J., McFadden, L., Tol, R. S. J., Hinkel, J., Spencer, T., Grashoff, P. S., Boot, G., & Klein, R. J. T. (2008). A new global database for assessing the vulnerability of coastal zones to sea-level rise. *Journal of Coastal Research, 24*(4), 917–924.

Walsh, C. L., Dawson, R. J., Hall, J. W., Barr, S. L., Batty, M., Bristow, A. L., Carney, S., Dagoumas, A., Ford, A., Tight, M. R., Watters, H., & Zanni A. M. (2011). *Assessment of climate change mitigation and adaptation in cities*. Proceedings of ICE: Urban design and planning, Special issue on Urban Development and Sustainability 164(DP2), 75–84. doi: 10.1680/udap.2011.164.2.75

Wilde, K. D. (2011). *Moving coastlines: Emergence and use of land in the Ganges-Brahmaputra-Meghna Estuary*. Dhaka: The University Press Limited (UPL).

Wong, P. P., Losada, I. J., Gattuso, J., Hinkel, J., Khattabi, A., McInnes, K. L., Saito, A., & Sallenger, A. (2014). Coastal systems and low-lying areas. In C. B. Field, V. R. Barros, D. J. Dokken, K. J. Mach, M. D. Mastrandrea, T. E. Bilir, M. Chatterjee, K. L. Ebi, Y. O. Estrada, R. C. Genova, B. Girma, E. S. Kissel, A. N. Levy, S. MacCracken, P. R. Mastrandrea, L. L. White (Eds.), *Climate change 2014: Impacts, adaptation, and vulnerability. Part A: Global and sectoral aspects*. Contribution of working group II to the fifth assessment report of the Intergovernmental Panel on Climate Change (pp. 361–409). Cambridge, UK/New York: Cambridge University Press.

World Bank. (2010). *Climate risks and adaptation in Asian coastal megacities*. A synthesis report. Washington, DC: World Bank. http://siteresources.worldbank.org/EASTASIAPACIFICEXT/Resources/226300-1287600424406/coastal_megacities_fullreport.pdf

Zanuttigh, B. (2011). Coastal flood protection: What perspective in a changing climate? The THESEUS approach. *Environmental Science and Policy, 14*, 845–863.

Zanuttigh, B., Bagli, S., Zagonari, F., Pescaroli, G., Bozzeda, F., Hoggarth, S., & Nicholls, R. J. (2013). THESEUS decision support system for coastal risk management. *Coastal Engineering*, THESEUS Special Issue *87*(1–2), 218–239.

Chapter 14
Integrated Coastal Assessment: The Way Forward

Robert J. Nicholls, Richard J. Dawson, Sophie A. Day (née Nicholson-Cole), Mike Walkden, Andrew R. Watkinson, Owen Tarrant, Jim W. Hall, and Peter Frew

Abstract The preceding chapters have laid out the range of challenges of coastal simulation to support future coastal management, with a particular focus on understanding erosion, flood and habitat changes and their links to coastal management. The detailed case study in Norfolk places the theory and generic principles in a

R.J. Nicholls (✉) • S.A. Day (née Nicholson-Cole)
Tyndall Centre for Climate Change Research, Faculty of Engineering and the Environment, University of Southampton, Southampton SO17 1BJ, UK
e-mail: r.j.nicholls@soton.ac.uk; day_soph@yahoo.co.uk

R.J. Dawson
School of Civil Engineering and Geosciences, Tyndall Centre for Climate Change Research, Newcastle University, Newcastle upon Tyne NE1 7RU, UK
e-mail: richard.dawson@newcastle.ac.uk

M. Walkden
WSP, Keble House, Southernhay Gardens, Exeter EX1 1NT, UK

WSP Group, Exeter, UK
e-mail: mike.walkden@wspgroup.com

A.R. Watkinson
School of Environmental Sciences, University of East Anglia, Norwich NR4 7TJ, UK
e-mail: a.watkinson@uea.ac.uk

O. Tarrant
Flooding and Communities, Environment Agency,
Horizon House, Deanery Road, Bristol BS1 5AH, UK
e-mail: owen.tarrant@environment-agency.gov.uk

J.W. Hall
Environmental Change Institute, Oxford University Centre for the Environment, University of Oxford, South Parks Road, Oxford OX1 3QY, UK
e-mail: jim.hall@eci.ox.ac.uk

P. Frew
Coastal Strategy for North Norfolk District Council, Coastal Management Consultant,
17 Main Street, Little Thetford, Ely CB6 3HA, UK
e-mail: pf@peterfrew.co.uk

R.J. Nicholls et al. (eds.), *Broad Scale Coastal Simulation*, Advances in Global Change Research 49, DOI 10.1007/978-94-007-5258-0_14

real-world management context that is particularly relevant to the UK, while international case studies (Chap. 13) have introduced a range of alternative challenges and perspectives.

The Tyndall Centre's coastal programme has highlighted how the management of the coastline needs to reflect the connectivity between the various geomorphological features, natural processes, engineered structures and socio-economic drivers. Moreover, it is crucial to understand the trade-offs that result from different management strategies. In the Norfolk case, the analysis strengthens the argument for a change in the widespread historic management approach of "hold the line" towards allowing as much of the coastline as possible to return to a more natural and dynamic configuration. However, this is challenging for coastal governance, raising questions about how to address the concerns of directly and indirectly affected landowners and householders on the eroding coast.

Given the large uncertainties in projections of future sea-level rise and other relevant climate factors, and the extent to which this influences the effectiveness of different management strategies, an adaptive (or learning) strategy that is regularly reviewed in the light of new information and an evolving governance/society context is most appropriate. To this end, the Tyndall Coastal Simulator approach of coupling visualisation tools with simulation models provides a powerful tool for analysing present and future trajectories of coastal evolution. This has been shown to motivate and engage policy makers and stakeholders in an on-going dialogue about the management choices ahead.

This Chapter reviews the key contributions of the Tyndall Coastal Simulator in the context of UK coastal management, and how coastal management and policy formulation may be informed by the evidence and engagement philosophy it encapsulates. The overall approach is distilled for wider application and a set of key issues that are important to consider are identified. The notion of living with uncertainty and the need for adaptive management is a core issue. How the future tools and process of UK coastal management might evolve in the light of the Tyndall Coastal Simulator results is examined, including incremental and more radical developments.

Keywords Integrated coastal assessment • Coastal simulation • Coastal governance • Coastal management

14.1 Introduction

This book has presented a detailed guide through a process of Integrated Assessment (IA) of coastal areas at a scale that is designed to support strategic coastal management: over tens of kilometres of coast and multiple decades. In this concluding chapter, we review the collective body of Tyndall Centre research on coasts and ask:

- what has been learned through the process of developing an integrated coastal assessment?

- what are the implications for managing coastal change?
- what are the research priorities?

We also reflect upon both the findings and experience of developing the Tyndall Coastal Simulator as applied to North Norfolk and the *approach* taken, which has great potential to be applied more widely (as shown in Chap. 13).

Preceding chapters have illustrated how IA, as illustrated by the Tyndall Coastal Simulator, can help us to understand better, how the coastal system functions and explore possible future trajectories for its evolution under different management regimes and in the face of other uncertainties such as the magnitude of climate change. Box 1.4 (Chap. 1) defines IA in this context: understanding coupled systems with the potential for non-linear or unexpected responses to inform their possible evolution and management. There have been extensive IA policy analyses of climate change at the global scale (e.g. Warren et al. 2008; Anthoff et al. 2010; Arnell et al. 2013). More spatially limited, Continental/Regional scale IAs – as described here – are less common. However, there are examples, including analysis of climate impacts (Harrison et al. 2013) and air quality at continental scales (Schöpp et al. 1998), cities (Rotmans et al. 2000; Hall et al. 2010; Walsh et al. 2013) and a few examples for coasts (Turner et al. 1998; de Kok et al. 2001; Bricker et al. 2003): we expect the nature of many societally-relevant problems will require this capacity to be enhanced in the coming years (e.g. Wilbanks 2003).

The research presented in this book is an example of a scientific study of a large coupled system with real-world application to support coastal management, and in particular flood and erosion risk management. In taking a systems perspective of the natural, physical and social environment at a scale that is relevant to livelihoods and the economy, it allows analysis of how the system as a whole might evolve in a changing physical and socio-economic environment. This highlights opportunities, challenges and trade-offs and emphasises that coastal governance should be both reflective and take a long-term view in order to promote effective adaptation to coastal change. Long-term policies and plans operate within a continuous process, and hence need to be regularly debated and reviewed to reflect improving knowledge and changing context. The research described in this book demonstrates that broad scale coastal simulation can provide a rich, on-going supporting evidence base and potentially provide an evolving platform for this adaptation process.

The application of the Tyndall Coastal Simulator to North Norfolk demonstrates that it is now feasible to explore long term (up to one century, or even more) integrated projections of coastal processes such as geomorphology, flood risk and land use change, greatly enhancing the evidence base available for coastal management decisions. Moreover, the methods are transferable to other coastal areas and other coastal issues/problems. In particular, this book has demonstrated the first quantification of the relationship between the risks of erosion and flooding at a scale relevant to coastal decision makers. These results were captured within the Tyndall Coastal Simulator interface allowing the technical results to be accessible to a wide range of stakeholders. The analysis emphasised trade-offs and the difficulties faced by coastal managers, who in reducing the risk of erosion may actually enhance flood risk (or the cost and viability of mitigating this risk) at downdrift sites (i.e., within the same coastal sub-cell). The approach more broadly shows that there is great potential for coastal management stakeholders to co-produce improved understanding of coastal futures, and for decisions to be based on a stronger evidence base (see Chap. 12). The book also relates the technical aspects of coastal change to the present debate around long term shoreline management and especially the strategic decision as to whether or not to 'hold the line', or allow a planned retreat, via 'managed realignment' or consider 'no active intervention' (Nicholls et al. 2013).

This concluding chapter is structured as follows. In Sect. 14.2, we synthesise the key findings from the preceding chapters. Section 14.3 then distils the Coastal Simulator experience into a generic process for wider application. Wider contextual issues and possible choices for future coastal management are considered in Sect. 14.4. In particular, the response to the 1953 storm surge event in Norfolk is revisited to consider how we might have responded had we the foresight from the methods available here. Section 14.5 presents an agenda for future research and application of integrated assessment for coastal areas considering technical, social and application perspectives, before Sect. 14.6 concludes.

14.2 Synthesis of Key Findings

Chapter 1 sets out the global context for the climatic and socio-economic changes that drive coastal change and need to be downscaled for strategic application at the management scale (in this case, a sub-cell covering tens of kilometres). Following Chap. 1 and Fig. 1.1 in particular, the chapters can be grouped into those concerning downscaling (Chaps. 2, 3 and 4), impact analysis (Chaps. 5, 6, 7, 8 and 9) and developing sustainable responses (Chaps. 10, 11 and 12) which are now discussed in turn.

14.2.1 Scenario Generation

Chapters 2, 3 and 4 consider the basic inputs necessary to examine climate impacts and risks that are analysed subsequently, especially Chaps. 7, 8 and 9. In these chapters, human-induced climate change due to greenhouse gas emissions and associated socio-economic changes are considered, following a consistent scenario framework. Importantly, climate change at the coast more than just sea-level rise. Chapter 2 examines how sea level, surge and wave climate might change as a result of wider climate change. It delivers appropriate scenarios, but large uncertainties are apparent for all the climate drivers and are considered in the downstream analysis. As the wave simulations, driven by climate models, are for offshore conditions, Chap. 3 considers their necessary onshore transformation to the coast. Off the coast of Norfolk and Suffolk, the bathymetry is complex with numerous offshore sand banks, including some that dry at low tide. Sensitivity to this process has been tested and shows that as sea levels rise and this shelter is reduced, wave heights at the shore can increase substantially. Morphodynamic changes may, however, counteract the increase in wave height – if banks can grow with sea-level rise. Hence, sand bank morphodynamics are an important topic for further research.

As well as climate drivers, changes in socio-economic conditions influence impacts and economic risk. In Chap. 4, scenarios of the future of the built environment are developed using two algorithm-based approaches, rather than expert judgement. This has the advantage of having transparent rules and the scenarios can easily be revised and updated, as required. In all these simulations, it was assumed that overall population increases in East Anglia, following national forecasts provided by the UK's Office for National Statistics. This drives an increase in the number of residential and non-residential buildings. However, the increase in economic risk of coastal erosion and flooding also depends on the distribution of buildings which varies between scenarios. Hence, Chaps. 2, 3 and 4 show how future changes to a wide range of coastal drivers can be explored and quantified in relation to their uncertainties.

14.2.2 Impact Analysis

Chapters 5 and 6 considered broad coastal geomorphic evolution, including changes in coastal habitats and ecosystems. This is a key element in assessing impacts and risks at the coast. Currently, we gain great benefit from geomorphic features such as offshore banks (e.g. Chap. 3), beaches, marshes and other geomorphic features, in terms of natural protection and shelter. These geomorphic and ecosystem services are often taken for granted but will change and generally decline with climate change and other erosive trends. However, even with state-of-the-art quantitative analysis, predicting future geomorphic and habitat evolution is challenging and subject to large uncertainties. Chapter 5 establishes a framework for geomorphic system analysis that integrates qualitative and quantitative knowledge and promotes stakeholder dialogue. This allows expert judgement to be captured and helps identify areas where there is consensus or not concerning geomorphic system functioning. This provides an improved understanding of the coastal systems, as well as a foundation for quantitative analysis as discussed in Chap. 7. Chapter 6 then considers coastal habitats and their future, most especially salt marsh and intertidal areas. It shows that while deterministic predictions are possible for a specific scenario of sea-level rise, they fail to capture the large uncertainties in habitat development.

Chapters 7, 8 and 9 consider quantified analysis of coastal erosion and flooding in North Norfolk, including the naturally eroding cliffs from Weybourne to Eccles and the extensive coastal lowland of the Norfolk Broads from Eccles to Winterton Ness. Chapter 7 applies the SCAPE (Soft Cliff And Platform Evolution) reduced complexity numerical model to this coast. A series of two-dimensional profile models are simulated and interact with each other via a one-line (longshore) model. Importantly, SCAPE can simulate responses to a range of drivers, including natural gradients in sediment transport (which we know has shaped the Norfolk coast over thousands of years), sea-level rise and the construction or removal of coastal engineering measures. SCAPE has been used to hindcast the last 100 years of shoreline evolution in Norfolk, including the effect of known engineering interventions, as well as forecast future evolution over a century for the scenarios developed in Chaps. 2 and 3, combined with a range of scenarios of management: from removal of all engineering intervention to complete protection and intermediate and more plausible strategies. These simulations reveal the large effect of historic engineering interventions on coastal morphodynamics, including starvation of the beaches in front of the Norfolk Broads. Hence, future management strategies that defend less of the cliffed coast will cause local losses where the defences are removed, but will benefit the downdrift coast in terms of increased sediment supply and larger beaches. Of fundamental interest is the fact that the SCAPE model displays unexpected emergent behaviour that is consistent with observations (e.g. migrating longshore sand waves). Aspects of the systematic approach in Chaps. 5, 6 and 7 are being taken forward in more recent work, including the *iCOASST Integrating Coastal Sediment Systems* Project (Nicholls et al. 2012).

Chapter 8 integrates the morphodynamic forecasts with the climate drivers and considers inundation probability in the Norfolk Broads. The migrating sand waves identified in Chap. 7 are important determinants of inundation probability; at the troughs the beach volume is lowered increasing the likelihood of defence failure (breaching). From a systemic perspective, this process might at first be seen as more detail than required, but it is important for realistic simulation of inundation probability along this frontage. Furthermore, an ensemble approach of multiple realisations of the future is vital to sample fully the range of potential future beach conditions, together with inundation probability. Inundation extent is dominated by water level and to a lesser extent wave conditions. Pre-failure beach levels also influence potential discharge into the floodplain and therefore flood extents. Rising sea level will notably increase the area at risk of flooding and the flood depths when defence failure occurs.

Chapter 9 integrates the outputs from the hazard models in Chaps. 7 and 8 to evaluate the impacts of coastal management and climate change scenarios on flooding and erosion risks, developing the analysis of Dawson et al. (2009). These, for the first time, have been presented quantitatively in terms of a *common currency* of expected annual damages. The evolution of future flood risk is sensitive to both climate and socio-economic drivers through sea-level rise and management decisions that affect the quantity of beach sediment from eroding cliffs and artificial beach nourishment – underlining the importance of strategic coastal management. Over the next 50 to 100 years, the potential increase in flood risk is an order of magnitude larger than erosion risk. However, in the presence of sufficient beach volume (i.e. an ample supply of beach sediment), the current level of flood risk can be maintained or even reduced in future under rising sea levels. This is a surprising result as the Norfolk Broads have been presented as a site where a large retreat may be inevitable and unstoppable. Our analysis shows that the key issue is whether a sand supply can be provided, either by artificial nourishment or by increased longshore supply. The latter approach has major implications for the future of the up-drift cliff coast and its residents.

14.2.3 Delivering Sustainable Responses

Chapters 10, 11 and 12 consider the process of communicating outputs from the Tyndall Coastal Simulator to stakeholders and how this can inform adaptive responses to coastal change. Chapter 10 highlights the challenges of communicating information about coastal change, including model simulations; the information changes through time, is highly uncertain and can impact on a diverse range of people, organisations and interests. A number of visualisation tools were explored as ways to communicate different coastal futures. While these tools showed great potential in improving communication that help to enable discussion and debate, concern was raised regarding communicating uncertainties. Hence, a suite of visualisation tools are likely to be required in any communication exercise. To maximise their effectiveness, these need to be tailored to the audience and the purpose and stage of the engagement process. Further to the methods explored here, gaming has also been shown to provide an effective means of enabling discussions over long-term planning issues, coastal change and sea-level rise (Lonsdale et al. 2008; Pontee and Morris 2011; Rijcken et al. 2012).[1]

Chapter 11 describes the development of an interface, based on a bespoke GIS, to manage and facilitate exploration by researchers and stakeholders of the wealth of data and visualisations from the Tyndall Coastal Simulator and the wider issues that are raised. It also allowed offline calculations via integrating datasets and simple models. The interface was developed in consultation with coastal managers whose input helped to guide its design and provide insight on its usability. The role of the Interface in promoting dialogue within the research consortium and more wide among stakeholders was an important activity which facilitated the overall Tyndall Coastal Simulator development process. However, the legacy of the Interface beyond the Tyndall Centre coastal project is an issue, as on-going maintenance and development is essential. In the future, alternative mechanisms such as web-services may be preferred. Embedding the operation and maintenance of these data and analysis tools within the relevant coastal operators and monitoring processes, such as the Environment Agency and Channel Coastal Observatory, would be a key step in the future.

Using the North Norfolk case study, Chap. 12 demonstrates that the governance arrangements that guide coastal management do not adequately bridge the gap between policy and application. Hence, coastal communities find it difficult to prepare proactively for coastal change. This is especially the case in Norfolk where coastal defences have been, or are proposed to be, abandoned, requiring a transition

[1] For example, RUFopoly: www.bcu.ac.uk/research/-centres-of-excellence/centre-for-environment-and-society/projects/relu/rufopoly

in coastal management. Nor do these governance arrangements presently reflect the connectivity and consequential trade-offs between the various geomorphological features that comprise the natural coastal system. At the local community level, for example, change in the coastal area in and around a village or small town is a complex and emotive subject for many reasons. Issues of trust, justice, economy and environment need to be considered, making integrated approaches to both the science and management of coasts essential. Shoreline Management Planning is questioning traditional hold the line strategies and considering retreat options in many locations, but our existing coastal governance mechanisms remain ill-prepared to deliver adaptive options suitable for living on a retreating coast. Stakeholder engagement is fundamental to the delivery of a successful coastal management policy. It must be part of the process from the outset so that the evidence base, of which tools such as the Tyndall Coastal Simulator form part, and management options are co-developed by policy makers and communities. Hence, IA tools as discussed here have great potential in analysing and visualising potential transitions in coastal change and management approach.

14.3 Reflections on the Process of Developing the Coastal Simulator

Collective learning has been an on-going aspect throughout this programme of research. Beyond the Norfolk case, more general lessons can be drawn out – relevant coastal and non-coastal applications of IA. There are four key factors, general to all IA studies, that we highlight here:

- Promote flexibility, learning and iteration within the assessment process, as the issues and questions being posed will become better formulated as the assessment develops.
- Select appropriate scales and boundaries for the phenomena of interest, generally drawing broad boundaries in the first iteration.
- Adopt a participatory approach to maximise its relevance and real-world application.
- Dedicate appropriate time and resources to the integration process.

These principles are discussed in turn in the following sections, followed by consideration of their application to other cases.

14.3.1 Promote Flexibility, Learning and Iteration

Our experience with the Tyndall Coastal Simulator showed us that IA is by its nature learning by doing – modelling advances reveal new insights, while interactions between the research team and stakeholders led to new understanding. These lessons need to be assimilated and acted upon within the research process. Hence, there is a need for regular reflection as well as sufficient flexibility and adaptability in the research programme to allow for this learning. As learning occurs and understanding develops, so subsequent research and application is improved, refined and focussed.

In the early stages of the research, conceptual development was undertaken in parallel with both qualitative and quantitative investigation and regular exchange amonge the whole research team. The research was much more than just plugging a set of models together in a consistent manner, though that alone is a complex and time-consuming process. The conceptual aspects identified the key issues across the dimensions of interest, including the views of key stakeholders. The qualitative/quantification investigations took existing methods that addressed these key issues and developed them as appropriate. Hence they assessed the feasibility and level of sophistication at which analysis would be possible: to some extent the conceptual and quantitative/qualitative aspects co-evolved. The multiple phases of the Tyndall Centre research were helpful. The understanding gained in Phase 1 of the Tyndall Centre coastal programme ensured that development in Phase 2 was greatly accelerated. These lessons have been applied in later IA research activities such as the Tyndall Centre cities programme (Walsh et al. 2011; 2013) and the Infrastructure Transitions Research Consortium[2] (Hall et al. 2012; 2013), where learning and iteration were included in the research design.

Consequently, the IA programme evolved as understanding of the specific research needs developed, data availability improved and technical capacity increased. Changes in the team of scientists and stakeholders presented both challenges and opportunities, bringing additional and complimentary perspectives and skills, but also an overhead associated with integrating new members – although this can promote team reflection. Over a multi-year project, it is also important to be aware of changes in regulation, institutional arrangements, data/information, stakeholder interest, priorities and awareness, technological advances and so on. Hence, any long-term IA research process must be dynamic to remain relevant and state-of-the-art.

14.3.2 Identifying System Scales and Boundaries

The starting point of an IA process is policy-driven research questions: this raises the issue of appropriate system scale and boundaries. Some of these boundaries may be at different scales, so there is a need to assess these issues as outlined in Fig. 1.11. Choosing the appropriate level of investigation influences the IA approach and what it can achieve. In Norfolk, the size of our main study area (SMP6) was strongly influenced by the scale at which management occurs on the coast. We could have selected a longer or a shorter stretch of coastline, but the Shoreline Management Plans in England provide an important management scale for which there is a physical basis and a recognised need (Nicholls et al. 2013). This includes physical coupling due to longshore transport as SMP6 is based on a sub-cell definition. Lastly, there

[2] http://www.itrc.org.uk/

was strong stakeholder interest in how the coast could be managed in the future with significant stretches of cliff defences identified for managed realignment (i.e. abandonment) in the current SMP, as well as concern about the effects of climate change for a coastal area with a long history of erosion and flooding.

There is a balance to be struck selecting appropriate modelling resolutions to gain insight, versus being swamped by detail (or creating computationally intractable models). This requires particular consideration for broad scale modelling, where the large spatial scale inevitably requires compromise between issues such as model resolution (spatial and temporal scales), the detail of processes modelled and the number of simulations required to understand the implications of key uncertainties.

We also learnt that it is important to consider the whole system, rather than focussing on the details of any particular system element. The main question is what are the key processes and their interactions and can they be represented appropriately in models. An element of pragmatism is necessary in constraining the scope of a study. This involved a willingness to experiment, including use of a range of approaches to conceptualise and model phenomena. For instance SCAPE, which is used for geomorphic predictions, represents all the relevant processes and their interactions in a relatively simple but comprehensive manner enabling rapid computation (see Chap. 7). Coupling models can also lead to the identification of emergent properties, which can be tested and explored and provide useful insights into the behaviour of coupled human-natural and engineered systems.

14.3.3 Stakeholder Engagement

The Tyndall Coastal Simulator project went beyond common practice with regards to stakeholder engagement. The on-going engagement of stakeholders at local to national levels was fundamental to its real-world relevance. This process is discussed in more detail in Chap. 12 and ranged from data provision to practical critiques. Ultimately, stakeholder engagement was critical in terms of data exchange, the conceptual development of the Tyndall Coastal Simulator and in testing and evaluating its function, relevance and potential application.

Having the means to generate a better understanding of the coastal system and better projections of how it may behave in the future are of limited value if the tools are not accepted by the potential user community. We found that the process of developing the Tyndall Coastal Simulator greatly benefitted from the genuine involvement of people who had practical expertise and an interest in the potential of the tool. This co-production of the Tyndall Coastal Simulator was important. Practitioners bring a level of insight and value into research of this kind that cannot be obtained theoretically, and engagement maximises the assessment's relevance. The Tyndall Coastal Simulator facilitated engagement with a wide range of stakeholders, including the general public, and the on-going debates on the future of the North Norfolk coastline. Conflict resolution may be a general application of this type of approach. Stakeholders were enthusiastic about the wider application of the Tyndall Coastal Simulator (or similar approaches and tools) and could even see opportunities for commercial application of this type of approach. Hence, a participatory appraisal is fundamental to applying this method elsewhere. This approach could be particularly beneficial in areas where a partnership model of management exists.

14.3.4 Dedicating Appropriate Time and Resources

A key issue affecting the speed with which developments made during the Tyndall Coastal Simulator project were able to pass into practice is the associated cost. These costs were more than that associated with a normal Shoreline Management Plan, Coastal Strategy Study or Project Appraisal. This is because the assessment required both dialogue and integration at the systems level. Although the timeframe of any research project could be reduced with additional resource, the development of shared methods and understanding still takes considerable time and investment. Human resources are especially important for IA and developing the appropriate skills is often counter to traditional training and hence hard to replace if staff move on.

Two integration activities are particularly important and potentially costly: (1) technical integration of models and (2) supporting dialogue and team working. Model integration requires human resources, expertise, time and technical capability. Key technical elements in the Tyndall Coastal Simulator project were relatively cost-effective. For example, the flood models used were relatively simple and quick to construct, while the original regional coastal geomorphic SCAPE model was constructed and calibrated within a few weeks. The value of these models comes not from their detail or precision, but rather their flexibility and rapid runtime which enable integration, long-term analysis and multiple simulations.

The provision of resources to support dialogue and team working to enable the development of shared understanding, knowledge and goals ensures that there is capability for integration. This is perhaps a less tangible element but of equal importance in an IA project. It is difficult to budget for the non-technical elements of integration. One of the key components is that studies of this kind need strong leadership and vision, perhaps an individual or small team who can guide the construction and development of an IA project. Specific skills and effort are necessary to drive integration as well as planning for and financing opportunities to enable shared understandings to develop. Through these opportunities and recognising the value of other disciplinary perspectives, new insights emerge.

As an example, future phases of Shoreline Management Planning could be supported by and benefit from dedicated IA, but at a cost: the additional resources required in terms of time, finance and personnel expertise could be significant and this needs to be balanced against the benefits. However, as IA methods improve and become more widely understood, they should become easier and less costly to apply. For example, the adoption of open standards in environmental modelling software, such as OpenMI (Gregersen et al. 2007), could greatly facilitate the process of integrating and linking models in the future. The on-going investment in consistent datasets and methods for analysis of coastal erosion and flooding by the Environment Agency and DEFRA (e.g. Burgess et al. 2002; Nicholls et al. 2013) means that widely applicable methods can be more easily developed and widely applied, including potential IA methods.

14.3.5 Application to Real-World Cases

The approach that we present in this book could be transferred to other locations. In some cases there is low complexity or few issues to consider – a formal IA may not be an appropriate tool for exploring and developing potential management options. This may be done effectively with expert analysis. IA is more appropriate to investigate systemic processes that we do not fully understand. Geomorphic coupling, as occurs in Norfolk, is one example where this approach is useful. More generally, coupling natural and human systems is an interesting research frontier where important insights could emerge (e.g. Lazarus et al. 2011; Murray et al. 2011).

Figure 14.1 illustrates a cyclical process that we propose as a generic approach to developing and implementing an IA to better understand a coastal system and to test long-term policy interventions. It involves ongoing engagement with relevant stakeholders who are involved with every step of the process, leading to co-production of the results and the process as a whole. Table 14.1 summarises how it was applied in the Tyndall Coastal Simulator for Norfolk.

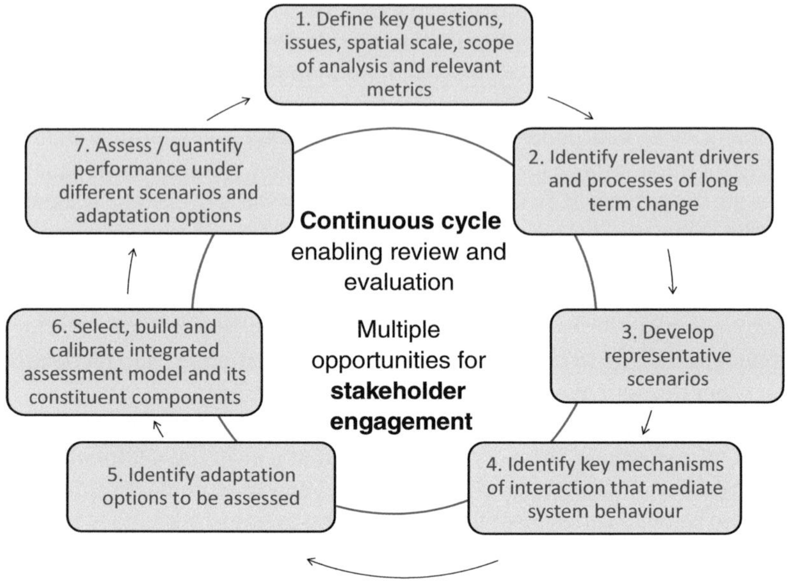

Fig. 14.1 Integrated assessment to test long-term policies: a generic seven-step approach

Table 14.1 The seven steps for developing the Tyndall Coastal Simulator for SMP6, following Fig. 14.1

Step		Example issues and/or actions
1.	Define key questions	On-going coastal erosion and flood risk in coastal sub-cell 3b (now SMP6) and cliff-top and flood plain communities
		Deterioration of defences; defences abandoned and removed at Happisburgh
		New Shoreline Management Plan (SMP) – which proposed significant reductions in cliff protection
		Stakeholders who felt that their views and needs were not adequately reflected in the SMP process
		Concerns about climate change, including faster or unexpected change
2.	Identify relevant drivers and processes	Climate change
		Geomorphic change
		Subsidence in the Norfolk Broads (*not considered due to insufficient data – in qualitative terms this would exacerbate flood area, depth and velocity*)
		Socio-economic change (urban expansion)
3.	Develop representative scenarios (see Appendix A)	UKCIP02 and UKCP09 climate scenarios
		Sensitivity testing (e.g. High (H++) scenario).
		Foresight socioeconomic futures and underpinning analysis (e.g. the national census).
4.	Identifying key mechanisms of interaction	Longshore transport
		Land use change
		Offshore bank morphodynamics
		Legacy of historical defences/SMP options
5.	Identify adaptation options	A range of cliff defence options from no protection to complete protection (informed by the current SMP)
		Beach nourishment
		Do nothing
		Raise or strengthen flood defences[a]
		Spatial planning (guide new development / urban expansion)
6.	Select, build and calibrate integrated assessment model	The model components were validated and / or calibrated as best as possible according to past observations. Validating the overall IA is probably impossible, so it is important to validate all the components possible and be explicit about uncertainties.
7.	Assess/quantify performance	Results are reported throughout the book.

[a]Our emphasis was on adaptation options that influence the mechanisms of interaction identified in step 4 and so this is not reported here

Figure 14.1 should not be confused with a risk-based *decision-making* cycle such as that identified by Willows and Connell (2003); rather Fig. 14.1 describes a process of defining the questions and process for IA that is consistent with the broader process of coastal management (Fig. 1.2). The cycle of continuous review is common to many policy-oriented modelling studies as it provides opportunity for reflection, iteration and learning. For example, the Thames Estuary (TE2100) project is based on regular review and monitored as conditions change (Box 14.1). Starting or entering the process will usually involve definition or refinement of the key issues and cross-cutting issues. Reflection and learning are a part of all human research and management, but making them explicit as we do here and throughout the book, reinforces and emphasises these important processes which are key to successful IA.

14.4 Key Conceptual Issues and Challenges

The previous section defined the process of an IA assessment. There are also some important conceptual and cross-cutting issues to be tackled when applying Fig. 14.1 to develop IA for coastal simulation. Consideration of these issues supports improved broad-scale analysis and management of complex coastal problems. Four issues are identified and discussed further in Sects. 14.4.1, 14.4.2, 14.4.3 and 14.4.4:

- Recognising trade-offs
- Considering uncertainties
- Responses to extreme events
- Integrated and adaptive governance

14.4.1 Recognising Trade-Offs

When considering the future, important trade-offs are likely to emerge and require analysis (Dawson 2011). Coasts are naturally dynamic and are widely retreating. At one level, this is simply coastal morphodynamics. The notion of permanent protection should be avoided as protection is a series of decisions which must be repeatedly reviewed (on a 50–100 year cycle) (cf. Penning-Rowsell et al. 2013a). However, from a societal perspective, retreat comes at the cost of coastal land and associated assets. Investments may be made to delay erosion in certain areas, but in addition to the direct tangible effects and costs of such intervention, unintended indirect effects on coasts may also occur due to a physical connection as in North Norfolk, or maybe a coupled human-physical connection (e.g. defences can often encourage further development).

Future costs are unavoidable on developed coasts due to their asset concentration and legacy of engineering interventions, regardless of management policy. Most options involve some sort of trade-off between a wide range of potential costs and benefits. For example, improved defences reduce risk but can hinder access and amenity, and it may be challenging to provide both at the same time. So before such interventions are made, these linkages need to be understood. Representation of such integrated 'whole' systems helps clarify the full costs and benefits of alternative management choices and how these can evolve in a wider context.

This is well recognised by experienced coastal managers, and shoreline management in England and elsewhere have been shaped to prevent or minimise such effects (cf. Nicholls et al. 2013). For such constraints to be effective, the linkages between and within coasts must be understood and ideally modelled, albeit recognising the significant uncertainties over long timeframes. The Tyndall Coastal Simulator has integrated appropriate tools for this task. Moreover, it shows that more sophisticated decisions can be made involving trade-offs between coupled hazards and their economic risks.

Critical economic trade-offs examined in this study include the risks from flooding and erosion, choices in future land use development and alternative coastal management adaptation policies. By quantifying these trade-offs, we established that the risk of flooding would increase without either allowing cliff erosion and/or substantial beach nourishment. Quantification of these two options in relation to their potential to reduce the risk of flooding could be used to inform the most appropriate management decision and potentially allow savings to be hypothecated to help communities at risk from erosion, i.e. rather than compensate them for their loss, share the benefits of that loss in terms of flood risk reduction.

Other decisions relate to societal well-being and the environment. Many coastal societies, for example face choices in terms of the trade-off between hard coastal defences and allowing a more natural coastline to evolve. The research presented in this book highlights the kinds of choices that society may have to make based on a sound scientific basis – without any judgement by the research team about what should be done. These choices include, for example, natural cliffs or defended cliffs and coastal grazing marsh or salt marsh as discussed in Chap. 6.

Broader questions arise from this work as illustrated below. How in a democracy could we develop an acceptable strategy that would allow the coast to revert to a more natural rather than protected coastal system? What options suggest the wisest use of resources? What constitutes a fair strategy? How can the gulf between national and local scales of decision making, budgets and needs be negotiated? Can management be holistic and make use of sound science when constrained by political boundaries? Additionally, there are issues of sustainability in decision-making, including ethics and fairness (e.g. private land values tied to defences which society often provides via general taxation). Different societies will weigh these and other questions differently, but they illustrate the real issues and trade-offs that need to be analysed.

14.4.2 Considering Uncertainties

Uncertainty is a key element of the simulation and analysis presented in this book. While improving technical capability and generating robust data can reduce uncertainty, it is an inescapable aspect of any simulation study or investigation of future choices. Hence, approaches which recognise and address uncertainty are critical to IA.

Future climate is uncertain – both because our scientific understanding of the workings of the climate system remains incomplete and because the evolution of future climate is (partly) contingent upon human activity (such as greenhouse gas emissions). This uncertainty about the course of future climate – on all scales – and uncertainty about how societies will respond to this change in climate presents new challenges for the way individuals, organisations and societies make decisions (Dessai et al. 2007). Similarly future socio-economic conditions are uncertain. Uncertainty is of particular importance in this analysis because of the complexity of the coastal system, the time-scale of appraisal and the role of human agency in mediating system behaviour. Long term projections of climate and socio-economic changes are fraught with uncertainty that propagate through the IA as the analysis moves to increasingly more local scales (Fig. 1.11). Furthermore, additional uncertainties are introduced by each modelling component.

Any IA must therefore be conducted within an appropriate uncertainty framework. Uncertainty analysis involves the identification of the sources of uncertainty in datasets and models, their implications in terms of outputs of interest (e.g. expected annual damages) and ultimately their use in decisions. Uncertainty analysis is required in order to understand how predictions of interest to a decision-maker (e.g. risk associated with a coastal management policy) may differ under plausible variations in assumptions. As an example, in Phase 2 of the Tyndall analysis we made use of the UKCP09 climate scenarios which quantify aspects of climate model uncertainties for given emissions scenarios. Specifically, we sampled from the probability distributions of future sea level rise trajectories provided in UKCP09.

Elsewhere, uncertainties are predominantly considered using a range of different scenarios. However, as with other scenario studies, these are plausible and internally consistent projections, conditional upon a clearly specified set of assumptions – they are not forecasts. For socio-economic change, we generated plausible bounds on quantities of interest and tested different adaptation policies in the context of these plausible bounds. Thus, for example, while we cannot specify the precise nature of future development in North Norfolk, we understand existing development and plausible land use change under different socio-economic and policy situations. We have calculated future flood risk taking account of these changing land use scenarios and the latest knowledge of potential sea level rise and other climate drivers.

Where a detailed analysis of the full range of uncertainties was computationally intractable (e.g. surge modelling) or the state of the art was unable to simulate the phenomena of interest (e.g. offshore sandbank evolution) we applied a 'guided sensitivity test'. This involved identifying realistic bounds on likely future changes to understand their implications (e.g. selecting a low and high sensitivity climate model variant to explore sensitivity of offshore waves).

Hence, our results capture a wide range of uncertainties, although computational tractability has limited the extent to which all of these can be explored. Formal approaches to handling uncertainty in modelling frameworks exist and should be adapted to address the particular challenge of dealing with uncertainty in IAs. New data sources, including remote sensing and pervasive sensors to monitor coastal change, have the potential to provide hitherto unavailable data streams which could be used to constrain the residual uncertainties in our analysis (and give new insights). These must be complemented by development and application of methods for decision analysis that can identify interventions and long term strategies that are less sensitive to uncertainty (see Box 14.1 and Sect. 14.4.4).

14.4.3 Responses to Extreme Events

Extreme events influence how coasts are managed, often shaping how they evolve over the decades which follow. History is peppered with examples of extreme coastal events (Kron 2013): this book has specifically mentioned the 1953 North Sea storm surge. Others, to name a few, include the 1970 and 1991 cyclones in coastal Bangladesh (Sect. 13.2.2), the floods in Hamburg in 1962, Hurricane Katrina in 2005 and more recently Hurricane Sandy in 2012 in the USA (Sect. 13.2.6) and the Great East Japan Earthquake and Tsunami in 2011 (Sect. 13.2.1). How we react and learn in response to extreme events is critical when we think about preparing for future coastal change.

One potential application of the Tyndall Coastal Simulator and similar results is retrospective learning of how the results might have changed the responses to such historic events. The 1953 North Sea flood has a major legacy. It led to the development of North Sea surge forecasts, better defences along the east coast, most especially the Thames Barrier and associated defences for London (Fig. 14.2), and continues to be a reference event for flood management. The 1991 cyclone in Bangladesh led to a system of cyclone shelters and a warning system which has reduced mortality to such events by two orders of magnitude. Hurricane Katrina in 2005 led to more sophisticated methods for assessing hurricane flood events and US$15 billion of new defences around New Orleans. The longer term outcome of Hurricane Sandy

Fig. 14.2 The Thames Barrier, London (Richard Dawson)

for Greater New York remains less clear, but large investments are being considered and important changes will inevitably result (PlaNYC 2013). Hallegatte et al. (2013) have argued that future flood events in coastal cities will become more severe under current trends of climate-driven rising sea levels and urban expansion, reinforced in some cases by subsidence. This raises questions about how to respond to these trends and future floods and disasters in particular? There is potential for policy to respond to virtual rather than real extreme events in the future. How events can be simulated in the context of other coastal drivers and at different time scales in order to help us prepare for future change remains an important question to build on tools like the Tyndall Coastal Simulator.

Responses to events of different magnitude should ideally be underpinned by good understanding of potential future vulnerability and a strategic plan. While we are always planning, extreme events offers windows of opportunity for change, which long-term strategies can proactively exploit. Rather than react to single real events, it is possible through coastal simulation to respond to many simulated disasters so that when an event happens we are better prepared. Equally, it is possible to over-prepare and a balance needs to be struck. At what point do we prepare realistically for extreme events, given their likelihood and at what expense? This leads us to a critical question for trade-offs (Sect. 14.4.2).

In Norfolk, as widely found around the UK, the coastal system is subject to a strong legacy of past events and coastal management decisions. This legacy is usually given limited consideration, with more attention focussed on future drivers of change such as sea-level rise. Past events, especially the 1953 surge (see Sect. 1.4.5) have heavily influenced how the UK coastline has since been managed. The Tyndall Coastal Simulator demonstrates that future management decisions are as important as climate change and the legacy of future management choices needs to be carefully considered as well.

The 1953 flood was the biggest flood event on the Norfolk coast in terms of flood damage and loss of life in the last 200 years. Figure 8.4 in Chap. 8 shows the aftermath of a breach at Sea Palling. There had been significant protection in North Norfolk prior to 1953, with some defences dating back to the mid nineteenth century. After the event, the main response was to expand and upgrade protection of both the cliffed and the flood-prone coasts and now most of the cliffs are protected (see Fig. 7.2). Figures 14.3a illustrates the sea wall at Horsey before 1953 and Figs. 14.3b and 14.3c show the concrete sea wall, which was built after the 1953 event in the same place.

After 1953, urban development continued both on the cliff-top and in the flood plain, and many of the residents believed that there was a permanent commitment from government to protect them. By limiting erosion, the creation of new beach material was also severely limited. Hence, downdrift beaches in Norfolk declined due to a lack of natural supply and continued coastal engineering intervention is required to keep the risk of breaching into the Norfolk Broads at an acceptable level. The most notable interventions to date have been the construction of nine detached breakwaters at Sea Palling and major nourishment (1.4×10^6 m^3) during the 1990s, followed by repeated renourishment when the beach falls below trigger levels

Fig. 14.3a The retaining sea wall at Horsey in Norfolk, taken on 4th February 1953 soon after the 31st Jan/1st February storm surge (Image courtesy of Norfolk County Council Library and Information Service)

Fig. 14.3b The sea wall at Winterton, near Horsey in Norfolk in 2012 (Sophie Day)

Fig. 14.3c The sea wall and groynes at Horsey in Norfolk in 2012 (Sophie Day)

(see Chap. 8). Spurred by these observations and rising costs, aspirations for the management of this coast have started to change, as exemplified by the current SMP and its proposal for widespread defence abandonment along the cliffed coast. Locally, some defences have already been abandoned and removed at the cliffs at Happisburgh. While this has a benefit of increasing sand supply and lowering protection costs, it also led to rapid erosion, the loss of a number of houses and access ramps to the beach between 1990 and 2010 (for example see Figs. 12.3a–d in Chap. 12), and importantly loss of trust in government by many coastal residents. More recently some of that trust has been re-gained by re-establishing a ramp and purchasing some houses that were most threatened. This work was funded through the DEFRA Pathfinder programme, which is promoting learning to live with a dynamic coast (See Chap. 12).

If the new knowledge explained in the earlier Chapters had been available in 1953, a different decision pathway might have been followed after this extreme event. With hindsight, we might have built fewer defences on the cliff coast, resulting in less expectation of cliff-top defence. This might have been reinforced with land use planning and policy instruments to create buffers on the cliff top (i.e. allowing the natural rates of cliff recession) and the risk associated with erosion would be lower. Along the cliff-top, appropriate land uses such as open space for recreation, or mobile buildings which could be moved with the eroding cliff-top could have been encouraged. The greater erosion would have resulted in larger natural beaches in front of the southern cliffs and the Norfolk Broads, and hence lower flood risk, less need for defences on the flood prone part of the coast and lower management costs. There would be less demand, certainly in frequency for beach nourishment and perhaps by the 1990s we would not have needed the breakwaters at Sea Palling. Hence, 60 years down the line, the Norfolk coast would require less intervention than we see in 2014. Recent chronic erosion at Happisburgh would not have occurred – maybe Happisburgh would not have been protected in the first place. We would have more scope for working with nature to achieve a coast that is sustainable in the long-term in the social, economic and environmental dimensions.

While this alterative pathway is speculative, it provides important lessons on future adaptation in many areas, where proactive rather than reactive responses to extreme events are needed. In addition, the recognition of learning is important and can be part of the process of moving to more proactive responses. The Thames Estuary 2100 project, which is planning London's flood management for the next 100 years and beyond, demonstrates this as discussed in Box 14.1. Defining adaptation pathways and exploring a learning approach is also compatible with the Tyndall Coastal Simulator method and this approach is expected to be widely applied in the future, as discussed further below.

Box 14.1 Thames Estuary 2100 Project and Adaptation Pathways

Like Norfolk, London has a significant legacy from the 1953 event in terms of a sophisticated flood defence system linked to a forecasting and warning system: perhaps the most iconic part of the defence system is the Thames Barrier (Fig. 14.2). Approximately 0.5 million homes and 1.25 million people are estimated to be located in the Thames Estuary floodplain. Although the Estuary is currently well protected by an integrated system of flood defence that provide a standard of protection estimated to be in excess of 1:1000 year, this standard is declining as sea level rises. Similarly, significant lengths of the 300 km of walls and embankment that protect the estuary floodplain are now reaching the end of their design life. At the same time, London is rapidly growing in an eastward direction along the Thames Estuary floodplain towards the coast.

Given this 'perfect storm' of drivers of future flood risk: sea level rise, socio-economic growth and a deteriorating asset base, it was recognised that a long-term plan for the future management of flood risk to London and the Thames Estuary was needed. This has been delivered by the Thames Estuary 2100 (TE2100) project (Environment Agency 2012).

When appraising decisions over the long life time of the new flood defence assets that might have to be built, issues of sea-level rise, storminess and other long-term uncertainties dominate. Importantly, sea-level rise could be rapid and it will continue to rise beyond the life of the new flood assets (Church et al. 2013), so a series of adaptation steps will be needed. Rather than adopt a highly precautionary strategy, it was recognised that a managed adaptive approach under which investment decisions are made in line with observa-

(continued)

Box 14.1 (continued)
tions of key processes such as sea-level rise is appropriate. Hence, a series of adaptation pathways as a function of the rise in sea level, rather than as the passage of time, were developed to guide future decision-making (see Fig. 14.4). It was found that for small rises in sea level, several different options were available following the managed risk pathway, but ultimately for larger rises a new downstream barrier would be needed to protect London. This is an adaptation plan that defines the options that are feasible as sea levels rise and links them together into a set of incremental steps, to be implemented when required. Hence, it prepares for sea-level rise and also avoids the issue of over-adapting. More information can be found in Tarrant and Sayers (2012), Penning-Rowsell et al. (2013b) and Ranger et al. (2013).

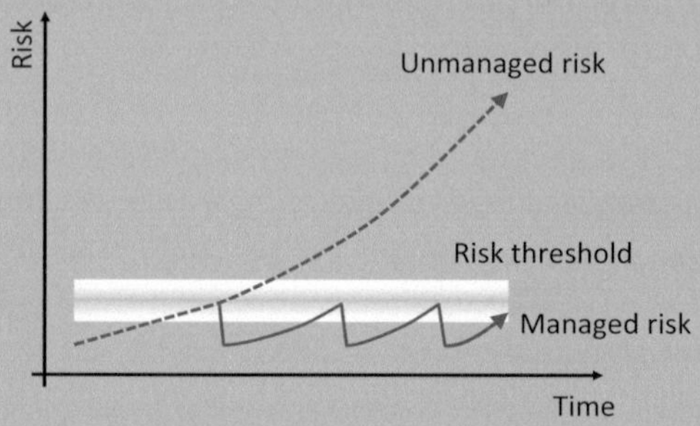

Fig. 14.4 An illustration of different risk pathways. Unmanaged risk reflects no intervention and risk grows to unacceptable levels. Managed risk reflects a number of interventions whose timing is related to the growth in risk and a desire to maintain risk within acceptable bounds. The definition of acceptable risk is a societal decision which may change over long time frames (Adapted from Environment Agency 2012)

14.4.4 Integrated and Adaptive Governance

Adaptation is the only way to deal with the inescapable impacts of climate change (Stern 2007).

Change at the coast is inevitable. To what degree and how this change is manifested is of course subject to a range of drivers and their interaction with management strategies that are pursued. While such uncertainties are inescapable, the Tyndall Coastal Simulator approach gives us improved insights into possible futures, according to how changes and decisions play out. Accordingly, we are in a better position to consider a range of options and policies in the context of long-term change, including the consequences of different decision pathways. In other words, the approach supports us in planning adaptively and with foresight for future change.

The approach has potential to play a much greater role in supporting future adaptation to coastal change. Firstly, by providing the capacity to integrate the many factors that influence and are affected by coastal change and secondly, through integration into decision-making and planning processes.

Engaging appropriate stakeholders in a meaningful debate is an important aspect of IA research. IA must be policy driven, by definition and conducting policy and issue-relevant research, with the potential to be adopted into practice was an important goal of the Tyndall Coastal Simulator research. There were critical links with practitioner and policy stakeholders which aimed to maximise the relevance and potential usefulness of the Simulator (Chaps. 11 and 12). If applied in practice, the Simulator or any IA tool would require continued improvement and development through use, maintenance and on-going investment and technological support.

While the technical aspects of an IA are necessarily expert driven, its conceptual development and outputs can and should be shared with stakeholders of many kinds, informed by different interests and used to support the communication and debate around possible futures and decision making about these. Tools to provide such capacity are much needed. For example, Chap. 10 discusses some of the potential for visualisations of coastal futures (as an output of a broad scale analysis or IA exercise) to be used as a communication tool in this respect.

There are different ways in which adaptation to coastal change can happen. In simple terms, adaptation can be driven either by the anticipation of or reaction to actual coastal change and events. Ideally, adaptation in practice is a mixture of both of these things: an on-going process of learning, reviewing decisions and responding to events as discussed in Sect. 14.3. Integrated approaches to analysing complex coastal problems have great potential to be supported by tools such as the Tyndall Coastal Simulator, but this requires appropriate institutional capacity.

As explained in more detail in Chap. 12, institutional arrangements and planning processes are not always set up in ways that enable adaptive planning, or which are capable of implementing adaptive responses when events happen. This may be, for example, because of a lack of in-built review mechanisms, inappropriate processes and short term financial arrangements. Long-term thinking is not necessarily mainstreamed, and hence such decisions can introduce unanticipated future vulnerability (e.g. locating new housing in areas that may become significantly more flood prone in future). Integrated planning across scales involves finding a balance between local issues and national strategic priorities.

14.5 An Agenda for Future Research and Application

This chapter has reviewed the earlier chapters contained within this book and has considered the lessons learnt. In this section, we identify key priorities for future research and challenges for the application of coastal integrated assessment. These reflections are useful for researchers and funders interested in developing future coastal simulation capacity in the UK and elsewhere and are grouped under three broad themes: (1) technical issues, (2) social issues and (3) application issues. The technical issues tend to be incremental in nature, while the social and application issues raise more fundamental questions and challenges.

14.5.1 Technical Issues

A number of technical improvements are readily possible, both in the components used in the assessment and in the overall assessment. The suggestions below are largely relevant to the Norfolk case study, but illustrate readily transferable lessons.

- *Spatial resolution and extent*: the approach could be applied at a finer scale, or equally at the same scale at a larger spatial extent, such as the East Coast of England, allowing easier linkage with the regional climate models (Chap. 2) and consideration of regional suspended sediment dynamics.
- *Wind, wave and surge climates*: more detailed information about how these might change in the future is needed, suggesting the need for many more simulations than the five ensembles considered in Chap. 2 and higher resolution to better resolve extremes.
- *Geomorphic coastal models*: there is a need to move from qualitative to quantitative geomorphic models (Chap. 5) building on the experience of reduced complexity models such as SCAPE (Chap. 7) and other less computationally intensive simulation models such as the estuary model ASMITA (Whitehouse et al. 2009; Nicholls et al. 2012). This should include consideration of the shallow seabed, especially sandbanks.
- *Integrated habitat assessment*: there is a need for habitats to be assessed with the flood and erosion (and other) components and also a broader consideration of changing ecosystem services (UK National Ecosystem Assessment 2011). For example, in Norfolk the cliff and beach habitats could be assessed together with the erosion and flood risk. This would increase the complexity of the problem, but provide a richer view of the possible futures.
- *Breach analysis*: the flood model components are fairly well developed, but more understanding of defence failure in general and specifically breach processes and how the failure of defences interact with natural beaches would improve analysis of inundation and flood risk.

14.5.2 Social, Economic and Political Issues

A host of different groups and organisations have an interest and role in managing the coast. Successful management of flood and coastal erosion risk requires the active engagement of local communities in developing, supporting and delivering coastal risk management. With a better understanding of physical processes and the resulting extreme events and long-term change that coastal societies will have to face, a wide range of social, economic and political aspects should also be considered:

- *Economic issues*: these could include the measures of risk (expected annual damages, as used here – Dawson et al. (2009) and Chap. 9) and also expand to the direct costs of different response options, as well as more complex economic effects and the indirect impacts around the issues of regional economic development at one level, planned retreat options and (property or land) blight at a more local level. Comprehensive analysis is to be encouraged and should include effects that are intangible and hard to measure (e.g. political costs, and economic viability), as well as appropriate methods (e.g. multi-criteria analysis).
- *Reconcile local and national priorities*: how do you analyse big-picture strategic issues versus local priorities where there may be a clash of goals and changing balance of power over time? For example, in the UK there is a recent focus on 'localism' and 'local' solutions that may be contrary to strategic needs and system perspectives as considered in this book.
- *Incorporate a wider range of impacts*: such as health effects of flooding. More substantial investment could develop more sophisticated impact analysis, including the consideration of possible social and economic feedbacks. For example, could more advanced agent-based models (Chap. 4) meaningfully represent processes such as coastal abandonment? What other methods might be suited to downscaling socio-economic change?
- *Managing transitions*: the Norfolk analysis has shown that implementing a successful transition in management has both technical and social dimensions and the associated analysis needs to reflect this. The Tyndall Coastal Simulator approach provides a way of analysing these issues and might effectively be combined with the concept of adaptation pathways to analyse and provide more effective routes for effective transitions towards more effective adaptation approaches.

14.5.3 Application Issues

Understanding whole-system behaviour presents us with a set of challenges as listed above, and linking this understanding to coastal management and policy and stakeholders in general raises further challenges. While the results of IA should by definition be policy relevant, policy processes do not necessarily have the capacity to absorb these results. From our experience, this understanding has to develop between the academics and stakeholders over time, as we have observed earlier in the chapter. Some factors that will support our ability to better develop whole-system models and their application to decision-making at a range of scales are as follows:

- *Linking to monitoring and management programmes*: Linking to national monitoring programmes could be a useful way to advance these methods and develop systems to assimilate and assess data as it is gathered. Present planning processes are rather poorly connected to these wider activities making adaptation difficult to achieve locally. Organisations such as the Channel Coastal Observatory (http://www.channelcoast.org/) could be important partners here.

- *Building ongoing relationships*: with coastal operating authorities is essential for the translation of IA results into policy. This would make IA application routine and promote integration into management systems. This includes training and education of both the academics and policymakers, and as IA becomes more routine, promotes the movement of these methods from academia into consultancy and application.
- *Relate outputs to existing procedures and standards*: The Tyndall Coastal Simulator methods readily fit into national policy processes such as the DEFRA Flood Defence Grant in Aid (FDGiA) calculation (Environment Agency 2014) and coastal datasets are consistent with data available across England and Wales. This provides institutional opportunities to set up suites of IAs for use on coastal flood and erosion risk management strategies. Funding of flood management in England and Wales is now moving to a partnership funding policy where the funds may come from a consortium of stakeholders. The outputs of an IA assessment explore the flow of benefits from such investment and inform existing and new partnerships about the costs and benefits of their investments. While it will always be context specific, similar national approaches could be considered elsewhere.
- *Exploitation of rapidly-evolving informatics*: to help change the way in which society (including policy makers) engages with coastal science (and science in general) – visualisation techniques and web-based delivery are examples of two exciting developments, and we anticipate enormous innovation in this area in the near future.
- *Learning from adaptation experience*: especially dealing with coastal change and the transition from defended to dynamic coasts. These are challenging issues which the book has addressed to some degree and the DEFRA Pathfinder projects are addressing across England. This research is important and needs to be continued. We understand how to defend a coast much better than how we can live with coastal change. This might be considered action research on coastal adaptation and the lessons captured and incorporated into coastal policy formulation.
- *Nurture and reward inter-disciplinary skills*: Cross-disciplinary research programmes such as the Tyndall Coastal Simulator cultivate the development of IA approaches, but supporting education programmes for early career researchers and practitioners would facilitate future development and the implementation of IA in practise. Promoting innovative funding is an important aspect.

14.6 Conclusions

The scale of the challenge facing the world's coast through the twenty-first century is immense (Chap. 1). Multiple drivers at multiple scales are in action, especially where there are large concentrations of people. Climate change is a growing factor and issues like sea-level rise will operate almost universally. Hence a range of existing problems will get worse and further new problems will emerge. Given the multiple drivers in operation, linear analysis will not diagnose many of the problems or their solution. As this book as argued throughout, there is a pressing need for the IA methods outlined here to be applied widely around the world's coast.

Generic concepts from this book that have relevance for many situations and locations have been identified. While these lessons are transferable, they must be contextualised on a case-by-case basis because coasts are heterogeneous. The nature of the physical system as well as the socio-cultural contexts and norms of each location mean that details will be distinct. Additionally, change will happen both gradually and incrementally – as well as suddenly during and in the aftermath of extreme events.

This chapter has reviewed detailed experience and generalized it to facilitate its wider application. We have shown that IA is a useful process for coupled systems, where unexpected results and outcomes may emerge that can challenge current management approaches. In coastal areas, this could be longshore transport of beach sediment and its influence on erosion and flood risk as in Norfolk. However, as we have illustrated other interactions can be significant, including those between socio-economic or management processes. The importance of these factors can only be understood by further investigation using the types of methods outlined in this book.

A set of principles has been developed for the detailed IA of coastal systems. The challenge of taking these ideas forward and further developing them should not be underestimated. Significant progress has been made by the Tyndall Coastal Simulator, although a number of research and application challenges remain. The challenges span a diverse range of disciplines and include major gaps in our knowledge about the functioning and services provided by the natural environment, key governing physical process and engineering and social sciences, including the link from science to management.

Despite these remaining challenges, we conclude that the Tyndall Coastal Simulator has demonstrated that IA is both technically feasible and fundamental to support sustainable coastal management policy. The rewards and benefits are potentially enormous as just halving the current rate of losses in coastal areas would have immense societal benefits (cf. Kron 2013). In areas with coastal problems, coalitions of stakeholders, including policymakers and academics across disciplines, need to start this process, including finding the resources to build these tools. This will not be easy, and there will inevitably be multiple routes to success. But these pioneers will be laying the foundation of an important dimension of future coastal management.

References

Anthoff, D., Nicholls, R. J., & Tol, R. S. J. (2010). The economic impact of substantial sea-level rise. *Mitigation and Adaptation Strategies for Global Change, 15*(4), 321–335. doi:10.1007/s11027-010-9220-7.

Arnell, N. W., Lowe, J. A., Brown, S., Gosling, S. N., Gottschalk, P., Hinkel, J., Lloyd-Hughes, B., Nicholls, R. J., Osborn, T. J., Osborne, T. M., Rose, G. A., Smith, P., & Warren, R. F. (2013). A global assessment of the effects of climate policy on the impacts of climate change. *Nature Climate Change, 3*(5), 512–519. doi:10.1038/nclimate1793.

Bricker, S. B., Ferreira, J. G., & Simas, T. (2003). An integrated methodology for assessment of estuarine trophic status. *Ecological Modelling, 169*, 39–60.

Burgess, K., Orford, J., Dyer, K., Townend, I., & Balson, P. (2002, July 1). FUTURECOAST – The integration of knowledge to assess future coastal evolution at a national scale. Paper presented at 28th International Conference on Coastal Engineering, Cardiff, pp. 3221–3233.

Church, J. A., Clark, P. U., Cazenave, A., Gregory, J. M., Jevrejeva, S., Levermann, A., Merrifield, M. A., Milne, G. A., Nerem, R. S., Nunn, P. D., Payne, A. J., Pfeffer, W. T., Stammer, D., & Unnikrishnan, A. S. (2013). Sea Level Change. In T. F. Stocker, D. Qin, G.-K. Plattner, M. Tignor, S. K. Allen, J. Boschung, A. Nauels, Y. Xia, V. Bex, & P. M. Midgley (Eds.), *Climate change 2013: The physical science basis. Contribution of Working Group I to the Fifth assessment report of the Intergovernmental Panel on Climate Change*. Cambridge/New York: Cambridge University Press.

Dawson, R. J. (2011). Potential pitfalls on the transition to more sustainable cities … and how they might be avoided. *Carbon Management, 2*(2), 175–188. doi:10.4155/cmt.11.8.

Dawson, R. J., Dickson, M., Nicholls, R. J., Hall, J., Walkden, M. J. A., Stansby, P. K., Mokrech, M., Richards, J., Zhou, J., Milligan, J., Jordan, A., Pearson, S., Rees, J., Bates, P. D., Koukoulas, S., & Watkinson, A. (2009). Integrated analysis of risks of coastal flooding and cliff erosion under scenarios of long term change. *Climatic Change, 95*, 249–288.

de Kok, J. L., Engelen, G., White, R., & Wind, H. G. (2001). Modeling land-use change in a decision-support system for coastal-zone management. *Environmental Modeling & Assessment, 6*(2), 123–132.

Dessai, S., Hulme, M., & O'Brien, K. (2007). Editorial: On uncertainty and climate change. *Global Environmental Change, 17*, 1–3. doi:10.1016/j.gloenvcha.2006.12.001.

Environment Agency. (2012). *Thames Estuary 2100. Managing flood risk through London and the Thames estuary*. Environment Agency, Thames Barrier, London, 226 pp. Available at http://webarchive.nationalarchives.gov.uk/20140328084622/http://cdn.environment-agency.gov.uk/lit7540_43858f.pdf

Environment Agency. (2014). *Calculate grant in aid funding for flood and coastal erosion risk management projects: Guidance for risk management authorities*. Version 1 updated February 2014, Environment Agency, Bristol. Available at: https://www.gov.uk/government/uploads/system/uploads/attachment_data/file/297377/LIT_9142_dd8bbe.pdf. Accessed 8 Aug 2014.

Gregersen, J. B., Gijsbers, P. J. A., & Westen, S. J. P. (2007). OpenMI: Open modelling interface. *Journal of Hydroinformatics, 9*(3), 175–191.

Hall, J. W., Dawson, R. J., Barr, S. L., Batty, M., Bristow, A. L., Carney, S., Dagoumas, A., Ford, A., Tight, M. R., Walsh, C. L., Watters, H., & Zanni, A. M. (2010). City-scale integrated assessment of climate impacts, adaptation and mitigation. In R. K. Bose (Ed.), *Energy efficient cities: Assessment tools and benchmarking practices* (pp. 43–64). World Bank: Washington, DC, USA.

Hall, J. W., Henriques, J. J., Hickford, A., & Nicholls, R. J. (Eds.). (2012). *A fast track analysis of strategies for infrastructure provision in Great Britain: Technical report*. Oxford: Environmental Change Institute, University of Oxford.

Hall, J. W., Henriques, J. J., Hickford, A., & Nicholls, R. J. (2013). Systems-of-systems analysis of national infrastructure. *Proceedings of ICE – Engineering Sustainability, 166*(5), 249–257.

Hallegatte, S., Green, C., Nicholls, R. J., & Corfee-Morlot, J. (2013). Future flood losses in major coastal cities. *Nature Climate Change, 3*, 802–806. doi:10.1038/nclimate1979.

Harrison, P. A., Holman, I. P., Cojocaru, G., Kok, K., Kontogianni, A., Metzger, M., & Gramberger, M. (2013). Combining qualitative and quantitative understanding for exploring cross-sectoral climate change impacts, adaptation and vulnerability in Europe. *Regional Environmental Change, 13*, 761–780.

Kron, W. (2013). Coasts: the high-risk areas of the world. *Natural Hazards, 66*(3), 1363–1382.

Lazarus, E. D., McNamara, D. E., Smith, M. D., Gopalakrishnan, S., & Murray, A. B. (2011). Emergent behavior in a coupled economic and coastline model for beach nourishment. *Nonlinear Processes in Geophysic, 18*, 989–999.

Lonsdale, K. G., Downing, T. E., Nicholls, R. J., Parker, D., Vafeidis, A. T., Dawson, R. J., & Hall, J. (2008). Plausible responses to the threat of rapid sea-level rise in the Thames Estuary. *Climatic Change, 91*(1–2), 145–169. doi:10.1007/s10584-008-9483-0.

Murray, A. B., Gopalakrishnan, S., McNamara, D. E., & Smith, M. D. (2011). Progress in coupling models of human and coastal landscape change. *Computers and Geosciences, Computers and Geosciences, 53*, 30–38.

Nicholls, R. J., Bradbury, A., Burningham, H., Dix, J., Ellis, M., French, J., Hall, J. W., Karunarathna, H. U., Lawn, J., Pan, S., Reeve, D., Rogers, B., Souza, A., Stansby, P. K., Sutherland, J., Tarrant, O., Walkden, M., & Whitehouse, R. (2012, July 1–6). *iCOASST – Integrating Coastal Sediment Systems*. Paper presented at the International Conference on Coastal Engineering, Santander, Spain. http://www.icce2012.com/

Nicholls, R. J., Townend, I. H., Bradbury, A., Ramsbottom, D., & Day, S. (2013). Planning for long-term coastal change: experiences from England and Wales. *Ocean Engineering, 71*, 3–16. doi:10.1016/j.oceaneng.2013.01.025.

Penning-Rowsell, E. C., Priest, S., Parker, D., Morris, J., Tunstall, S., Viavattene, C., Chatterton, J. B., & Owen, D. (2013a). *Flood and coastal erosion risk management: A manual for economic appraisal*. London: Routledge.

Penning-Rowsell, E. C., Haigh, N., Lavery, S., & McFadden, L. (2013b). A threatened world city: The benefits of protecting London from the sea. *Natural Hazards, 66*, 1383–1404. doi:10.1007/s11069-011-0075-3.

PlaNYC. (2013). *A Stronger, More Resilient New York*. New York: City of New York.

Pontee, N. I., & Morris, K. (2011). CoastRanger MS: A tool for improving public engagement. *Journal of Coastal Research, 27*(1), 18–25.

Ranger, N., Reeder, T., & Lowe, J. A. (2013). Addressing 'deep' uncertainty over long-term climate in major infrastructure projects: Four innovations of the Thames Estuary 2100 Project. *European Journal on Decision Processes, 1*(3–4), 233–262.

Rijcken, T., Stijnen, J., & Slootjes, N. (2012). SimDelta – Inquiry into an internet-based interactive model for water infrastructure development in The Netherlands. *Water, 4*(2), 295–320.

Rotmans, J., van Asselta, M., & Vellinga, P. (2000). An integrated planning tool for sustainable cities. *Environmental Impact Assessment Review, 20*, 265–276.

Schöpp, W., Amann, M., Cofala, J., Heyes, C., & Klimont, Z. (1998). Integrated assessment of European air pollution emission control strategies. *Environmental Modelling & Software, 14*, 1–9.

Stern, N. (2007). *The economics of climate change: The stern review*. Cambridge: Cambridge University Press. 712 pp.

Tarrant, O., & Sayers, P. (2012). Managing flood risk in the Thames Estuary – The development of a long term robust and flexible strategy. In P. Sayers (Ed.), *Flood risk management – The design and management of supporting infrastructure*. London: Thomas Telford.

Turner, R. K., Lorenzoni, I., Beaumont, N., Bateman, I. J., Langford, I. H., & McDonald, A. L. (1998). Coastal management for sustainable development: Analysing environmental and socio-economic changes on the UK coast. *Geographical Journal, 164*, 269–281.

UK National Ecosystem Assessment. (2011). *The UK national ecosystem assessment: Technical report*. Cambridge: UNEP-WCMC.

Walsh, C. L., Dawson, R. J., Hall, J. W., Barr, S. L., Batty, M., Bristow, A. L., Carney, S., Dagoumas, A., Ford, A., Tight, M. R., Watters, H., & Zanni A. M. (2011). Assessment of climate change mitigation and adaptation in cities, *Proceedings of the ICE: Urban Design and Planning, Special issue on Urban Development and Sustainability, 164*(DP2), 75–84. doi:10.1680/udap.2011.164.2.75.

Walsh, C. L., Roberts, D., Dawson, R. J., Hall, J. W., Nickson, A., & Hounsome, R. (2013). Experiences of integrated assessment modelling in London and Durban. *Environment and Urbanization, 25*(2), 257–376.

Warren, R., de la Nava Santos, S., Arnell, N. W., Bane, M., Barker, T., Barton, C., Ford, R., Füssel, H.-M., Hankin, R. K. S., Klein, R., Linstead, C., Kohler, J., Mitchell, T. D., Osborn, T. J., Pan, H., Raper, S. C. B., Riley, G., Schellnhüber, H. J., Winne, S., & Anderson, D. (2008). Development and illustrative outputs of the Community Integrated Assessment System (CIAS), a multi-institutional modular integrated assessment approach for modelling climate change. *Environmental Modelling & Software, 23*, 592–610. doi:10.1016/j.envsoft.2007.09.002.

Whitehouse, R., Balson, P., Beech, N., Brampton, A., Blott, S., Burningham, H., Cooper, N., French, J., Guthrie, G., Hanson, S., Nicholls, R., Pearson, S., Pye, K., Rossington, K., Sutherland, J., & Walkden, M. (2009). *Characterisation and prediction of large-scale, long-term change of coastal geomorphological behaviours: Science report* (SC060074/SR1). Bristol: Environment Agency.

Wilbanks, T. J. (2003). Integrating climate change and sustainable development in a place-based context. *Climate Policy, 3*(Suppl 1), S147–S154.

Willows, R. I., & Connell, R. K. (Eds.). (2003). *Climate adaptation: Risk, uncertainty and decision-making* (UKCIP technical report). Oxford: UKCIP. http://www.ukcip.org.uk/wordpress/wp-content/PDFs/Risk.pdf. Last accessed 15 Oct 2012.

Appendix A: Scenarios Framework

The Scenarios Framework Used in the Tyndall Coastal Simulator

This appendix summarises the scenarios framework used in the Tyndall Coastal Simulator including a quantification of the scenarios used in Phase 1 (2001–2006) and Phase 2 (2006–2009) activities. It addresses climate change, socio-economic change and management responses.

Introduction

Scenarios, or plausible futures, are a recognised technique for investigating long-term change where there are many complex and interacting variables and where the future is very uncertain (e.g. Carter et al. 2007). The extreme complexity of evolving coastal erosion and flooding systems and other uncertainties surrounding climate and socio-economic change make scenarios a valuable tool for the development of long-term shoreline management policies (Nicholls et al. 2008, 2012).

We have used pre-existing frameworks to construct a number of alternative coastal scenarios and assessed the scale and nature of the resulting erosion and flood risks. In so doing, it is possible to gain a broad appreciation of the scale of future risks that may need to be addressed and the extent to which policies can be developed to manage them.

Two different types of scenario have been combined in our analysis:

1. Climate change projections driven by global greenhouse gas emissions scenarios and their subsequent impact on environmental loadings such as sea-level rise and storm characteristics
2. Socio-economic scenarios which describe changes to population, assets and land use as well as the context in which shoreline management policy and practice might be enacted

R.J. Nicholls et al. (eds.), *Broad Scale Coastal Simulation*, Advances in Global Change Research 49, DOI 10.1007/978-94-007-5258-0

This combination of scenarios is important: climate change will tend to affect the nature of coastal hazards, whilst socio-economic factors will largely determine the severity of resulting impacts. Both are needed to provide a complete picture of future flood and erosion risks. There is no exact causal link between the climate and socio-economic scenarios considered, in part because they were developed for different purposes but also because the climate scenarios are based upon global greenhouse gas emissions, whilst the socio-economic scenarios consider local or national issues. However, some combinations of climate and socio-economic scenarios exhibit trends that are more coherent than others – for example, the highest national growth in UK population and global growth in emissions are often considered to be more plausible than a situation where the UK bucks global trends in greenhouse gas emissions. However, to understand the sensitivity and relative contribution of socio-economic and climate factors, we analysed all scenario combinations. The climate change scenarios focus on key drivers of global climate change such as energy consumption and greenhouse gas emissions, whilst the socio-economic trajectories were developed to consider UK social, political and economic trajectories. Whilst global socio-economic scenarios have been developed (e.g. Carter et al. 2007), there are so many local factors to consider that are not ideal for the scale of analysis in this book. Furthermore, a low-carbon UK does not necessarily imply a decarbonised planet, or vice versa, so testing a wide range of combinations of these scenarios provides the most representative range of possible future flood and erosion risks.

Many of the results reported do not account for socio-economic change and therefore consider the impact of a changing climate on today's development and vulnerability. This is useful to understand the relative contribution to changing erosion and flood risks of climate and socio-economic drivers.

A coastal manager has a range of options by which they can respond to these climate and socio-economic pressures. The range of alternative management policies are also considered across the range of possibilities.

Climate Change Scenarios

The UKCIP02 (Hulme et al. 2002) and, subsequently, UKCP09 (Lowe et al. 2009) national climate simulations were used to drive the Tyndall Coastal Simulator in Phase 1 and Phase 2, respectively. These both follow the SRES (Special Report on Emissions Scenarios) scenario framework shown in Fig. A.1 (Nakicenovic et al. 2000) typically adopted in international climate change studies. The relationship between SRES, UKCIP02 and UKCP09 scenarios is shown in Table A.1. The process of downscaling from these scenarios to extract climate variables over the Coastal Simulator's model domain is explained in Chap. 2 of this book. Whilst we have focused on the SRES framework, our analysis is compatible with other scenario frameworks such as the emerging RCP/SSP (Reference Concentration Pathway/Shared Socio-economic Pathway) framework being developed for the IPCC AR5 (e.g. Moss et al. 2010; Kriegler et al. 2010).

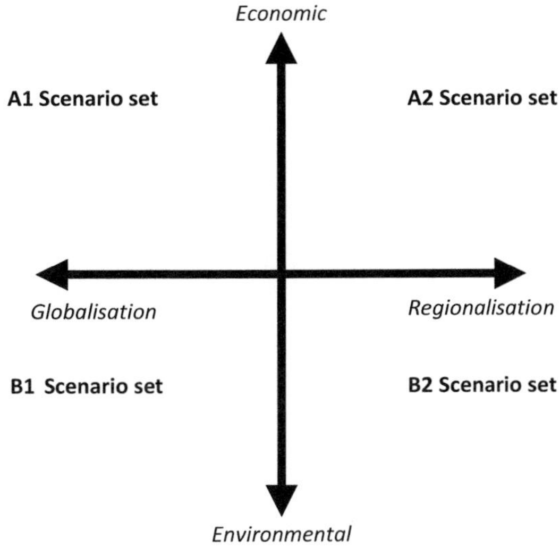

Fig. A.1 Axes of the four SRES storylines and the resulting sets of scenarios

The four SRES scenario sets are distinguished by two global trends: one varies between strong economic values and strong environmental values, the other trend between increasing globalisation and increasing regionalisation. The scenario sets are summarised as follows (Nakicenovic et al. 2000):

- A1 scenario set – a future world of very rapid economic growth, declining global population after the middle of the century and rapid introduction of new and more efficient technologies. Variants include A1FI (FI–fuel intensive) with high emissions and A1B (B–balanced) with medium emissions.
- A2 scenario set – a heterogeneous world with continuously increasing global population and regionally oriented economic growth that is more fragmented and slower than the other sets of scenarios.
- B1 scenario set – the same global population as the A1 storyline, but with rapid changes in economic structures towards a service and information economy, with reductions in material intensity and the introduction of clean and resource-efficient technologies.
- B2 scenario set – a world in which the emphasis is on local solutions to sustainability issues, with continuously increasing population, although lower than A2, and intermediate economic development.

In all 40 scenarios are defined by SRES, and five of the scenarios typically used and considered as representative 'markers' form the basis of UKCIP02 and UKCP09 scenarios (Table A.2). A summary of all the climate scenarios considered over both phases of the work is shown in Table A.3. Further detail is provided in the following sections.

Table A.1 Summary of the labels used throughout the book to refer to climate change scenarios and management strategies.

		Phase 1 scenarios (based on UKCIP02)									Phase 2 scenarios (based on UKCP09)			
	Sea-level rise	Low				Med	High				Low	Med	High	H++
	Offshore wave (H_s)	Low	High	High+	High–	Med	Low	High	High+	High–				
Management strategy (% cliff coast protected)	M1 (100 %)	1	2	3	4	5	6	7	8	9	101	102	103	104
	M2 (71 %)	10	11	12	13	14	15	16	17	18	105	106	107	108
	M3 (34 %)	19	20	21	22	23	24	25	26	27	–	–	–	–
	M4 (34 %)	–	–	–	–	–	–	–	–	–	109	110	111	112
	M5 (34 %, no nourishment)	–	–	–	–	–	–	–	–	–	113	114	115	116
	M6 (16 %)	28	29	30	31	32	33	34	35	36	–	–	–	–
	M7 (0 %)	37	38	39	40	41	42	43	44	45	117	118	119	120

When reporting results that also consider socio-economic change, the same numbering scheme is used but prefixed by WM, GS, LS and NE to correspond to the world markets, global sustainability, local stewardship and national enterprise socio-economic scenarios, respectively. Full descriptions of the climate, socio-economic and management scenarios are provided in this Appendix

Table A.2 Relationship between the SRES, UKCIP02 and UKCP09 scenarios

SRES	UKCIP02	UKCP09
A1FI	High	High
A2	Medium high	–
A1B	–	Medium
B2	Medium low	–
B1	Low	Low

Table A.3 Climate scenarios considered during Phases 1 and 2 of the Tyndall Coastal Simulator study

Phase 1	
Relative sea-level rise	
Low	+0.2 m by 2100 (no change in rate of SLR)
Medium	+0.45 m by 2100
High	+1.2 m by 2100
Offshore wave conditions	
Low	No change
Medium	7 % increase in winter wave height by 2100
High	10 % increase in winter wave height by 2100
High+	High, plus clockwise rotation of wave rose 10°
High *surge conditions*	High, plus anticlockwise rotation of wave rose 10° No change
Phase 2	
Relative sea-level rise	5th–95th percentiles
Low	0.223–0.614
Medium	0.239–0.751
High	0.264–0.913 m
H++	0.925–2.1 m
Offshore wave conditions	No change
Surge conditions	No change

Phase 1 Climate Scenarios

Three representative relative sea-level rise scenarios were selected to encompass the range of uncertainty in global, regional and local sea levels. The 'low' scenario of relative SLR represented little or no anthropogenic influence, i.e. a continuation of the recent historic rate of 1.5 mm year^{-1} (see also Miller and Douglas 2004). The 'medium' scenario follows the UKCIP02 medium-high scenario and results in a twenty-first century increase of 0.45 m by 2100. The 'high' scenario is based on the IPCC high limit plus an additional regional (North Atlantic scale) sensitivity of 50 % (following Hulme et al. 2002) to allow for spatial variability in thermal expansion (see Gregory et al. 2001). This scenario results in an increase in sea level of 1.2 m by 2100 over the twenty-first century. All three SLR scenarios used include a regional subsidence rate based on long-term geological observations of 0.7 mm $year^{-1}$ (Shennan and Horton 2002).

Long-term predictions of changes in wind, and hence wave, speed and direction from climate models remain uncertain – although as shown in Chap. 2, our capability in this area evolved alongside the research programme. However, climate model predictions are supported by observations in the North Sea and Atlantic Ocean, which suggested that extreme wave heights may be increasing (Bouws et al. 1996; Kushnir et al. 1997). Given the concern about increasing wave heights including from coastal practitioners and the lack of credible scenarios, the *sensitivity* of erosion and flood risk to a range of possible future conditions rather than downscaling climate model outputs directly was explored. Wave heights were unchanged in the 'low' scenario, whereas in the 'medium' and 'high' scenarios, offshore winter wave heights were increased up to a maximum of 7 % and 10 %, respectively, by 2100. These increases were applied linearly over the twenty-first century. In addition, the 'high' scenarios included +10° (clockwise) and –10° (anticlockwise) rotations of the offshore wave rose. Potential changes to wave period were not considered.

In Phase 1 of the research programme, it was not feasible to test all the combinations of sea-level rise, wave height and wave direction scenarios as well as testing all the management options. Therefore, there was a deliberate focus on the low and the high scenarios of relative sea-level rise, to provide potential boundary limits on the future outcomes. Table A.1 summarises the combinations of scenarios and management options that were tested in the coupled analysis.

Phase 2 Climate Scenarios

As described in Chap. 2, climate scenarios that were more detailed, and included probabilistic elements, became available during Phase 2 of the Tyndall programme (these were provided by UKCP09, see Lowe et al. 2009). This coincided with the availability of enhanced computing capacity, which allowed more extensive exploration of their consequences and the associated uncertainties. This included detailed analysis of the influence of climate ensemble members on marine climate. This analysis, and the Phase 1 risk analysis (see Chap. 9), showed negligible future (climate-driven) change in surge magnitude and wave height of the Norfolk coast. Consequently in the second phase of analysis, these elements of climate were assumed to be stationary (i.e. to be statistically constant and neither grow nor decay over time). For the erosion and flood impacts analysis, 250 realisations of each of four sea-level rise scenarios were explored. Figure A.2 shows the 5th and 95th percentiles of the low, medium and high scenarios (which correspond to those in Table A.2) and a high-end 'H++' scenario (Lowe et al. 2009; Nicholls et al. 2011). This H++ scenario was interpreted from the IPCC Fourth Assessment Report and subsequent literature that responded to a lack of scientific understanding on the behaviour of ice sheet failure. The lower limit of the H++ scenario was based on the

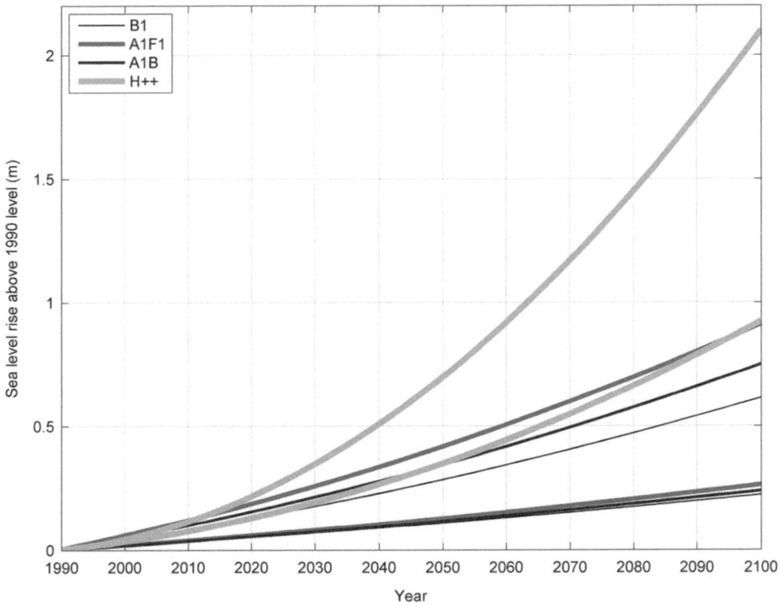

Fig. A.2 Projected sea-level rise around North Norfolk (Lowe et al. 2009)

maximum global mean sea-level rise given by the IPCC Fourth Assessment Report, whilst the upper limit was inferred from several lines of evidence, including indirect observations of sea-level rise during the last interglacial. This upper limit is believed to be highly unlikely and is intended to inform certain types of contingency planning.

All sea-level rise scenarios are assumed to be normally distributed about the central value (which is at the midpoint of the 5th and 95th percentiles) and the H++ trend is treated similarly to the other three projections.

Socio-economic Change

To account for potential changes in socio-economic conditions, we use the ForeSight programme[1] scenarios (DTI 2002) which have been applied in a range of future analyses such as the Foresight Future Flooding Study (Thorne et al. 2007). The ForeSight scenarios were developed to explore alternative directions in which social, political, economic and technological changes may evolve over coming

[1]The UK government launched the ForeSight programme in 1994 to prepare for the future by bringing together representatives from business, government, academia and others to identify potential future threats and opportunities. It is currently based in BIS within the UK government.

Fig. A.3 The ForeSight scenario axis and the four scenarios used in this analysis (SPRU 1999; DTI 2002)

decades in the UK. Importantly, they describe a whole range of drivers and issues, not just those related to climate change, and these feed through the entire integrated assessment (cf. Nicholls et al. 2008, 2012).

Typically the four ForeSight scenarios are presented along two axes which define fundamental different ways in which society could evolve (Fig. A.3). Along the vertical axis is the system of governance, ranging from autonomy where power remains at the national level to interdependence where power increasingly moves to other institutions, e.g. up to the EU or down to regional government. Along the horizontal axis are social values, ranging from individualistic values to more community-oriented values.

The national scenarios are summarised in Table A.4, and their interpretation for East Anglia is summarised in Table A.5. This 'qualitative downscaling' to consider the local implications of the ForeSight scenarios was a precursor to enumerating key variables for use in the quantitative analysis of land use in Chap. 4. This subsequently enabled the estimation of the change in the number of properties in the floodplain (Table A.6). Contrary to the climate scenarios, the same socio-economic scenarios were used throughout both phases of the research programme. However, as described in Chap. 4, a more sophisticated agent-based modelling approach was used to simulate land use in Phase 2 of the programme, so the details of these scenarios changed from Phase 1 to Phase 2.

Table A.4 Summary of the ForeSight socio-economic scenarios for the UK (DTI 2002)

	World markets (WM)	Global sustainability (GS)	Local stewardship (LS)	National enterprise (NE)
Description	*People aspire to personal independence, material wealth and mobility to the exclusion of wider social goals. Integrated global markets are seen as the best way to deliver this. Internationally coordinated policy sets framework conditions for the efficient functioning of markets. Wherever possible, the provision of goods and services is privatised, under the principle of minimal government. Rights of individuals to personal freedoms are enshrined in law*	*People aspire to high levels of welfare within communities with shared values, more equally distributed opportunities and a sound environment. These objectives are thought to be best achieved through active public policy and international cooperation within the EU and at the global level. Social objectives are met through public provision, increasingly at an international level. Markets are regulated to encourage competition amongst national players. Personal and social behaviour is shaped by commonly held beliefs and customs*	*People aspire to sustainable levels of welfare in federal and networked communities. Markets are subject to social regulation to ensure more equally distributed opportunities and a high quality local environment. Active public policy aims to promote economic activities that are small scale and regional in scope and acts to constrain large-scale markets and technologies. Local communities are strengthened to ensure participative and transparent governance*	*People aspire to personal independence and material wealth within a nationally based cultural identity. Liberalised markets together with a commitment to build capabilities and resources to secure a high degree of national self-reliance and security are believed to best deliver these goals. Political and cultural institutions are strengthened to buttress national autonomy*
Social values	Internationalist, libertarian	Internationalist, communitarian	Localist, co-operative	Nationalist, individualist
Governance	Weak, dispersed, consultative	Strong, coordinated, consultative	Strong, local, participative	Weak, national, closed
Role of policy	Minimal, enabling markets	Corporatist, political, social and environmental goals	Interventionist, social and environmental	State-centred market regulation to protect key sectors
Economic development	High growth, high innovation, capital productivity	Medium-high growth, high innovation, resource productivity	Low growth, low innovation, modular and sustainable	Medium-low growth, low innovation, maintenance economy
Structural change	Rapid, towards services	Fast, towards services	Moderate, towards regional systems	More stable economic structure
Fast-growing sectors	Health and leisure, media and information, financial services, biotechnology, nanotechnology	Education and training, Large systems engineering, new and renewable energy, information services	Small-scale manufacturing, food and organic farming, local services	Private health and education, domestic and personal services, tourism, retailing, defence
Declining sectors	Manufacturing, agriculture	Fossil fuel energy, traditional manufacturing	Retailing, tourism, financial services	Public services, civil engineering
Income	High	Medium high	Low	Medium low
Equity	Strong decline	Improvement	Strong improvement	Decline

Table A.5 Summary of the four storylines and how they are interpreted for East Anglia, including detailed consideration of demography, household location preferences and planning strategies

Storyline for East Anglia	Demography and migration	Location preferences	Planning strategy
World market (WM)			
The integrated trading systems lead to more development of the main urban centres in East Anglia, which are still shadowed by London. They attract more workforces, at the expense of more remote locations. The increasing economy and wealth boosts private consumption and materialistic personal well-being, testified by larger new houses and more second homes. There will be a continued pressure to build more houses, and these are attracted to coastal and floodplain locations. Consequently, significant developments would occur in all areas including the floodplain and cliff areas	With medical advances, life expectancy rises significantly. Today's decrease of the birth rate accelerates. Inflow migrations are happening for two distinct reasons. On the one hand, the developing urban centres are attractive to the workforce. On the other hand, an increasing number of London workers seek cheaper homes farther from the city, increasing the resident commuters group. There is also an increased individual desire for more permanent leisure location at the coast for both retirees and second-home owners as their purchasing power is increasing	Households have more interest for locations with (very) good connection to the rest of the region and the country. Therefore, it is vital for them to be close to main roads. They are also more interested in locations with a good accessibility to jobs and services in the larger cities, to the expense of remote, small and medium-sized urban areas. Therefore, the current ranking of urban areas is reinforced. Residents' desire for environmental amenities is also increasing. Also, the importance to be located close to the coast is getting critical for retirees and second-home owners	Regulations are going down as both urban and rural rule out the necessity for defining priority areas. There is a lot less planning intervention in the residential development process so restrictions are not enforced and people choose their location at their own risk
Global sustainability (GS)			
The 'think global, act local' motto encourages people to reduce their ecological footprint and to invest in their social network rather than in material and goods. The economy grows at a lower path than today. The main urban centres tend to delegate to smaller towns part of the activities so services are closer to everyone and size differences are lessened. Population and household numbers increase; new developments will be mostly confined to existing urban areas. Less new development on coasts and floodplains arises partly through more planning and also through smaller growth in the number of households. Floodplain occupancy is discouraged under this scenario	With medical advances, life expectancy rises and the number of today's average births reduce, but to a lesser extent than for WM. The 'think global, act local' motto encourages most households to not purchase second homes and to buy existing buildings within urban areas where they work and of reasonable size as it creates economies of scale for the community services and diminish commuting. Migration flows are reduced as economic activities target local workforce and sustainability over high profits	Households are aware of the risks attached to hazardous zones. They prefer locations where they have less impact on the environment. Thus, existing urban areas are favoured to the countryside	Regulations are more restrictive. Both urban and rural planners impose a strict priority hierarchy amongst development areas in order to increase urban density. Risk prevention is enforced throughout the region with safety buffers applied around hazardous locations

Local stewardship (LS)			
Political actions are put in place to boost the local economy which increases jobs provision. The concern about natural resources and the environment pushes political actions to be taken to better inform on the negative consequences of new development in general and second homes ownership in particular. Most people are receptive to these concerns and information. The local economic activities maintain a sustainable viability of every community which are as autarchic as possible and mainly maintain contacts with the surrounding communities. A conservationist ethic and lower demand for new development contributes to the continuing dominance of traditional housing. Floodplain occupancy is kept stable. The urbanisation of the countryside comes to a halt as planning controls are tightened and strong land use control avoids flood-prone areas	With medical advances, life expectancy rises. Today's birth rate trend is inverted as people are encouraged to contribute to the local age pyramid for maintaining the local community alive. Nevertheless, the increased jobs provision still attracts new migrants who settle permanently. External retirees tend to be attracted as today by the region because of its mild weather and its welcoming local community attitude	Autarchy values encourage households to go and remain in areas of origin. Also, they tend to avoid protected and hazardous areas to preserve the community safety as suggested by the politics. Thus, secondary urban areas are growing more than cities. Local economy growth focuses on active neighbourhood activities and local exchanges, which reduces the commuting loads and thus the importance to reside close to main roads	Rural planners impose development priorities to better organise and structure the local landscape around the local community, as they are freed from regional and national strategies. Restrictions are applied firmly for community safety in hazardous zones only. Urban planners' strategies remain the same as the community logic does not prevail as much in larger built areas
National enterprise (NE)			
Liberalised markets do not greatly shift the gravity centre of economy towards the region but maintain their activities to today's levels. Population, households, urban areas and second homes increase largely in East Anglia. However, the increase will not be as high as that of the world markets scenario since there are lower levels of investment in housing. As well as the issue of urban sprawl, the increase in the number of households may be exacerbated by a coastal attraction and people relocating in coastal regions from large urban areas like London. Planning controls are relaxed, allowing more buildings and more people to live in floodplains	With medical advances, life expectancy rises. The birth rate keeps at today's level. Personal independence and material wealth encourages people to invest in second homes at the coast. London keeps its influence and migrations occur similarly as today	Households devote their attention on medium-sized urban areas because they are perceived as the right balance between social needs and open-space will. Also, they need to stay close enough to the main jobs centres, represented by the cities and large towns. At the same time, residents have more desire for environmental amenities. Retirees and second-home owners' attraction to the coast is increasing	The planning control is getting weaker, particularly in the countryside, where the economical focus prevails over landscape conservation. Restrictions on hazardous zones remain as today's and less attention is given to the conservation of natural areas

Table A.6 Summary of the new and percentage increase of residential (RP) and non-residential (NRP) properties in the floodplain and within 400 m of the cliff top under the four socio-economic scenarios to the 2080s (Chap. 4, see also Mokrech et al. 2012)

Scenario	Type of property	Local stewardship		Global sustainability		National enterprise		World markets	
		N	%	*N*	%	*N*	%	*N*	%
New properties in the coastal floodplain	RP	1,343	8	10,882	64	19,550	114	19,719	116
	NRP	80	2.5	1,082	34	1,908	60	2,127	66
New properties within 400 m of cliff top	RP	330	7	511	11	654	14	787	17
	NRP	17	2	44	5	52	6	85	9

Management Strategies

This coast has been shaped by human management for hundreds of years as shown in Chap. 1. These strategies comprised increasing lengths of hard defences on the cliffed coast and a mixture of toe protection, offshore reefs and beach nourishment on the flood-prone coast. Here we consider different options of protection for the cliffed coast and beach nourishment for the flood-prone coast assuming the other defences on the flood-prone coast are maintained.

A number of different scenarios of coastal management policies were represented to explore how human interventions may change future coastal evolution and flood risk and also influence the system's sensitivity to climate change. As with the exploration of climate and socio-economic scenarios, our approach has been to explore a sufficiently broad range of management policies to ensure we develop a rich understanding of the system behaviour. For example, to understand the importance of cliff erosion, and the sensitivity of flood risk to sediments from the cliffs, we tested policies with no artificial beach nourishment, which at the moment is undertaken from Sea Palling to Winterton to manage flood risk (see Figs. 1.4 (b) and 1.8 in Chap. 1).

Over the timeframe of the project, the Shoreline Management Plan (SMP) was proposed and developed through several consultation periods (see Box A.1). Several of the management strategies correspond to realisations of SMP policies. Management strategy M4+N (with nourishment) is the most closely aligned to current policy; however, shoreline management planning is a dynamic process, and consequently none of the management strategies reported here provide a precise representation of the current SMP (NNDC 2012) policies. M4-N (no nourishment) was intended to reveal the value of beach nourishment in reducing flood risk. M3 and M5 represent policies closely aligned to those considered during the first phase of modelling (M3 is very similar to M4-N), with M5 corresponding to a more radical realignment than M3.

Management policy M2 was included with the intention of informing public debate around the implications of maintaining historic defence policies. M1 and M6 were included to explore the most extreme changes in coastal management (i.e. almost a sensitivity analysis).

In those management scenarios where structures were gradually removed from the coast (M3, M4, M5), effort was made to simulate realistically managed retreat, in that structures were assumed to be removed first from areas with the lowest economic value and allowed to expire without renewal at the end of their expected lives, generally before 2030. This was implemented deterministically for management policies tested during Phase 1 (M3, M5) and probabilistically during Phase 2 (M4). In addition to representing uncertainty in structure failure years, the probabilistic treatment also provided a more progressive representation failure, rather than a sudden simultaneous collapse of extended lengths of structures as was assumed for M3 and M5.

Box A.1 A Brief History of SMPs in North Norfolk[2]

In 1996, North Norfolk District Council adopted the first generation Shoreline Management Plan along the Sheringham to Lowestoft SMP (Subcell 3b, now SMP6) frontage. In 2002, in advance of the preparation of new guidance from Defra, a review commenced for a modified frontage from Kelling to Lowestoft (SMP6) as one of three national case studies. That revised plan appeared in draft in 2004 to much public disquiet and concern because of the proposal not to maintain or renew large sections of cliff defence (a nonactive intervention policy), where before the policy was to 'hold the line'. Waveney District Council adopted the original plan in 2007. Subsequently an amended version (similar to North Norfolk's – see below) was approved by Great Yarmouth Borough Council. North Norfolk District Council's Cabinet agreed at its meeting of September 2007 that a revised SMP is accepted conditionally and that the government be urged to include a wider economic appraisal, implement integrated coastal zone management and undertake research into the effects of intermittent defences.

In 2009, a revised version of the SMP and a strategic environmental assessment (SEA required by legislation) was subject to consultation. Following the consultation and review, a slightly amended SMP was finally adopted by The Anglian (Eastern) Transitional Regional Flood and Coastal Committee at its meeting on 15 April 2011. As some changes were made to the SMP since the first strategic environmental assessment was undertaken on the 2005 version of the SMP, there was a statutory requirement to reassess and reconsult on an amended SEA. The consultation was completed between February and April 2012, and the consultation responses are now being considered.

North Norfolk District Council's Cabinet resolved to adopt SMP6 alongside its partner organisations at a meeting in November 2011. All partners finally adopted SMP6 at the end of 2012 (NNDC 2012).

---

[2]Adapted from North Norfolk District Council's webpages: http://www.northnorfolk.org/coastal/9871.asp accessed on 1st December 2012.

Table A.7 Management strategies tested during the development and application of the coastal simulator[a]

	Description	Percentage of cliffed coast with coastal protection (%)	Nourishment	
			No (−N)	Yes (+N)
M1	Construction of seawalls along the entire coast in the year 2000	100	Phase 1	Phase 2
M2	Maintenance of all existing structures	71	Phase 1	Phase 2
M3	Structures removed from less inhabited locations (similar to early draft of the SMP)	34	Phase 1	–
M4	Implementation of the policies recommended by the North Norfolk Shoreline Management Plan (SMP, North Norfolk District Council 2010)	34		Phase 2
M5	Implementation of the policies recommended by the North Norfolk Shoreline Management Plan (SMP, North Norfolk District Council 2010) but with no beach nourishment at Sea Palling		Phase 2	
M6	Structures removed from larger towns/industrial sites, i.e. Overstrand, Mundesley, Bacton Gas Terminal (most radical proposals from an early draft of the SMP)	16	Phase 1	–
M7	Removal of all coast protection structures (i.e. those intended to prevent erosion) in the year 2000	0	Phase 1	Phase 2

[a]In M3 to M6 in Phase 1, structures are removed in 2030, whilst in Phase 2, their failure was simulated probabilistically based on a reliability assessment (Chap. 7)

References

Bouws, E., Jannink, D., & Komen, G. J. (1996). The increasing wave height in the North Atlantic Ocean. *Bulletin of the American Meteorological Society, 77*(10), 2275–2277.

Carter, T. R., Jones, R., Lu, X., Bhadwal, S., Conde, C., Mearns, L., O'Neill, B., Rounsevell, M., & Zurek, M. (2007). New assessment methods and the characterisation of future conditions. In M. L. Parry, O. F. Canziani, J. P. Palutikof, P. J. van der Linden, & C. E. Hanson (Eds.), *Climate change 2007: Impacts, adaptation and vulnerability* (pp. 133–171). Cambridge, UK: Cambridge University Press.

DTI. (2002). *Foresight futures 2020: Revised scenarios and guidance*. London: Department of Trade and Industry.

Gregory, J. M., Church, J. A., Boer, G. J., Dixon, K. W., Flato, G. M., Jackett, D. R., Lowe, J. A., O'Farrell, S. P., Roeckner, E., Russell, G. L., Stouffer, R. J., & Winton, M. (2001). Comparison of results from several AOGCMs for global and regional sea-level change 1900–2100. *Climate Dynamics, 18*, 225–240.

Hulme, M., Jenkins, G. J., Lu, X., Turnpenny, J. R., Mitchell, T. D., Jones, R. G., Lowe, J., Murphy, J. M., Hassell, D., Boorman, P., McDonald, R., & Hill, S. (2002). *Climate change scenarios for the United Kingdom: The UKCIP02 scientific report*. Norwich: Tyndall Centre for Climate Change Research, School of Environmental Sciences, University of East Anglia. 120 pp.

Kriegler, E., O'Neill, B. C., Hallegatte, S., Kram, T., Lempert, R., Moss, R. H., & Wilbanks, T. J. (2010, October). *Socio-economic scenario development for climate change analysis* (CIRED Working Paper DT/WP No 2010–23). Available for download at: http://www.centre-cired.fr/IMG/pdf/CIREDWP-201023.pdf

Kushnir, Y., Cardone, V. J., Greenwood, J. G., & Cane, M. A. (1997). The recent increase in North Atlantic wave heights. *Journal of Climate, 10*(8), 2107–2113.

Lowe, J. A., Howard, T., Pardaens, A., Tinker, J., Holt, J., Wakelin, S., Milne, G., Leake, J., Wolf, J., Horsburgh, K., Reeder, T., Jenkins, G., Ridley, J., Dye, S., & Bradley, S. (2009). *UK climate projections science report: Marine and coastal projections*. Exeter: Met Office Hadley Centre.

Miller, L., & Douglas, B. C. (2004). Mass and volume contributions to twentieth-century global sea level rise. *Nature, 428*, 406–409.

Mokrech, M., Nicholls, R. J., & Dawson, R. J. (2012). Scenarios of the future built environment for coastal risk assessment of climate change using a GIS-based multi-criteria analysis. *Environment and Planning B, 39*(1), 120–136. doi:10.1068/b36077.

Moss, R. H., Edmonds, J. A., Hibbard, K. A., Manning, M. R., Rose, S. K., van Vuuren, D. P., Carter, T. R., Emori, S., Kainuma, M., Kram, T., Meehl, G. A., J. Mitchell, F. B., Nakicenovic, N., Riahi, K., Smith, S. J., Stouffer, R. J., Thomson, A. M., Weyant, J. P., & Wilbanks, T. J. (2010). The next generation of scenarios for climate change research and assessment. *Nature, 463*(7282), 747–756.

Nakicenovic, N., Alcamo, J., Davis, G., de Vries, B., Fenhann, J., Gaffin, S., Gregory, K., Grübler, A., Jung, T. Y., Kram, T., La Rovere, E. L., Michaelis, L., Mori, S., Morita, T., Pepper, W., Pitcher, H., Price, L., Riahi, K., Roehrl, A., Rogner, H.-H., Sankovski, A., Schlesinger, M., Shukla, P., Smith, S., Swart, R., van Rooijen, S., Victor, N., & Dadi, Z. (2000). *Special report on emissions scenarios: A special report of working group III of the Intergovernmental Panel on Climate Change*. Cambridge, UK: Cambridge University Press, 599 pp. Available online at: http://www.grida.no/climate/ipcc/emission/index.htm

Nicholls, R. J., Wong, P., Burkett, V., Woodroffe, C. D., & Hay, J. (2008). Climate change and coastal vulnerability assessment: Scenarios for integrated assessment. *Sustainability Science, 3*(1), 89–102.

Nicholls, R. J., Hanson, S. E., Lowe, J. A., Warrick, R. A., Lu, X., Long, A. J., & Carter, T. R. (2011). *Constructing sea-level scenarios for impact and adaptation assessment of coastal area: A guidance document.* Supporting Material, Intergovernmental Panel on Climate Change Task Group on Data and Scenario Support for Impact and Climate Analysis (TGICA), 47 pp. http://www.ipcc-data.org/docs/Sea_Level_Scenario_Guidance_Oct2011.pdf

Nicholls, R. J., Woodroffe, C. D., Burkett, V., Hay, J., Wong, P. P., & Nurse, L. (2012). Scenarios for coastal vulnerability assessment. In E. Wolanski & D. S. McLusky (Eds.), *Treatise on Estuarine and coastal science Vol. 12 Ecological economics of estuaries and coasts*. Maryland Heights: Elsevier.

NNDC (North Norfolk District Council). (2012). *Shoreline Management Plan 6. Kelling Hard to Lowestoft Ness*. Available at http://www.northnorfolk.org/smp6/. Last downloaded 24 June 2015.

Shennan, I., & Horton, B. (2002). Holocene land- and sea-level changes in Great Britain. *Journal of Quaternary Science, 17*, 511–526.

SPRU, CSERGE, CRU, & PSI. (1999). *Socio-economic futures for climate impacts assessment: Final report*. Brighton: Science and Technology Research, University of Sussex.

Thorne, C. R., Evans, E. P., & Penning-Rowsell, E. C. (Eds.). (2007). *Future flood and coastal erosion risks*. London: Thomas Telford.

Index

R.J. Nicholls et al. (eds.), *Broad Scale Coastal Simulation*, Advances in Global Change Research 49, DOI 10.1007/978-94-007-5258-0

MIX
Papier aus verantwortungsvollen Quellen
Paper from responsible sources
FSC® C105338

If you have any concerns about our products,
you can contact us on
ProductSafety@springernature.com

In case Publisher is established outside the EU,
the EU authorized representative is:
Springer Nature Customer Service Center GmbH
Europaplatz 3, 69115 Heidelberg, Germany

Printed by Libri Plureos GmbH
in Hamburg, Germany